Inertial
Confinement Fusion

Inertial
Confinement Fusion

JAMES J. DUDERSTADT
Department of Nuclear Engineering
The University of Michigan

GREGORY A. MOSES
Department of Nuclear Engineering
The University of Wisconsin

A Wiley-Interscience Publication

JOHN WILEY & SONS

New York   Chichester   Brisbane   Toronto   Singapore

Library of Congress Cataloging in Publication Data:

Duderstadt, James J., 1942-
  Inertial confinement fusion.

"A Wiley-Interscience publication."
  Includes bibliographical references and index.
  1. Pellet fusion.   I. Moses, Gregory A.
II. Title.

TK9204.D82   621.48'4   81-11472
ISBN 0-471-09050-6   AACR2

Printed in the United States of America

10 9 8 7 6 5 4 3 2 1

To our parents

Preface

Controlled thermonuclear fusion may someday provide a clean, safe, and abundant source of energy. But the difficulties involved in demonstrating the scientific feasibility of this process, in igniting and controlling the fusion fire, are formidable. Controlled fusion has proven to be an elusive goal. Today, after some 30 years of effort, we stand only on the threshold of demonstrating its feasibility. The challenges of developing a viable fusion technology capable of massive implementation will almost certainly require an even greater effort.

Most fusion research to date has been directed at confining a dilute fusion fuel using cleverly designed magnetic fields. However, during the past decade an alternative approach known as *inertial confinement fusion* has begun to receive considerable attention. In this approach intense laser or charged particle beams are used to rapidly compress a tiny pellet of fusion fuel, typically from 1 to 5 mm in diameter, to the enormous densities and temperatures required for efficient thermonuclear burn. If the fuel pellet is compressed to sufficient densities, then it will burn so rapidly that appreciable fusion energy will be released before it can blow apart; it will be "confined" during the fusion burn by its own inertia.

The effort directed toward the development of inertial confinement fusion has grown to the point where it now rivals that of the more traditional magnetic confinement fusion approach. But whereas there exist several excellent texts on the physics of magnetic confinement fusion, the literature concerned with inertial confinement fusion remains relatively diffuse. It therefore seems an appropriate time to attempt to pull together the many disciplines involved in inertial confinement fusion research into an introductory text.

It is important to recognize that the relevant subject matter for inertial confinement fusion differs quite significantly from that for magnetic confinement fusion. In the latter field, the primary emphasis is on plasma physics and electromagnetic theory. In contrast, any introduction to inertial confinement fusion should include material concerned with the physics of inertially confined thermonuclear fusion reactions, hydrodynamics and shock waves, transport processes in dense plasmas, and the interaction of laser or charged particle beams with plasmas. In addition, material of a more applied nature should be included such as laser and charged particle driver beam development, target design and fabrication, experiment-diagnostic methods, and possible applications of the inertial confinement fusion process.

This text has resulted from our attempt over the past several years to develop a course on this subject. It is aimed at advanced undergraduate or graduate students in engineering and physics, as well as at practicing engineers and scientists seeking an introduction to inertial confinement fusion. Only the usual undergraduate background in mathematics and physics has been assumed. Although some additional exposure to plasma physics would prove useful to those intending to enter this field, it is not essential to understanding most of the material presented in this text.

Very little of the material presented in such a broad treatment can claim originality. We have drawn heavily from the technical literature in developing material for this text. Although we have attempted to include a comprehensive bibliography, the wide range and rapidly changing nature of the present literature on inertial confinement fusion makes this a difficult task. Therefore we have provided particular reference to a number of excellent review articles on various aspects of this subject. Our effort has also benefited greatly from the knowledge, experience, and assistance of a number of colleagues. Of particular note is the influence of James Shearer and Ray Kidder (LLL); Eldon Linnebur, Bill Varnum, Paul Rocket, and David Bach (LASL); Fred Mayer (KMSF); Stephen Bodner (NRL); and Richard Osborn and Rudi Ong (Michigan). We would also like to acknowledge the comments and suggestions concerning the manuscript provided by Barry Ripin, John McMahon, David Mosher, Shyke Goldstein, and Jerry Cooperstein (NRL); Mary Ann Sweeney and Thomas Mehlhorn (Sandia); David Berwald (TRW); Thomas Sutton and Shin Takeshita (Michigan); and Donald Kania (LASL). Finally, we express our appreciation to Todd Spindler (Wisconsin) for his help on the references.

JAMES J. DUDERSTADT
GREGORY A. MOSES

Ann Arbor, Michigan
Madison, Wisconsin
October 1981

Contents

Inertial
Confinement Fusion

ONE

Introduction

As the limitations of the Earth's resources of conventional fuels have become more apparent, scientists have turned their attention toward the stars for a new source of energy. It has been known for several decades that *nuclear fusion* reactions are a major energy source in stars. In this process the nuclei of light elements are fused together at very high temperatures to produce more tightly bound, heavier nuclei, releasing energy in the process.

An example of such a reaction is that which occurs when the two heavier isotopes of hydrogen, deuterium (D) and tritium (T), combine to produce helium plus a neutron. This fusion reaction releases 17.6 MeV of energy, which is carried off as kinetic energy by the reaction products. The energy content of such fusion fuels is truly enormous. A thimbleful of deuterium would release as much energy from fusion as the combustion of 20 tons of coal. The natural deuterium contained in one liter of water would produce the fusion energy equivalent of 300 liters of gasoline.

The potential of such reactions for generating large amounts of energy is evident. We need only look at any star to see a massive example of fusion energy release. In a sense, nuclear fusion can be regarded as the most primitive form of solar power, since it is also the energy source of our sun. Hence it was natural for scientists to wonder whether fusion might be employed as a terrestrial energy source. The awesome potential of this quest was demonstrated by the development of nuclear fusion weapons—the hydrogen bomb—in the early 1950s. Since that time, proponents of fusion power have predicted that someday this nuclear process would provide us with a safe, clean, and abundant source of energy.[1-7]

But the difficulties involved in igniting and controlling a fusion reaction are formidable. The light nuclei that must fuse together are positively charged and strongly repel one another. To overcome this repulsion, we must slam the two nuclei together at very high velocities. One way of doing this is to take a

1

mixture of deuterium and tritium and heat it to such high temperatures that the velocities of thermal motion of the nuclei are sufficient to overcome charge repulsion and initiate the fusion reaction. Such a scheme is referred to as a *thermonuclear fusion reaction*. The temperature required is quite high—roughly 100 million degrees (or 10 keV, where 1 keV corresponds to 1.16×10^7 K). Until quite recently scientists had imitated the sun only in a rather violent fashion by using a nuclear fission explosion to create temperatures high enough to ignite the fusion reaction in the hydrogen bomb.

But simply heating the fusion fuel to enormous temperatures is not enough to ignite the fusion reaction. For most of the time, when the nuclei run into each other, they simply bounce off or scatter without fusing together. Indeed, such scattering collisions are a million times more probable than fusion events. So somehow we have to hold the high temperature fusion fuel together long enough to allow the nuclei to collide the millions and millions of times necessary to induce the fusion reactions.

Therefore to achieve thermonuclear fusion energy we must solve two problems: (1) produce and heat a plasma fuel to thermonuclear temperatures, and (2) confine it long enough to produce more fusion energy than we have expended in heating and containing the fuel. These twin requirements are usually quantified by a mathematical relation known as the *Lawson criterion*,[8] which essentially reflects the balance between thermonuclear energy production and heating energy. This criterion can be expressed as a condition on the product of the fuel density n and the time of fusion fuel containment τ. If we express n in units of number of nuclei per cm^3 and τ in seconds, then the Lawson criterion demands that the product $n\tau$ exceed roughly 10^{14} s/cm^3 for a D-T fusion reaction (and 10^{16} s/cm^3 for the D-D reaction).

But how are we to accomplish the twin goals of heating and confinement in such a way as to satisfy the Lawson criterion? In a star the enormous mass causes gravitational forces that confine the reacting fuel, compressing it and heating it to the necessary temperatures. Certainly we cannot expect gravity to do that job here on Earth.

In thermonuclear weapons no attempt is made to confine the reacting fuel. Instead one attempts to heat the fuel to thermonuclear temperatures so fast that an appreciable number of fusion reactions occur before it is blown apart ("explosively disassembles"). This scheme is known as *inertial confinement*, since it is the inertia of the reacting fuel that keeps it from blowing apart prematurely. But to heat an appreciable mass of fuel to such high temperatures requires an extremely large energy source, and the source used in thermonuclear weapons is an explosive fission chain reaction. That is, an atomic bomb is used to heat the thermonuclear fuel to ignition temperatures. Again this approach is highly unsuited for a controlled application.

The approach to fusion power that has been most extensively studied to date works with far smaller quantities of thermonuclear fuel. In particular, it takes advantage of the fact that at the high temperatures necessary for fusion to occur, the fuel becomes an ionized or charged gas known as a *plasma*. Since such charged particles have difficulty moving across magnetic-field lines (instead tending to spiral along them), the primary approach has been to design a "magnetic bottle" composed of strong magnetic fields to contain the fuel. Traditionally these *magnetic confinement fusion* schemes have worked with very low fuel densities ($\sim 10^{14}$ cm^{-3}) and have attempted to achieve confinement times of the order of a second to satisfy the Lawson criterion. After two decades of intensive research, magnetic confinement fusion has reached the threshold of achieving the goal of scientific breakeven, in which the Lawson criterion is satisfied and the fusion energy produced by the fuel exceeds the energy necessary to heat and confine it.

Recently, however, scientists have become excited about an alternative approach to controlled thermonuclear fusion based on inertial confinement.[9-11] In this approach intense laser or charged particle beams would be used to rapidly compress a tiny pellet of deuterium-tritium fuel to tremendous densities and temperatures and ignite a thermonuclear fusion reaction or burn. If the fuel pellet is compressed to sufficient densities, then it will burn so rapidly that appreciable fusion energy will be released before it can blow apart.

More precisely, the intense laser or charged particle beams (the "driver" beams) would strike the pellet surface, ionizing this surface and ablating it off into the vacuum surrounding the pellet (see Figure 1.1). As the outer surface of the pellet blows away, an enormous pressure is generated (much as by a rocket exhaust) that would compress the core of the fuel pellet to densities as high as 1000 to 10,000 times solid-state density. This compression would also raise the temperature of the core of the pellet to fusion temperatures so that a thermonuclear burn is ignited. This burn would then propagate outward through the

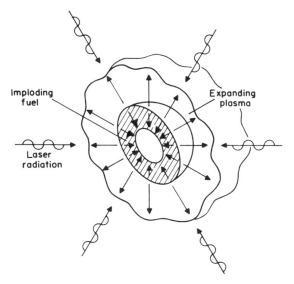

Figure 1.1. Implosion of an inertial confinement fusion target.

rest of the fuel pellet, igniting and burning it, to result in the explosive release of fusion energy. The process of compression and thermonuclear ignition and burn would occur in a time much shorter than the time required for the pellet to blow apart ($\sim 10^{-9}$ s). Hence a premium is placed on developing driver beams capable of delivering large quantities of energy onto tiny targets (1 to 10 mm in diameter) in a very short pulse (0.1 to 20 ns).

In a sense, the inertial confinement approach to controlled fusion represents a scaling down of the hydrogen bomb over a millionfold to a tiny microexplosion. For a brief instant, the driver beams compress or implode the fuel pellet to produce conditions similar to those found in stars. A tiny sun is produced, which bursts in an instant, releasing its fusion energy. If we can capture this energy, then we can convert it to useful purposes.

Edward Teller[12] has noted that inertial confinement fusion (ICF) is essentially the internal combustion engine approach to fusion. To make the analogy more precise, recall that the internal combustion engine of a car is based on a four-stage combustion cycle (see Figure 1.2): (1) injection of fuel (gas and air) into the cylinder, (2) compression of the fuel mixture by a piston, (3) ignition of the compressed fuel by a spark plug, and (4) combustion of the fuel mixture in a small explosion that drives the piston and hence the crankshaft (converting chemical energy into mechanical energy).

Inertial confinement fusion schemes are based on the following analogous sequence: (1) a tiny pellet of deuterium-tritium isotopes is injected into a blast chamber, (2) the pellet is compressed to very high density with intense laser or charged particle beams, (3) the high density and compression heat induce the ignition of a thermonuclear reaction, producing a microscopic thermonuclear explosion, and (4) the thermonuclear energy carried by the reaction products, including neutrons, X rays, and charged particles, is deposited as heat in a blanket that then acts as a heat source in a steam thermal cycle to produce electricity (conversion of nuclear energy into electric energy). The inertial confinement fusion internal combustion engine would use a series of micro-thermonuclear explosions (from 1 to 100 per second, each generating the energy equivalent of several kilograms of high explosive) to generate power.

The applications of inertial confinement fusion fall into several categories: power production,[13] weapons applications,[11, 14, 15] and fundamental physics studies. Much of the funding for research activities in this area has been stimulated by the recognition that the environment created by the implosion and thermonuclear burn of a tiny fuel pellet is similar in some respects to that of a thermonuclear weapon. Hence there has been considerable interest in using inertial confinement fusion targets to simulate weapons physics and effects on a microscopic scale.

Perhaps the most immediate application of inertial confinement fusion will be in basic physics studies. The imploded fuel pellet produces conditions of temperature and pressure that are quite unusual (at least on a terrestrial scale). Inertial confinement fusion implosions can be used to study properties of matter under extreme conditions, the interaction of intense radiation with matter, and aspects of low energy nuclear physics. Indeed, inertial confinement

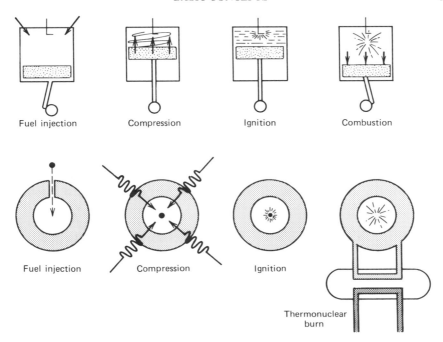

Figure 1.2. A comparison of internal combustion engines: the top sequence for a gasoline engine of an automobile, the bottom sequence for an inertial confinement fusion reactor.

fusion presents us with a unique opportunity to study certain aspects of astrophysics such as stellar interiors on a laboratory scale.

But perhaps the most significant application of inertial confinement fusion will be to the production of energy that can then be used for a variety of purposes such as the generation of electricity, the production of process heat or synthetic fuels, or propulsion. The importance of this application is apparent when it is recognized that there are only three major inexhaustible energy source options available to our civilization at the present time: the nuclear fission breeder reactor, solar energy, and nuclear fusion.[16, 17] Serious social and political questions threaten to stall breeder reactor development.[18] Solar energy faces major challenges of both a technical and economic nature.[19] Hence the importance of an aggressive nuclear fusion research effort appears evident.

Controlled thermonuclear fusion in general, and inertial confinement fusion in particular, present us with a staggering technological challenge. But the potential benefit of such an abundant energy source compels us to address this challenge with a determined effort.

1.1. BASIC CONCEPTS

We have noted that the basic requirements of the fusion game involve heating a plasma fuel (e.g., D-T) to thermonuclear temperatures (approximately 10 keV) and then confining this high temperature fuel for a sufficiently long time that it produces more fusion energy than the energy invested in its heating and confinement. The scoreboard for this game is the Lawson criterion,[8] which demands a certain minimum value of the product of number density n and confinement time τ—for example, the scientific feasibility criterion for a D-T fuel is $n\tau > 10^{14}$ s/cm³. As we have also noted, the traditional approach to fusion has been to attempt to confine a very low density plasma fuel (at $n \sim 10^{14}$ cm⁻³) for a relatively long time ($\tau \sim 1$ s) in a suitably shaped magnetic field (e.g., toroidal fields such as in the Tokamak).

The inertial confinement fusion scheme takes the opposite approach. The aim is to heat a dense fuel to thermonuclear temperatures extremely rapidly so that an appreciable thermonuclear reaction energy will be produced before the fuel blows itself apart. To see what we are up against, consider a small pellet of radius 1 mm. The "disassembly time" τ_d required for the heated pellet to blow itself apart is roughly the time required for a sound wave to traverse the pellet. Since the speed of sound in a 10-keV D-T plasma is roughly 10^8 cm/s, the disassembly time $\tau_d \sim 0.1/10^8 = 10^{-9} = 1$ ns. Hence to satisfy the Lawson criterion, we must use a fuel density in excess of $n \sim 10^{14}/\tau_d \sim 10^{23}$ cm⁻³ which is roughly liquid-state density.

Therefore the new game we must play in inertial confinement fusion is to heat a small, liquid-density D-T pellet to thermonuclear temperatures before it has a chance to expand—that is, in 1 ns. Actually the energy required is not too great—roughly 1 MJ or about 0.28 kWh—about the energy consumption in one evening's operation of a television set. But when this energy is delivered in 10^{-9} second, it corresponds to a power level of $10^6/10^{-9} = 10^{15}$ W. The staggering demands of such a power level become apparent when it is noted that the present electrical generating capacity of all the power plants in the United States is somewhat under 10^{12} W.

Hence we are faced with the task of generating enormous powers and focusing these down on a tiny pellet, roughly 1 mm in radius. But this is what a laser is very good at doing. For not only can a laser focus large amounts of energy onto very tiny spots, but it can also zap this energy in a very short time —easily within 1 ns (indeed, laser pulses as short as 10^{-12} s have been achieved).

So if we use the laser just like a very big flashlight to zap the fuel pellet to fusion temperatures very rapidly, we can visualize that a laser fusion system might work something like that shown in Figure 1.3. The laser light is focused on the pellet, heating it rapidly to thermonuclear temperatures and thereby inducing a thermonuclear fusion microexplosion. The energy from this explosion is then captured and converted to electricity through a steam thermal cycle. After using part of this energy to reenergize the laser, the remaining energy is then distributed to the electrical power grid.

So far, so good! And this was essentially the "public image" presented by the inertial confinement fusion effort in the B.D.C. (before declassification) days prior to 1972. But this simple-minded scheme had a fatal flaw, which

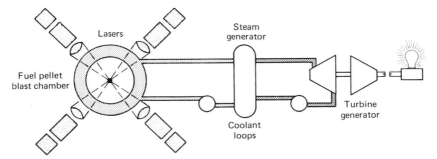

Figure 1.3. A simple schematic of an inertial confinement fusion reactor.

became apparent when one tried to estimate the laser energy required to produce such a microexplosion.

Suppose (because of laser and thermal cycle inefficiencies) we require the thermonuclear energy produced by the pellet to be M times the incident laser energy. Suppose further that only a fraction ε_D of the incident driver energy can be coupled into the target. Then we can factor these expressions into the Lawson criterion (in a manner that will be made more explicit in Chapter 2) to arrive at an estimate of the required laser energy for inertial confinement fusion as[21]

$$E_{\text{laser}} = \frac{M^3}{\varepsilon_D^4} \left(\frac{n_L}{n} \right)^2 \qquad \text{MJ}$$

where n_L is the number density for liquid D-T (4.5×10^{22} cm^{-3}). Let us now apply this estimate to calculate the laser energy required for scientific feasibility, that is, for $M=1$. If we take $\varepsilon_D=1$ and $n=n_L$, we find a laser energy requirement of 1 MJ $=10^6$ J. To place this number in perspective, the largest laser in the world today, the Shiva-Nova laser at the Lawrence Livermore Laboratory, produces a pulse of only 100 kJ—a factor 10 times too small. For a reactor, we would have to require $M=10$ (at least) which would imply a hopelessly large laser energy of 10^9 J. Viewed in this light, laser fusion is clearly a fool's quest.

Or is it? We mentioned that this was the "naïve" or B.D.C. approach. We must be a bit more sophisticated in our analysis. Let us begin by reexamining the criterion for achieving net fusion energy release in a somewhat different light. We can identify two times of major significance for inertial confinement fusion schemes: the *disassembly time*, which scales as

$$\text{disassembly time} = \tau_d \sim \frac{R}{c_s}$$

and the *thermonuclear burn time*

$$\text{burn time} = \tau_b = \frac{m}{\rho} \langle v\sigma \rangle^{-1}$$

Here R is the pellet radius, ρ is its mass density, m is the ion mass, c_s is the speed of sound, v is the relative speed of the colliding nuclei, and σ is the fusion reaction cross section.

If we regard τ_b as a measure of the time required to burn a fuel pellet of density ρ, and τ_d as the time during which the thermonuclear reaction will occur, then we can define a "thermonuclear burn efficiency" as just the ratio of these two times[22]:

$$\text{thermonuclear burn efficiency} = \varepsilon_b = \frac{\tau_d}{\tau_b}$$

The quantity $(\langle v\sigma \rangle / mc_s)$ is essentially constant in the temperature range of interest (20 to 80 keV). Hence we find that the thermonuclear burn efficiency scales as the product of fuel density and radius, ρR. If we insert the appropriate numerical constants, we find an alternative to the Lawson criterion that is far more appropriate for inertial confinement fusion schemes. This new criterion becomes

$$\rho R > 1 \text{ g/cm}^2$$

(Actually, if we are a bit more careful and take into account fuel depletion,[23] we find that the burn efficiency becomes

$$\varepsilon_b = \frac{\rho R}{(6.3 + \rho R)}$$

Hence for $\rho R = 3$, roughly one third of the pellet fuel would be burned.)

To understand the implications of this result, note that for a 1-mm pellet, $\rho R = 1$ implies a fuel density of $\rho = 10 \text{ g/cm}^3$. But since the liquid density of D-T is only $\rho_L = 0.2 \text{ g/cm}^3$, we find that this implies a compression of the fuel pellet to at least 50 times its initial density. Hence the key to inertial confinement fusion is apparently high compression.

More generally, if we note that ρ scales with fuel radius as R^{-3} (for fixed fuel mass), we find that the thermonuclear burn efficiency scales as $\rho R \sim R^{-2}$. The more we compress the fuel, the larger ρR becomes, and the more efficient the thermonuclear burn becomes. For example, a compression of 1000 would reduce the requirements for scientific breakeven to only 1 J and those for a reactor to 1000 J. Actually, these simple scaling arguments are still too naïve since they predict breakeven requirements several orders of magnitude below those suggested by more complex models (which tend to cluster about 1 MJ). However, they do illustrate the strong dependence of the required driver beam energy on the compression factor.

The only remaining question, then, is "how." How do we achieve such tremendous compressions? Certainly not by normal mechanical forces. Nor will chemical explosives do the job (since they are limited to compressions of roughly 10 by the strength of interatomic forces). Densities as large as 1000 times liquid-state density are not common even on an astronomical scale, occurring only in very dense white dwarf stars.

The trick involves using the driver beams themselves.[24-28] Suppose we can focus a number of beams of intense laser light onto the fuel pellet surface (see Figure 1.4). As the pellet absorbs this intense light energy, its surface is rapidly vaporized, ionized, and heated to high temperature, blowing off into the vacuum surrounding the pellet. This blowoff or ablation of the pellet surface drives a shock wave back into the pellet (recall Newton's third law—or better yet, imagine the ablation as you would the thrust from a rocket exhaust). As this shock wave implodes in toward the center of the pellet, it compresses the fuel in a small central region to high density and thermonuclear temperatures so that ignition occurs. At these very high densities (large ρR), the energetic alpha particles produced in the D-T fusion reactions are absorbed in this central region "spark," heating it to still higher temperatures and causing the fuel to burn even more rapidly. As the central spark burns, alpha particles are deposited in the adjacent fuel, bringing it to ignition temperatures. This process continues, leading to a thermonuclear burn wave that propagates outward into the cold, compressed fuel surrounding the ignited pellet core, consuming the fuel in a very rapid thermonuclear microexplosion. After only a few picoseconds a significant fraction of the imploded pellet fuel has burned, and the very high energy release blows the pellet apart, thereby terminating the reaction.

Hence the key idea is to use the laser beam to bring the central region of the pellet to ignition densities and temperatures, but in such a way that the rest of

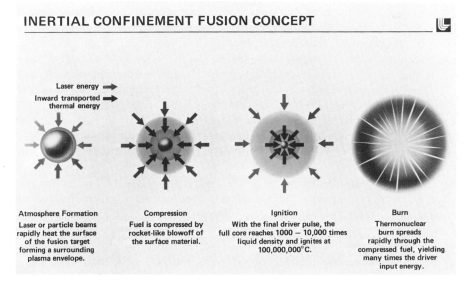

INERTIAL CONFINEMENT FUSION CONCEPT

Laser energy ➡
Inward transported thermal energy ➡

Atmosphere Formation
Laser or particle beams rapidly heat the surface of the fusion target forming a surrounding plasma envelope.

Compression
Fuel is compressed by rocket-like blowoff of the surface material.

Ignition
With the final driver pulse, the full core reaches 1000 – 10,000 times liquid density and ignites at 100,000,000°C.

Burn
Thermonuclear burn spreads rapidly through the compressed fuel, yielding many times the driver input energy.

Figure 1.4. The scenario for the implosion of an inertial confinement fusion target. (Courtesy of Lawrence Livermore National Laboratory.)

the compressed fuel remains cold (so that the required compression energy is minimized). That is, one wishes to isentropically compress the fuel. Only a central spark is produced in the compressed fuel to light the fusion fire. In this way, one lowers the laser energy requirements to roughly 10,000 to 100,000 J.

This simple picture is complicated somewhat by the fact that laser light cannot penetrate very far into a very dense plasma without being reflected. In fact, if the plasma density is above 10^{21} cm^{-3}, the incident laser light (from the Nd glass lasers commonly used in some inertial confinement fusion experiments) will not penetrate. Hence during the actual laser irradiation, a low density cloud or atmosphere ablates off and surrounds the pellet core, shielding it from direct laser radiation. The laser energy absorbed in this atmosphere or corona is then transported into the denser regions of the pellet by processes such as electron thermal conduction to drive the imploding shock wave.

The general features of the pellet implosion scheme were first confirmed in laboratory experiments performed in 1974.[29] Laser beams were focused by specially shaped mirrors onto the surface of tiny pellets consisting of glass shells (from 50 to 100 μm in diameter and 1 to 5 μm in thickness) containing D-T gas up to 100 atm in pressure. Such glass microballoons were imploded to densities roughly 100 times that of the initial fill gas, and the first thermonuclear neutrons were detected. Subsequent targets utilizing multiple layers of materials to provide for the efficient absorption of incident light and energy transfer have led to still higher implosion densities and neutron yields.

However, the success of such implosion experiments should not be interpreted as a demonstration of the scientific feasibility of the inertial confinement fusion scheme. Even the most advanced experiments have demonstrated a target energy gain (the ratio of fusion energy produced to driver beam energy) of only 10^{-4} to 10^{-3}. The high gains required for most applications (roughly 100) will require highly efficient implosions in which driver beam absorption and energy transfer are maximized, the compression process is nearly isentropic, and the fuel is compressed to 10^3 to 10^4 times liquid density so that only a small core region need be ignited to trigger burn propagation to the remainder of the pellet. Present estimates are that the achievement of break-even gain (corresponding to $\rho R \sim 0.3$ to 1 g/cm^2) will require drivers in the 100 to 500 TW range. An inertial confinement fusion reactor faces even more severe requirements ($\rho R \sim 3$ to 5 g/cm^2) corresponding to drivers in the 1 to 10 MJ, 1000 TW level.

These target gain requirements present a very difficult challenge for the design of inertial confinement fusion drivers. Four classes of drivers have been considered to date: lasers, relativistic electron beams, light ion beams, and heavy ion beams.

Laser drivers typically consist of a source or oscillator that feeds light into a number of trains or beamlines of successively more powerful laser amplifiers (see Figure 1.5). Large laser systems may consist of dozens of these beamlines and hundreds of amplifiers.

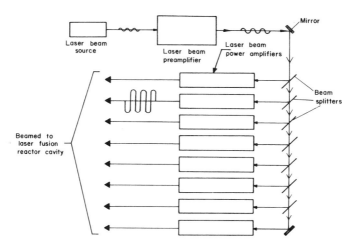

Figure 1.5. Configuration of a large laser driver.

Most high energy laser facilities designed for laser fusion research utilize large Nd glass lasers that emit infrared light at a wavelength of 1.06 μm.[30] To date these lasers have been restricted by glass damage considerations to energies less than 100 J per beam. Several laboratories in the United States and abroad have Nd laser systems operating or under development that approach 10 TW or greater in power level. Since these lasers must be pumped using flashlamp techniques, they are intrinsically very inefficient (less than 0.25%) and therefore would not be suitable for reactor applications. However the advanced state of Nd laser development has led to the extensive use of this driver type in inertial confinement fusion research.

To achieve the high efficiencies and power levels required by reactor applications, it will probably be necessary to use gas lasers. For example, CO_2 lasers have been operated at efficiencies of several percent at high power levels.[31] Furthermore, pulsed CO_2 laser technology appears capable of achieving the necessary power levels at efficiencies of 8 to 10% and repetition rates of 10 to 100 shots per second. Unfortunately, the long-wavelength light emitted by CO_2 lasers (10.6 μm) may not couple effectively to drive the pellet implosion.

An advanced gas laser design based on a krypton-fluorine mixture appears capable of high power, high efficiency (4 to 7%) operation at a shorter wavelength (0.2 μm) that should improve beam-target coupling. However the very early nature of KrF laser development and the lack of target interaction experiments at this wavelength (aside from some early experiments using frequency-quadrupled Nd laser light at 0.26 μm) make any major commitment to this driver type somewhat premature at this point.

Another gas laser that shows promise is the hydrogen fluorine chemical laser. The HF laser has a broad band of wavelengths between 2.6 and 3.4 μm.

This broad band characteristic may possibly ameliorate the problem of plasma instabilities leading to hot electron generation. The HF laser, because it is driven by chemical processes, has an electrical efficiency that can exceed 100%. However, when the energy required to reprocess the spent lasing gas is included the net efficiency is expected to be about 5%.

The projected laser driver requirements for both breakeven experiments and reactor applications are given in Table 1.1. These goals are compared with both present and projected capabilities of several major laser types. It should be apparent from this comparison that the development of suitable laser drivers for inertial confinement fusion applications is a matter of considerable uncertainty.

There are several alternatives to using high-powered lasers as the "pistons" to drive the pellet implosion. Charged particle acceleration is an attractive technology because high energy electron or ion beams can be efficiently produced, possibly at high repetition rate. If short pulses of such beams can be focused onto tiny ICF fuel pellets, they should be able to deposit energy quite effectively to drive the implosion process.

Early charged particle drivers used pulsed diodes to produce relativistic electron beams.[32,33] These accelerators basically consist of a high voltage source that stores energy in capacitor banks and then rapidly switches this electrical energy into an insulated pulse-forming line and thence into a diode. Electrons are accelerated to the anode from a dense plasma that forms on the cathode surface. These electrons can then be passed through a foil and focused onto a target. Such relativistic electron beams have been used to implode fusion targets both in the United States and the Soviet Union.

However, early experiments indicated that such high energy electrons couple very inefficiently to the target. The range of relativistic electrons is too large to create adequate ablation pressure. Furthermore, the relativistic electrons produce hard X rays through breamsstrahlung that can penetrate into the target

Table 1.1. A Comparison of Projected Requirements Versus Actual Capabilities of Various Laser Drivers

	Projected Requirements	Present Capabilities			
		Nd	CO_2	I	KrF
Energy	300 to 500 J	30 kJ	20 kJ	2 kJ	100 J
	3 to 10 MJ (reactor)				
Focal spot size	1 mm	100 μm	100μm	100 μm	100 μm
Pulse length	0.1 to 10 ns	0.1 to 1 ns	1 ns	1 ns	1 ns
Repetition rate	1 to 10 Hz	10^{-3} Hz	1 to 10 Hz	10^{-3} Hz	10^{-3} Hz
Laser efficiency	10%	0.2%	5%	0.1%	1 to 5%
Wavelength	0.3 to ?	1.06 μm (0.26 to 0.53 μm with frequency multiplication)	10.6 μm	3 μm	0.3 μm

core, preheating the fuel. It is also difficult to focus electron beams on a target located at some standoff distance from the diode. The small mass of the electrons leads to strong space charge effects that tend to prevent the tight focusing necessary for high beam intensity on target.

Therefore in recent years the polarity of pulsed diode accelerators has been reversed so that they can be used to produce instead beams of high energy (1 to 10 MeV) light ions (ranging from protons to carbon ions).[34, 35] Such light ion beams couple relatively strongly to the target. Their much larger mass overcomes many of the focusing difficulties caused by space charge effects in relativistic electron beams. Furthermore, pulsed power accelerators appear to be capable of scaling at modest cost to the high power levels and potentially high efficiencies (20 to 30%) required for inertial confinement fusion applications. The major uncertainties in such light ion beam drivers involve beam transport and power concentration on the target.

The coupling of light ion beams to the target is certain to be nearly 100% efficient. However, this may still be insufficient to achieve high target gains. Furthermore, space charge and self-generated magnetic fields can still cause focusing problems. Hence recent interest has been directed at developing heavy ion beam accelerators based on RF or induction linear accelerators and storage rings.[36] Beams of heavy ions (Xe to U ions) with energies as large as 10 GeV can be focused very easily because their large mass (inertia) overcomes space charge repulsion. Heavy ion beams should be absorbed quite effectively by the target. The large energy of the heavy ions (in the GeV range) would permit the necessary power to be delivered to the target at much lower beam currents than those that characterize light ion beam drivers. Furthermore, the technology of high energy accelerators is quite highly developed, although experience with producing high beam currents of very heavy ions is essentially nonexistent. As with pulsed diode accelerators, present technology seems capable of scaling to the requisite power levels, efficiencies, and pulse rates. The major uncertainties involve beam transport and focusing. Furthermore, the large size and high cost of heavy ion accelerators have prevented the performance of target experiments in the absence of a major funding commitment. This stands in sharp contrast to light ion beam accelerators, which can be built relatively inexpensively. We have compared several of the advantages and disadvantages of various inertial confinement fusion driver types in Table 1.2.

The key concept in these inertial confinement fusion schemes is to use the driver (whether laser or charged particle beam) to ablate off the surface of the fuel pellet, thereby driving a rocketlike implosion of the fuel to high density. However, we will demonstrate in the next chapter that even with strong coupling of the beam to the target, most (90%) of the incident energy goes into the thermal and kinetic energy of the ablated material rather than the compressed fuel.

Hence there has been some interest in alternative inertial confinement fusion schemes that avoid the surface ablation process. One such scheme involves the use of hypervelocity particles as the "drivers" to produce high compressions.[37]

Table 1.2.   A Comparison of Advantages and Disadvantages of Various Inertial Confinement Fusion Driver Types

Driver Type	Advantages	Disadvantages
Lasers	Very high intensity Focusable to small spot size Advanced technology Versatile pulse length, frequency doubling	Low energy per pulse Inefficient Hot electron generation Energy absorption efficiency
Light ions	Large energy per pulse Classical deposition in matter No hot electrons Uses existing technology High efficiency Inexpensive—can be made small	Low intensity Uncertain focusability Uncertain beam propagation at required current
Heavy ions	Classical energy deposition Low current beams compared to light ions High efficiency High repetition rate Large energy per pulse	Very costly—unlikely to be small system Uncertain focusing and beam transport Transport requires hard vacuum

In such an approach, macroscopically-sized projectiles (typically several milligrams in size) would be accelerated to velocities of 100 to 300 km/s (perhaps by laser-driven ablation) and then allowed to collide with one another or on a target block in a reactor chamber to produce the densities and temperatures necessary for a thermonuclear fusion reaction.

A different approach involves a hybrid inertial/magnetic confinement scheme known as the imploding liner.[38,39] The general scheme is to discharge a very large current through a thin metal cylinder, liner, or array of wires (roughly 0.2 m initial radius, 3 mm initial thickness, and 0.2 m in length). The self-magnetic fields resulting from the current implode the liner at high velocities ($\sim 10^4$ m/s) onto a 0.5-keV, 10^{18} cm^{-3} D-T plasma that is initially formed or injected into the liner. As the liner implodes (in 20 to 40 μs), adiabatic compression raises the plasma to thermonuclear burn temperatures. During the implosion and subsequent burn, the fuel is confined inertially by the metal liner and endplug walls. The embedded magnetic field acts as an insulator against radial and axial thermal conduction. Between implosions the liner and several meters of adjacent electrical leads are replaced.

Many questions surround the imploding liner concept. The hydrodynamic stability of the liner implosion is one. Furthermore, there is a large impedance mismatch between the liner and the power source input line. Finally, the development of a refueling scheme in which both the liner and electrical leads could be rapidly replaced after every shot is a difficult challenge.

Let us set aside for the moment the question of driver type and turn to a consideration of how such thermonuclear microexplosions can be used to produce useful energy in some kind of reactor device. In a typical design the pellet implosion might be assumed to yield some 10^8 J (about 20 kg of high explosive worth of energy). If such explosions are repeated 30 times per second, then such a reactor would yield 3000 MW of thermal power corresponding to a steam thermal cycle electricity output of 1000 MW.[40-42]

The thermonuclear explosion energy appears as various types of radiation emitted from the exploding pellet. Typically the energy will appear as fast 14 MeV neutrons, energetic charged particles, and X rays. Surprisingly enough, it is relatively easy to design a blast chamber that can withstand the force of such a blast. The principal concern is the damage that the incident radiation can do to the chamber wall. However, by careful design—for example, by shielding the wall surface with a flowing liquid lithium curtain to absorb the X rays and charged particle debris—it should be possible to design a blast chamber to contain such pellet microexplosions.

Most of the explosion energy would be carried by fast neutrons, and therefore the blast chamber would be surrounded by a blanket, such as lithium, designed to absorb the neutron energy (and produce tritium for further refueling as well). This blanket could then be cooled using conventional techniques, and the heat withdrawn by a coolant would be used to produce steam for a turbine-generator. (See Figure 1.6.)

Inertial confinement fusion reactors can be contrasted with magnetic fusion systems in several important respects. First, the pulsed repetitive nature of the radiation from the microexplosions produces radiation environments and cyclic stresses that place particularly severe requirements on first-wall and blanket designs. However, balanced against this is the advantage that the driver is decoupled from the reactor environment. Furthermore, fusion chamber vacuum requirements are much less demanding in inertial confinement fusion systems, thereby allowing the use of liquid metals and/or buffer gases in first-wall protection schemes.

Two of the most important parameters influencing reactor designs are driver efficiency and target gain. These parameters are strongly coupled when applications of inertial confinement fusion to electric power generation are considered. For example, the 2 to 5% efficiency anticipated with short wavelength laser drivers such as the KrF laser would require very large target gains of 200 to 500. More efficient drivers such as light or heavy ion beams (10 to 50%) would reduce target gain requirements considerably to 20 to 100.[11]

Since inertial confinement fusion systems should be capable of producing large quantities of neutrons, it has been suggested that alternative uses of these devices may be of interest. (See Figure 1.7). The neutrons might be used to convert fertile material (e.g., uranium-238 or thorium-232) into fissile material (plutonium or uranium-233).[43-46] Or perhaps the neutrons could be used to

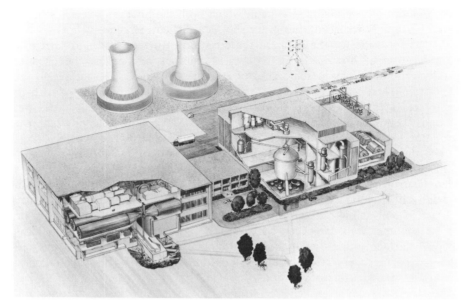

Figure 1.6. A conceptual laser fusion reactor. (Courtesy of the Lawrence Livermore National Laboratory.)

transmute long-lived radioactive waste (actinides) into shorter-lived or stable isotopes.[47] Yet another application would be to use the neutrons to radiolytically decompose water into hydrogen and oxygen, and then use the hydrogen in chemical processes to produce methane that can supplement our vanishing natural gas reserves.[48,49]

Whatever the application, it should be evident that the successful development of a viable inertial confinement fusion reactor is still many years down

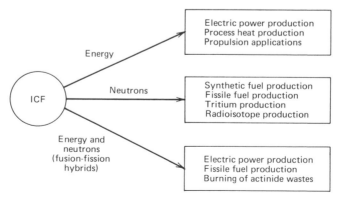

Figure 1.7. Various applications of inertial confinement fusion. (Courtesy of the Lawrence Livermore National Laboratory.)

the road. Indeed, we are still several years away from the demonstration of scientific feasibility (just as we are with magnetic confinement fusion approaches). But the promise of this particular approach for controlled thermonuclear fusion cannot be denied. The potential of inertial confinement fusion demands that we direct a major effort toward its development as a future energy source.

1.2 HISTORICAL DEVELOPMENT

The roots of inertial confinement fusion might be traced back to Bethe's recognition in 1931 that nuclear fusion was a primary energy source in stars, or perhaps to the development of the basic theory of thermonuclear fusion reactions by Teller, Fermi, Tuck, and others at Los Alamos during the 1940s.[51] Actually, the foundation for inertial confinement fusion was laid many years earlier by the hydrodynamic analysis of spherical bubble implosions (cavitation) by Besant in 1859 and Rayleigh in 1917.[52] Of particular interest was a self-similar solution to the problem of an imploding shock wave given by Guderley[53] in 1942. These ideas were applied to the design of nuclear fission weapons by Neddemeyer, Von Neumann, Teller, Tuck, Christy, and others at Los Alamos during the days of the Manhattan Project.[54] Moderately high compressions were achieved by using high explosives to drive spherical implosions. However, as we will demonstrate later, the maximum compressions that could be achieved using chemical explosives fall far short of those needed for inertial confinement fusion microexplosions.

As early as 1961 a Livermore scientist, John Nuckolls, realized that the powerful light beam of a pulsed laser could be used to achieve the energy densities necessary to produce very high compressions.[55] His early calculations (based on the laser pellet coupling physics developed by Ray Kidder[56,57] and Sterling Colgate) suggested that carefully tailored laser light pulses could produce ablatively driven implosions of D-T pellets to compressions as high as 10,000 times liquid-state density. Similar calculations were performed by others during the mid-1960s, including Kidder and Zabawski at Livermore, Dawson[58] at Princeton, Lubin at Rochester, Hertzberg, Daiber, and Wittcliff[59] at the Cornell Aeronautical Laboratory, Brueckner at the University of California, and Tuck and others at Los Alamos.

The calculations of Nuckolls and Kidder led to the initiation of a classified experimental laser fusion program at Livermore in 1963. By the mid-1960s Kidder and Mead had constructed a 12-beam ruby laser system to test the implosion calculations. During the late 1960s the development of high-powered neodymium glass lasers by the French and the rapid progress in CO_2 laser development by the Department of Defense accelerated the interest in laser fusion.

Experimental and theoretical analyses of laser-driven fusion continued to appear, both within the classified weapons program and the open literature. Of particular note were the activities of Lubin[60,61] at Rochester, Haught et al.[62-64] at United Aircraft, Basov et al.[65-71] at the Lebedev Institute, Bobin et al.[72] at Limeil, Caruso[73,74] in Italy, Witkowski et al.[75] in Germany, and Yamanaka et al.[76] in Japan. The Russian group reported the first indication of laser-produced fusion temperatures and neutrons in 1968.[77] Nuckolls and his colleagues continued to develop the theory of laser-driven implosions under the cloak of security classification at Livermore. These calculations suggested that adiabatic implosions might yield scientific breakeven at 1000 J of absorbed laser energy.

In 1972 the first major declassification of the implosion scheme occurred. Stimulated by references in the Russian literature suggesting inertial confinement fusion implosions to super high densities, Brueckner[21] at KMS Fusion, Nuckolls[24] and colleagues at Livermore, and Clark et al.[27] at Los Alamos simultaneously presented papers detailing the concept of using ablatively driven compression to produce implosions to superhigh density in D-T pellets.

The first major experimental results involved the implosion of a 100-micron-diameter CD_2 microsphere with a few-hundred-joule, nine-beam laser system by Basov's group at the Lebedev Institute.[70] This experiment yielded roughly 3×10^6 neutrons.

A second major milestone was reached in 1974 when KMS Fusion irradiated D-T-gas-filled glass microballoon targets using a two-beam laser system capable of 200 J in 100 ps and achieved detectable thermonuclear burn at compressions of roughly 100 times the gas-fill density.[29,78] The 10^4 neutrons produced in these experiments were verified to be of fusion origin.[79] Although only a few hundred ergs of energy were released in these implosions, a Lawson number of roughly 2×10^{12} and a D-T fuel temperature of 1 keV were achieved. By late 1974 KMS scientists were routinely producing pellet implosions yielding 10^6 to 10^7 neutrons per shot. In December of that year, Livermore began similar experiments on their JANUS laser system using a single beam at a power level of 0.2 TW. Subsequent experiments on the 2-beam, 0.4-TW JANUS system increased neutron yields by several orders of magnitude during 1975.[80] The 4-TW, 2-beam ARGUS laser system increased neutron yields to 10^9 to 10^{10} and ion temperatures to 10 keV by early 1976.

These early experiments were performed with D-T gas-filled glass microballoons which behaved in an exploding pusher mode. That is, the glass shell was heated and exploded by electron thermal conduction from the laser-heated plasma surrounding the target. The inward-moving shell or pusher acted as a piston which compressed the D-T gas to nearly the original pusher density (1 g/cm³) and produced high ion temperatures. However it rapidly became apparent that this type of target could never achieve the fusion energy gains needed for breakeven performance. Rather, the experiments would have to be redirected toward laser pulses and targets suited to the isentropic compression required for very high fuel compressions.

In 1976 the first experiments were begun on the ARGUS system at Livermore with impulsively driven targets in order to produce high fuel density (although at low fuel temperatures).[81] By 1978 with ARGUS operating at 2 kJ

Livermore had achieved 10 times liquid density, and by 1979 had announced compressions as high as 100 using the 10-kJ SHIVA laser system. The fuel temperatures in these fuel compressions were kept low (approximately 0.5 keV) to maximize fuel compression and provide only sufficient numbers of thermonuclear reactions for diagnostic purposes.[82-85]

The CO_2 laser program at Los Alamos followed a parallel track of success. By early 1977 implosions to fusion conditions were achieved using the 0.2-TW, two-beam GEMINI laser system. In 1978 the 10-kJ HELIOS system came on line, and experiments with impulsively driven targets were begun. Compressions as high as 30 have been reported to date.[86-88]

Although the first attempts to demonstrate inertial confinement fusion utilized high-powered lasers, interest in electron and ion beam drivers grew rapidly during the 1970s. The Russian electron beam fusion effort under Rudakov produced thermonuclear neutrons in 1976. A variety of electron and ion beam experiments were conducted at Sandia Laboratory during the late 1970s using the Proto I and Proto II pulsed diode accelerators. The presence of fusion neutrons was detected in these experiments.[32,33,34] The large Particle Beam Fusion Accelerator (PBFA-1) went into operation in 1980 at the 30-TW level.

During this same time period significant work was also underway at the Naval Research Laboratory in both the laser and light ion approaches to ICF. The very important laser-plasma coupling problem was studied with the two beam PHAROS Nd laser. Thin foils were accelerated using long laser pulses in the first experimental attempt at truly ablative acceleration. In the ion beam area, pioneering work was done in the analysis and eventual explanation of electron and ion diode behavior. Intense beams of light ions were produced and focused into plasma channels where they were propagated for over a meter in length.

A significant theoretical effort directed at heavy ion beam fusion was begun at a number of laboratories during the late 1970s, including the Lawrence Berkeley Laboratory, Argonne National Laboratory, and Brookhaven National

Table 1.3. The Parameters Characterizing Ignition of an Inertial Confinement Fusion Target

Definition of ignition requirements: thermonuclear energy deposited in fuel exceeds initial fuel thermal energy

Typical Parameters

D-T density	$\sim 1000 \times$ liquid density
ρR	> 0.3 g/cm^2
$n\tau$	$> 10^{14}$ cm^{-3} s
Temperature	5 keV $\rightarrow$ 10 keV (bootstrapping)
Burn efficiency	$> 1\%$
DT gain	~ 5
Target gain	~ 0.1

Laboratory.[36] A variety of approaches were studied that aimed at extrapolating existing high energy physics accelerator technology to the high beam intensities necessary for inertial confinement fusion driver applications.

A number of large laser and particle beam systems are now coming on line, both in the United States and elsewhere. Livermore, Los Alamos, and the University of Rochester have under development large laser systems at the 50 to 100 TW level (NOVA, ANTARES, and OMEGA). The Sandia Laboratory plans an upgrade of the PBFA machine (PBFA-II) to the 100-TW level by the mid-1980s. And the Soviet Union is continuing its electron beam fusion effort using the Angara 5 accelerator.[90]

The next step in the experimental programs is to achieve greater than 1000 times liquid density compressions coupled with the production of fusion temperatures to initiate thermonuclear burn (see Table 1.3). Scientific break-even experiments are projected for the middle to late 1980s at several laboratories.

1.3. STATUS

The inertial confinement fusion effort has evolved to the point that many laboratories, both in the United States and abroad, are now conducting vigorous research programs using a variety of driver and target designs.[91-93]

During the late 1970s a number of laboratories imploded D-T targets to high density using lasers or particle beam drivers operating in the several kJ, TW range. Most of these early implosion experiments used exploding pusher targets based on simple glass microballoons filled with D-T gas at high pressure. These targets have the advantage that they can yield relatively large numbers of neutrons (because of the high ion temperatures produced in the nonadiabatic implosion) with moderate scale drivers.

Unfortunately, exploding pusher target experiments do not address the primary technical questions of high compression implosions necessary for appreciable gain. (See Figure 1.8). Although thermonuclear ignition could be obtained with a sufficiently large exploding pusher target, the energy required to drive such a target is beyond the capabilities of any projected driver. Thus later experiments have turned instead to ablative targets in which the fuel is compressed to high density at relatively low temperatures (adiabatic compression). This requires careful driver pulse shaping, minimizing fuel preheat, and high symmetry implosions.

Four laboratories in the United States (Livermore, Los Alamos, KMS Fusion, and the University of Rochester) have performed laser-driven high density implosion experiments achieving compressions of from 1 to 100 times liquid D-T density (0.2 g/cm^3). These measurements are compared in Figure 1.9. Both the KMS Fusion and Rochester experiments were performed with modest energy (100 J or less), but took great care to achieve spherically symmetrical target illumination. The KMS Fusion experiments used glass microballoon targets filled with D-T gas that was then solidified into a thin

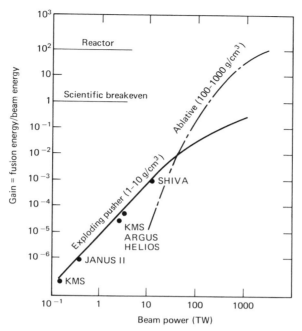

Figure 1.8. Target gains (both observed and projected) for exploding pusher and ablative designs. (After Brueckner, Ref. 93.)

shell using cryogenic methods. Compressions of up to 35 times liquid density have been achieved. Rochester also used glass microballoons as targets with their four-beam DELTA laser system to obtain compressions of several times liquid density. Compressions of as high as 100 have been achieved with targets of a classified design using the SHIVA laser at Livermore. At Los Alamos results have been achieved by irradiation of plastic-coated glass microballoons using the eight-beam HELIOS CO_2 laser system operating at 2300 J (1-ns pulse width).

Such experiments demonstrate that spherically symmetrical implosions to high density can be achieved, and provide some confidence for future experiments on larger 100-TW laser systems designed to initiate thermonuclear burn (NOVA and ANTARES). On a longer time scale, several laser systems are being proposed for the 300 to 500 TW level felt to be necessary to achieve scientific breakeven by the middle to late 1980s. Based on these experiments, more detailed plans can be made for the high gain experiments necessary for reactor applications.[92,93]

The laser fusion experiments of the 1970s achieved significant milestones. They demonstrated that high fuel compressions (50 to 100 times liquid density) and high implosion velocities can be achieved. They were able to exhaustively study the physics of a simple target design, the exploding pusher (D-T-filled glass microballoons). Unfortunately, these experiments also uncovered some

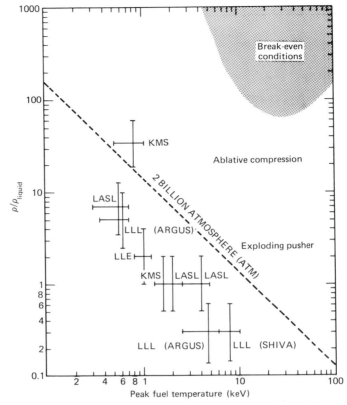

Figure 1.9. A comparison of target compressions. (After Ref. 91.)

serious problems in the area of coupling laser beam energy into the target. In particular, they found that at the high intensities required for laser fusion applications, a significant fraction of the incident light energy is coupled into high energy electrons. These fast electrons not only degrade target performance by preheating the fuel core, but also greatly complicate the theoretical analysis and understanding of the pellet implosion process. Other problems have included stimulated scattering of the incident beam, the hydrodynamic stability of pellet implosions, and a variety of problems in laser development. Although it appears that many of these problems are mitigated at shorter laser wavelengths (such as would characterize the KrF laser), they have nevertheless stimulated a gradual increase in the projections of laser-driver energies required for breakeven or reactor applications. At the present time, breakeven requirements are set at 300 to 500 kJ while significant gain (~ 100) is projected to require a driver energy of 2 MJ (to within a factor of 3).

Concern about inadequate coupling between incident laser light and the target has stimulated a rapid growth in the development of ion beams as ICF drivers. The advanced state of pulsed diode accelerator development has led to a rapid buildup in the light ion beam fusion program. In 1979 Sandia achieved proton beam power densities of 1 TW/cm^2 in experiments using the Proto-I accelerator. This provided confidence that the pellet ignition requirement of 100 TW/cm^2 could be achieved by overlapping a large number of such ion beams.[34] The Particle Beam Fusion Accelerator started target experiments with 36 light ion beam modules in 1981 at the 30-TW level (with several hundred kilojoules on target). This machine's output is scheduled to be increased to 100 TW (72 beams) by the mid-1980s.

Serious studies of heavy ion drivers are also under way. Both RF linear accelerators with charged particle storage rings and linear induction accelerators are being examined. Unfortunately, the cost of even a modest heavy-ion-driven fusion experiment is very high[93] (about 10 times that of a comparable light ion beam experiment), and a major commitment to this driver type must await further detailed studies.

Any major change in driver, from lasers to light or heavy ion beams, necessitated by target interaction physics is likely to have a serious impact on the progress of the inertial confinement fusion program, possibly delaying the demonstration of scientific feasibility (breakeven) until the late 1980s or early 1990s. An inertial confinement fusion test facility capable of demonstrating feasibility is now projected to cost upward of $500 million, regardless of driver type. The uncertainty of target designs and possible driver configurations also complicates an assessment of practical applications of this technology. In fact, the interrelation of driver, target design, and reactor design is so complex that it is impossible to say whether or not an economical inertial confinement fusion reactor can be designed, even if driver-target physics problems are solved.

Despite this uncertainty, there is still strong support for continued research and development on inertial confinement fusion for energy applications. In 1979 the Foster Committee[92] concluded that "inertial confinement fusion shows excellent promise of succeeding eventually in civilian (energy) applications. We can see no insurmountable technical difficulties."

Table 1.4. A Summary of Inertial Confinement Fusion Driver Status

	Lasers	Electron Beams	Light Ion Beams	Heavy Ion Beams
Beam particles	Photons	Electrons	p, α, C^{+4}	Xe-U
	(0.2–10 μm)	(1–10 MeV)	(1–10 MeV)	(1–50 GeV)
Power	30 TW	30 TW	30 TW	2 TW
	(Shiva)	(PBFA-I)	(PBFA-I)	(ISR)
Energy	20 kJ	1000 kJ	1000 kJ	6 MJ
	(Shiva)	(PBFA-I)	(PBFA-I)	
Seriously proposed	300 TW/300 kJ	100 TW/4000 kJ	100 TW/4000 kJ	600 TW/10 MJ
	(Nova)	(PBFA-II)	(PBFA-II)	

1.4. SCOPE OF THE TEXT

This text is intended to serve as an introduction to inertial confinement fusion. In any such field that is as yet so far removed from practical applications, there are a great many uncertainties that tend to influence a selection of topics. Primary among these is a choice of driver. Although early investigations have emphasized laser fusion, more recent efforts have shifted attention to ion beams. We have chosen to deal with the uncertainty surrounding a final choice of driver by keeping the first half of the text as general as possible, describing inertial confinement fusion physics in a manner independent of driver type.

A second uncertainty is somewhat more difficult to deal with. This concerns the mists of security classification that tend to veil certain aspects of inertial confinement fusion because of the presumed similarity to nuclear weapons physics. Fortunately, the classification veil is rather narrow and involves only certain aspects of target design. Hence most of the physics of inertial confinement fusion, including the details of driver design, driver energy deposition, target dynamics, and many aspects of inertial confinement fusion target design have appeared in the open literature and will be discussed in this text.

In line with our effort to keep our development of inertial confinement fusion concepts as independent of driver type as possible, we will adopt a pedagogy of working backward: from the details of inertially confined thermonuclear burn, to hydrodynamic implosions and compression, to energy transport in dense plasmas, and finally to a detailed consideration of driver energy deposition in the target. It is only at this last stage that we will need to introduce particular details of driver types. We will devote particular attention to the theoretical, computational, and experimental tools used to analyze the behavior of inertial confinement fusion targets.

We then turn our attention to the inertial confinement fusion drivers and consider laser and charged particle beam drivers in detail. Here we not only discuss present driver types, but also attempt to look ahead at possible future driver types intended for advanced applications.

Our final topic is applications of inertial confinement fusion. After a brief discussion of target design, we turn to a detailed discussion of applications, with particular attention devoted to power production (including the production of process heat and synthetic fuels).

REFERENCES

1. A. S. Bishop, *Project Sherwood: The United States Program in Controlled Fusion* (New York, Doubleday, Anchor Books, 1960).

2. S. Glasstone and R. H. Lovberg, *Controlled Thermonuclear Reactions* (New York, Van Nostrand, 1960).

3. L. A. Arzimovich, *Elementary Plasma Physics* (Waltham, Mass., Blaisdel, 1963).

4. R. F. Post and R. L. Ribe, "Fusion Power," *Science* **186**, 397 (1974).

5. J. P. Holdren, "Fusion Energy in Context: Its Fitness for the Long Term," *Science* **200**, 168 (1978).

6. G. L. Kulcinski, G. Kessler, J. P. Holdren, and H. Hafele, "Energy for the Long Run: Fission or Fusion?" *Am. Sci.* **67**, 78 (1979).

7. H. P. Furth, "Progress Toward a Tokamak Fusion Reactor," *Sci. Am.* **241**, 50 (August, 1979).

8. J. D. Lawson, "Some Criteria for a Power Producing Thermonuclear Reactor," *Proc. Phys. Soc. London* **B70**, 6 (1957).

9. J. L. Emmett, J. Nuckolls, and L. Wood, "Fusion Power by Laser Implosion," *Sci. Am.* **231**, 24 (June, 1974).

10. C. H. Stickley, "Laser Fusion," *Phys. Today* 50 (May, 1978).

11. J. A. Maniscalco, "Inertial Confinement Fusion," *Ann. Rev. Energy* **5**, 33 (1980).

12. E. Teller, "A Future ICE (Thermonuclear, That Is!)," *IEEE Spectrum* 60 (January, 1973).

13. L. A. Booth, D. A. Freiwald, T. G. Frank, and F. T. Finch, "A Laser Fusion Reactor Design," *Proc. IEEE* **64**, 1460 (1976).

14. "Laser Fusion: An Energy Option, But Weapons Simulation Is First," *Science* **188**, 30 (1977).

15. A. J. Toepfer and L. D. Posey, "The Applications of Inertial Confinement Fusion to Weapons Technology," Sandia Laboratories Report 77-0913 (1978).

16. E. T. Hayes, "Energy Resources Available to the United States, 1985–2000," *Science* **203**, 233 (1979).

17. K. M. Siegel, "The Energy Crisis and a Potential Laser-Fusion Solution," *J. Appl. Sci. Eng.* **1**, 3 (1975).

18. J. Duderstadt and C. Kikuchi, *Nuclear Power: Technology on Trial* (Ann Arbor, University of Michigan Press, 1979).

19. "Energy in Transition, 1985–2010," Report of the Commission on Nuclear and Alternative Energy Sources (CONAES), National Research Council, National Academy of Sciences, Washington, D. C. (1980).

20. A. Fraas and M. Lubin, "Fusion by Laser," *Sci. Am.* **225**, 21 (June, 1971).

21. K. Brueckner and S. Jorna, "Laser Driven Fusion," *Rev. Mod. Phys.* **46**, 325 (1974).

22. K. Boyer, "Laser Fusion," *Aero. and Astro.* **11**, 44 (1973).

23. G. S. Fraley, E. J. Linnebur, R. J. Mason, and R. L. Morse, "Thermonuclear Burn Characteristics of Compressed Deuterium-Tritium Microspheres," *Phys. Fluids* **17**, 474 (1974).

24. J. Nuckolls, L. Wood, A. Thiessen, and G. Zimmerman, "Laser Compression of Matter to Super-High Densities: Thermonuclear (CTR) Applications," *Nature* **239**, 139 (1972).

25. J. L. Emmett, J. Nuckolls, and L. Wood, "Fusion Power by Laser Implosions," *Sci. Am.* **231**, 24 (June, 1974).

26. K. A. Brueckner, "Introduction to Laser Driven Fusion," *Energy*, Vol. 3, *Nuclear Energy and Energy Policies*, Ch. 23, ed. by S. S. Penner (Reading, Mass., Addison-Wesley, 1976), p. 349.

27. J. S. Clarke, H. N. Fisher, and R. J. Mason, "Laser Driven Implosion of Spherical DT Targets to Thermonuclear Burn Conditions," *Phys. Rev. Lett.* **30**, 89 (1974).

28. J. Nuckolls, J. Emmett, and L. Wood, "Laser Induced Thermonuclear Fusion," *Phys. Today* 46 (August, 1976).

29. P. M. Campbell, G. Charatis, and G. R. Montry, "Laser-Driven Compression of Glass Microspheres," *Phys. Rev. Lett.* **34**, 74 (1975).

30. *Laser Program Annual Reports*, Lawrence Livermore National Laboratory, UCRL-50021, 1973 to present.

31. *Laser Fusion Program Progress Reports*, Los Alamos National Scientific Laboratory, 1974 to present.

32. G. Yonas, "Fusion Power With Particle Beams," *Sci. Am.* **239**, 48 (November, 1978).

33. G. Yonas, Ed., *Particle Beam Fusion Program—Publications and Related Reports: A Bibliography*, Sandia National Laboratory Report SAND-80-0809 (1980).

34. *Particle Beam Fusion Program Reports*, Sandia Laboratory, Albuquerque, 1976 to present.

35. G. Cooperstein et al., "NRL Light Ion Beam Research for Inertial Confinement Fusion," NRL Memo Report 4387, Nov. 1980.

36. R. C. Arnold, "Heavy Ion Beam Inertial Confinement Fusion," *Nature* **276**, 19 (1978).

37. A. I. Peaslee, Ed., *Proc. Impact Fusion Workshop*, Los Alamos, NM (July 1979), U.S. Department of Energy Document (NTIS) CONF-79074 (1979).

38. A. Robson, "A Conceptual Design for an Imploding-Linear Fusion Reactor (LINUS)," NRL Memo Report 3861, Sept. 1978.

39. S. I. Anisimov et al., "Generation of Neutrons as a Result of Explosive Initiation of the D-D Reaction in Conical Targets," *JETP Lett.* **31**, 61 (1980).

40. L. A. Booth, D. A. Freiwald, T. G. Frank, and F. T. Finch, "A Laser Fusion Reactor Design," *Proc. IEEE* **64**, 1460 (1976).

41. G. A. Moses, R. W. Conn, and S. I. Abdel-Khalik, "The SOLASE Conceptual Laser Fusion Reactor Study," *Proc. Third Topical Meeting on the Technology of Controlled Nuclear Fusion*, Santa Fe, NM (1978).

42. J. Maniscalso, J. A. Blink, J. Hovingh, W. R. Meier, M. J. Monsler, and P. E. Walker, "A Laser Fusion Power Plant Based on a Fluid Wall Reactor Concept," *Proc. Third Topical Meeting on the Technology of Controlled Nuclear Fusion*, Sante Fe, NM (1978).

43. B. R. Leonard, "A Review of Fission-Fusion Hybrid Concepts," *Nucl. Technol.* **20**, 161 (1973).

44. L. Lidsky, "Fission-Fusion Systems: Hybrid, Symbiotic, and Augean," *Nucl. Fusion* **15**, 151 (1975).

45. W. D. Metz, "Fusion Research. III. New Interest in Fusion Assisted Breeders," *Science* **193**, 307 (1976).

46. Robert McGrath, "A Suppressed Fission Hybrid Reactor for Producing Fissile Fuel," submitted for publication to *Nucl. Technol.*, 1980.

47. D. R. Berwald and J. J. Duderstadt, "Preliminary Design and Neutronic Analysis of a Laser Driven Fusion Actinide Waste Burning Hybrid Reactor," *Nucl. Technol.* **42**, 34 (1978).

48. H. J. Gomberg and W. W. Meinke, "Production of Synthetic Fuels: An Important Civilian Application of Laser Fusion," *Miami International Conference on Alternative Energy Sources*, ed. by T. N. Veziroglu, Miami, FL (1977).

49. H. I. Avci, K. D. Kok, R. G. Jung, and R. C. Dykheizer, "Production of High Temperature Process Heat in Pebble Beds in ICTR Blankets," *Trans. Am. Nucl. Soc.* **32**, 39 (1979).

50. H. Bethe, "Energy Production in Stars," *Phys. Rev.* **55**, 434 (1939).

51. R. G. Hewlett, *The Atomic Shield* (University Park, Pennsylvania State University Press, 1972).

52. Lord Rayleigh, "Investigation of the Character of the Equilibrium of an Incompressible Heavy Fluid of Variable Density," *Scientific Papers*, II, Cambridge, England (1900), pp. 200–207.

53. G. Guderley, "Starke Kugelige und Zylindriche Verdichtungsstösse in der Nahe des Kugel-mittelpunktes bzw. der Zylinderachse," *Luftfahrforschung* **19**, 302 (1942).

54. R. G. Hewlett and O. E. Anderson, Jr., *The New World, 1939/1946* (University Park, Pennsylvania State University Press, 1962).

55. L. Wood and J. Nuckolls, "Prospects for Unconventional Approaches to Controlled Fusion," AAAS Philadelphia Meeting (December, 1971).

56. R. E. Kidder, "Some Aspects of Controlled Fusion by Use of Lasers," UCRL-73500, Symposium on Fundamental and Applied Physics, Esfahan, Iran, August, 1971.

57. R. E. Kidder, "Interaction of Intense Photon Beams With Plasmas (II)," UCRL-74040, presented at the Japan–United States Seminar on Laser Interaction With Matter, Kyoto, Japan, September, 1972.

58. J. M. Dawson, "On the Production of Plasma by Giant Pulse Lasers," *Phys. Fluids* **7**, 981 (1964).

59. J. W. Daiber, A. Hertzberg, and C. E. Wittliff, "Laser-Generated Implosions," *Phys. Fluids* **9**, 617 (1966).

60. M. Lubin and A. Fraas, "Fusion by Laser," *Sci. Am.* **225** (June, 1971).

61. J. Soures, L. M. Goldman, and M. Lubin, "Short Pulse Laser Heated Plasma Experiments," *Nucl. Fusion* **13**, 829 (1973).

62. A. F. Haught and D. H. Polk, "Plasmas for Thermonuclear Research Produced by Laser Beam Irradiation of Single Solid Particle," *Proc. Conference on Plasma Physics and Controlled Nuclear Fusion Research*, Culham (1965) (Vienna, International Atomic Energy Agency, 1966) p. 219.

63. W. J. Fader, "Hydrodynamic Model of Spherical Plasma Produced by a Q-Spoiled Laser Irradiation of a Solid Particle," *Phys. Fluids* **10**, 2200 (1968).

64. H. F. Haught and D. H. Polk, "Formation and Heating of Laser Irradiated Solid Particle Plasmas," *Phys. Fluids* **13**, 2825 (1970).

65. N. G. Basov and O. H. Krokhin, "The Conditions of Plasma Heating by the Optical Generation of Radiation," *Proc. Third International Congress on Quantum Electronics* (New York, Columbia University Press, 1964) p. 1373.

66. O. N. Krokhin, "Self-Regulating Regime of Plasma Heating by Laser Radiation," *Z. Angew. Math. u. Phys.* **16**, 123 (1965).

67. Yu. P. Raizer, "Heating of a Gas by a Powerful Light Pulse," *Sov. Phys. JETP* **21**, 1009 (1965).

68. N. G. Basov, V. A. Boiko, V. A. Demetev, O. N. Krokhin, and G. V. Sklizkov, "Heating and Decay of Plasma Produced by a Giant Laser Pulse Focused on a Solid Target," *Sov. Phys. JETP* **24**, 659 (1967).

69. N. G. Basov, V. A. Gribkov, O. N. Krokhin, and G. V. Sklizkov, "High Temperature Effects of Intense Laser Emission Focused on a Solid Target," *Sov. Phys. JETP* **27**, 575 (1968).

70. N. G. Basov, V. A. Boiko, S. M. Zakharov, O. H. Krokhin, and G. V. Sklizkov, "Generation of Neutrons in a Laser CD-2 Plasma Heated by Pulses of Nanosecond Duration," *ZhETF Pis. Red.* **13**, 691 (1971).

71. N. Basov, O. H. Krokhin, and G. V. Sklizkov, "Heating of Laser Plasmas for Thermonuclear Fusion," *Proc. Second Workshop on Laser Interaction and Related Plasma Phenomena*, ed. by H. Hora and H. Shwarz (1971).

72. F. Floux, D. Cognard, L. G. Denoeud, G. Piar, D. Parisot, J. L. Bobin, F. Delobeau, and C. Fauquignon, "Nuclear Fusion Reactions in Solid Deuterium Laser-Produced Plasma," *Phys. Rev.* **A1**, 821 (1970).

73. A. Caruso, B. Bertotti, and P. Guipponi, "Ionization and Heating of Solid Material by Means of a Laser Pulse," *Nuovo Cimento* **B45**, 176 (1966).

74. A. Caruso and R. Gratton, "Some Properties of the Plasmas Produced by Irradiating Solids by Light Pulses," *Plas. Phys.* **10**, 867 (1968).

75. P. Mulser and S. Witkowski, "Numerical Calculations of the Dynamics of a Laser Irradiated Solid Hydrogen Foil," *Phys. Lett.* **A28**, 703 (1969).

76. C. Yamanaka, T. Yamanaka, T. Sasaki, K. Yoshida, M. Waki, and H. B. Kang, "Anomalous Heating of a Plasma by Lasers," *Phys. Rev.* **A6**, 2335 (1972).

77. N. G. Basov, P. G. Kryokov, S. D. Zakharov, Yu. V. Senatskiy, and S. V. Chekalin, "Experiments on the Observation of Neutron Emission at the Focus of High-Power Laser Radiation on a Lithium Deuteride Surface," *IEEE J. Quantum Electron.* **QE-4**, 864 (1968).

78. G. Charatis, J. Downward, R. Goforth, B. Guscott, T. Henderson, S. Hildum, R. Johnson, K. Moncur, T. Leonard, F. Mayer, S. Segall, L. Siebert, D. Solomon, and C. Thomas, "Experimental Study of Laser Driven Compression of Spherical Glass Shells," *Plasma Physics and Controlled Thermonuclear Fusion* (Vienna, International Atomic Energy Agency, 1974).

79. G. F. McCall, F. Young, A. W. Ehler, J. F. Kephardt, and R. P. Godwin, "Neutron Emission from Laser-Produced Plasmas," *Phys. Rev. Lett.* **30**, 1116 (1973).

80. V. W. Slivinsky, H. G. Ahlstrom, K. G. Tirsell, J. Larsen, S. Glaros, G. Zimmerman, and H. Shay, "Measurement of the Ion Temperature in Laser Driven Fusion," *Phys. Rev. Lett.* **35**, 1083 (1975).

81. D. R. Speck et al., "The Performance of Argus as a Laser Fusion Facility," UCRL-79816, presented at the Eleventh European Conference on Laser Interaction With Matter, Oxford, England (1977).

82. D. R. Speck et al., "Performance of the Shiva Laser Fusion Facility," Lawrence Livermore Laboratory Report UCRL 82117, presented at the 1979 IEEE Conference on Laser Engineering and Applications, Washington, D.C. (1979).

83. T. J. Gilmartin, "Nova, the Laser Fusion Scientific Feasibility Experiment," Lawrence Livermore Laboratory Report UCRL-82094, presented at the 1979 IEEE conference on Laser Engineering and Applications, Washington, D.C. (1979).

84. H. G. Ahlstrom, "Progress of Laser Fusion at Lawrence Livermore Laboratory," Lawrence Livermore Laboratory Report UCRL-82835 Rev 1 (1979).

85. K. R. Manes and J. A. Glaze, "Recent Inertial Confinement Fusion Results From the SHIVA Target Irradiation Facilities," Lawrence Livermore Laboratory Report UCRL-83274 (1979).

86. C. A. Fenstermacher, M. J. Nutter, W. T. Leland, and K. Boyer, "Electron Beam Controlled Electrical Discharge as a Method of Pumping Large Volumes of CO-2 Laser Media at High Pressure," *Appl. Phys. Lett.* **20**, 56 (1972).

87. K. B. Mitchell, D. B. Van Hulsteyn, G. H. McCall, P. Lee, and H. Greim, "Compression Measurements of Neon Filled Glass Microballoons Irradiated by CO_2 Laser Light," *Phys. Rev. Lett.* **42**, 232 (1979).

88. R. B. Perkins, "Recent Progress in Inertial Confinement Fusion Research at the Los Alamos Scientific Laboratory," Los Alamos Scientific Laboratory Report LA-UR-78-1629 (1978).

89. W. D. Metz, "Energy Research: Accelerator Builders Eager to Aid Fusion Work," *Science* **194**, 307 (1976).

90. S. V. Basenkov et al., "Accelerator Module of ANGARA-5," IEEE Pulsed Power Conference Lubbock, Texas (1979), U.S. Department of Energy Document (NTIS) CONF-790622 (1979).

91. "Report of the Panel on High Density Compression Experiments," W. J. Shafer Associates, United States Department of Energy Report WJSA-78-6-SR7 (1978).

92. "Final Report of the Ad Hoc Experts Group on Fusion" (The Foster Committee), United States Department of Energy Report DOE/ER-0008 (Washington, 1978).

93. K. Brueckner, "An Assessment of Drivers and Reactors for Inertial Confinement Fusion," K. A. Brueckner Associates, prepared for the Electric Power Research Institute, EPRI-AP-1371 (1980).

TWO

Inertially Confined Thermonuclear Fusion Reactions

The dynamical behavior of a burning thermonuclear fuel depends on many complex phenomena. These include charged particle collision processes, nuclear fusion reaction kinetics, the hydrodynamic behavior of the fuel (including both its motion and its temperature), the production and transport of radiation, and the transport of nonthermal particles such as reaction products. The relative importance of these phenomena differs significantly for high density, inertially confined fusion fuels compared to the relatively dilute thermonuclear plasmas characterizing magnetic fusion systems.

In this chapter we focus our attention on the dynamics of the nuclear fusion reaction itself, postponing a detailed consideration of hydrodynamic and energy transport processes in inertial confinement fusion to later chapters. Of particular importance is a discussion of the various nuclear fusion processes of most interest in ICF applications, the criteria for efficient thermonuclear ignition and burning of fuels, and the important processes that determine the fuel gain (i.e., the ratio of the fusion energy produced by the fuel and the driver energy necessary to ignite the fusion burn).

2.1. FUSION REACTION PHYSICS

At sufficiently high temperatures there are many possible candidates for fusion fuels. Although most first generation fusion concepts are based on D-T or D-D fusion reactions, other fuel cycles based either on deuterium fuels (e.g., D-^{3}He and D-^{6}Li) or proton fuels (p-^{6}Li and p-^{11}B) become attractive alternatives for advanced fusion systems.

All such fusion reactions are binary in the sense that two-body collision processes are involved. The fusion reaction represents a barrier penetration phenomenon in which the colliding nuclei penetrate the repulsive Coulomb barrier to within the range of attractive nuclear forces.[1] For example, the nuclear fusion cross section for the D-T reaction has a resonance near zero energy, reaching a peak of about 5 barns at 125 keV. More generally, the energy dependence of the fusion cross section can be roughly described by Gamow's barrier penetration model as

$$\sigma(E) \sim \frac{A}{E} \exp\left(\frac{-B}{E^{1/2}}\right) \tag{2.1}$$

where E is the energy available to the collision in the center of mass frame, and A and B are constants characterizing the particular fusion reaction species.

In a thermonuclear fusion process, the fuel ions are confined in such a way that they collide with one another millions and millions of times at high temperature, scattering about until a fusion reaction occurs. Hence the rate at which such fusion reactions occur involves an average over the velocity distributions of the participating species. More precisely, if n_A is the ion density of species A while n_B is the density of species B, then the rate at which fusion reactions occur is given by

$$R_{AB} = n_A n_B \langle v \sigma_{AB} \rangle \tag{2.2}$$

where $\langle \cdots \rangle$ indicates an average over the velocity distributions of both species[2, 3]:

$$\langle v\sigma \rangle = \frac{1}{n_A n_B} \int d^3 v_A \int d^3 v_B v_{\text{rel}}\, \sigma(v_{\text{rel}}) N_A(\mathbf{v}_A) N_B(\mathbf{v}_B) \tag{2.3}$$

where $v_{\text{rel}} = |\mathbf{v}_A - \mathbf{v}_B|$. In most cases, we assume that these distributions are those characterizing a plasma fuel in thermal equilibrium at a temperature T, that is, by a Maxwell-Boltzmann distribution

$$N(\mathbf{v}) = M(\mathbf{v}) = n\left(\frac{1}{2\pi mkT}\right)^{3/2} \exp\left(-\frac{mv^2}{2kT}\right) \tag{2.4}$$

We can then use analytic expressions for the cross section such as the Gamow formula Eq. 2.1 or more precise tabulated data to perform the integration and determine the Maxwellian-averaged reaction rate parameter $\langle v\sigma \rangle$ for various fuel temperatures T. Plots of this parameter versus temperature have been provided for the reactions of most interest in fusion applications in Figure 2.1.[4-9]

Two comments concerning the Maxwellian-averaged fusion reaction rates are appropriate here. Although the fusion cross sections peak at relatively high particle collision energies, the Maxwellian-averaged collision rate parameters

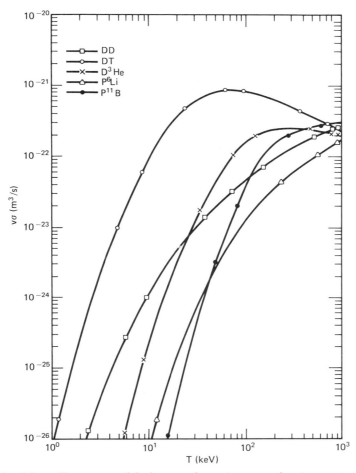

Figure 2.1. Maxwellian-averaged fusion reaction rates versus ion temperature for the principal isotopes of interest to inertial confinement fusion applications.

rise to appreciable values at far lower temperatures (e.g., 10 keV in the case of the D-T reaction). The reason for this is that the high energy particles in the Maxwellian distribution provide most of the contribution to the reaction rate. That is, the particles in the tail of the distribution function with energies many times that of thermal energy control the rate of the fusion reaction.

The second important comment concerns the relative magnitude of the reaction rates for various fusion reactions. It is apparent from Figure 2.1 that for lower fuel temperatures (10 keV), the D-T reaction proceeds at a rate almost two orders of magnitude larger than that characterizing the D-D reaction (or any other species, for that matter). Hence it is not surprising that this particular reaction has received most attention in fusion research.

2.1.1.   D-T AND D-D FUSION REACTIONS

Of most interest in early fusion concepts are the deuterium-based fusion reactions, D-T, D-D, and D-^{3}He, since these are characterized by appreciable reaction rates at temperatures below 100 keV. The largest reaction rate at low temperatures is provided by the D-T reaction:

$$D + T \rightarrow \alpha(3.5 \text{ MeV}) + n(14.1 \text{ MeV})$$

where the 17.6-MeV reaction energy (the "Q value") is partitioned between a 3.5-MeV alpha particle ($^4_2$He) and a 14.1-MeV neutron. From Figure 2.1 we find that the reaction rate for the D-T process has a broad maximum between temperatures of 20 and 80 keV.

The next most probable reaction at low temperatures is the D-D fusion reaction. It is characterized by two branches of approximately equal probability:

$$D + D \underset{\searrow}{\overset{\nearrow}{}} \begin{array}{l} {}^3_2\text{He}(0.82 \text{ MeV}) + n(2.45 \text{ MeV}) \\ {}^3_1\text{T}(1.01 \text{ MeV}) + p(3.02 \text{ MeV}) \end{array}$$

The reaction products, $^3_2$He and $^3_1$T, have a high probability of reacting with deuterium as they slow down:

$$^3_1\text{T*} + {}^2_1\text{D} \rightarrow {}^4_2\text{He}(3.5 \text{ MeV}) + n(14.1 \text{ MeV})$$

$$^3_2\text{He*} + {}^2_1\text{D} \rightarrow {}^4_2\text{He}(3.6 \text{ MeV}) + p(14.7 \text{ MeV})$$

where the asterisk denotes a nonthermal particle. In fact, the largest fraction of the overall Q-value is a result of such secondary reactions. The total reaction

$$6D \rightarrow 2\, {}^4_2\text{He} + 2p + 2n$$

has a Q-value of 43.25 MeV. This gives a specific yield of 345 MJ/mg for the D-D reaction as compared to 339 MJ/mg for D-T.

It should be noted that a D-T fuel mixture will always be accompanied to some degree by D-D and D-^{3}He reactions. More specifically, we can write down the rate equations for a D-T thermonuclear fuel mixture as

$$\frac{dn_\text{D}}{dt} = -n_\text{D}n_\text{T}\langle v\sigma_\text{DT}\rangle - n_\text{D}^2\langle v\sigma_\text{DD}\rangle$$

$$\frac{dn_\text{T}}{dt} = -n_\text{D}n_\text{T}\langle v\sigma_\text{DT}\rangle \tag{2.5}$$

Since the D-D reaction produces one triton T for every two fusion reactions, we might expect that the equation for n_T should also contain a source term of the form $n_\text{D}^2\langle v\sigma_\text{DT}\rangle$. However here we must remember that the tritium ion appears with an energy of roughly 1 MeV. Thus it is not in thermal equilibrium with the tritium fuel, and its contribution to the reaction rate must be treated separately (in effect, as a "beam plasma" interaction with an enhanced reaction rate).

It should be apparent from Figure 2.1 that the D-T reaction will dominate if the fuel contains more than a few percent tritium. However, if the fuel can be brought to temperatures where the D-D reaction will contribute, then the products T and ^{3}He will react promptly with the D nuclei, yielding an overall energy release per unit burned mass that is the same as that for the D-T reaction, namely 3.5 MeV per nucleon (assuming that the neutron reaction product escapes the reaction region without appreciable energy loss).

Since tritium is a radioactive nuclide (with a half-life of 12.3 years), it is not naturally occurring but rather must be produced artificially to fuel a fusion system. Fortunately, tritium can be produced quite easily by using neutron capture reactions in lithium:

$$n + {}^6_3\text{Li} \rightarrow \text{T} + {}^4_2\text{He} + 4.8 \text{ MeV}$$

$$n + {}^7_3\text{Li} \rightarrow \text{T} + {}^4_2\text{He} + n - 2.4 \text{ MeV}$$

The first process is a capture reaction involving neutrons of any energy. The second process is effectively an inelastic scattering reaction requiring a neutron energy in excess of the 2.4-MeV threshold. Since each D-T reaction produces one neutron, it is possible to breed tritium by surrounding the fusion reaction region with a lithium blanket.

2.1.2. ADVANCED FUSION FUELS

The first fusion devices to achieve energy break-even and then commercialization as reactors will undoubtedly be based on the D-T fuel cycle. However, the ultimate goal is to eventually move toward advanced fuels that minimize radioactivity and are based on naturally occurring fuel nuclides. Examples include the deuterium-based fuel cycles D-D, D-^{3}He, and D-^{6}Li and the proton-based fuel cycles p-^{11}B and p-^{6}Li. (See Table 2.1.)[10-12]

More specifically consider the dominant reactions characterizing the proton-based fuel cycles:

$$p + {}^{11}\text{B} \begin{cases} \nearrow \alpha + {}^8\text{Be}(8.590 \text{ MeV}) \\ \searrow 3\alpha(8.682 \text{ Mev}) \end{cases}$$

and

$$p + {}^6\text{Li} \rightarrow \alpha + {}^3\text{He}(4.023 \text{ MeV})$$

In both cases the fuel isotopes are naturally occurring. Furthermore, the reaction products are charged and therefore would be confined in the reaction region of the fuel, unlike the high energy neutrons produced in the D-T reactions (and to a lesser extent in the D-D reaction). Note that the p-^{11}B reaction might almost be termed a thermonuclear "fission" reaction since it

Table 2.1. Thermonuclear Reactions for Advanced Fuels

Fuel	Reaction Products (MeV)	Q Value (MeV)
(D, T)	$_2^4$He(3.5), $_0^1 n$(14.1)	17.6
(D, D)	$_2^3$He(.82), $_0^1 n$(2.45)	3.65
	$_1^3$T(1.01), $_1^1$H(3.02)	
(D, He^3)	$_2^4$He(3.6), $_1^1$H(14.7)	18.3
(D, Li^6)	$2\,_2^4$He(22.4)	22.4
(P, Li^7)	$2\,_2^4$He(22.4)	22.4
(P, B^{11})	$3\,_2^4$He(8.682)	8.682

breaks the ^{11}B nucleus into three alpha particles emitted with a continuum energy spectrum, much as in beta decay.

The advantages of such advanced fuel cycles are significant. The fuels involved in the p-^{11}B or p-^{6}Li cycles are sufficiently abundant to ensure an inexhaustible fuel supply. The absence of tritium in the fuel cycle eliminates the difficulties involved in breeding and managing this radioactive material. The effective elimination of neutron production from the fuel cycle eliminates the difficulties posed by radiation damage. Hence the major environmental and safety drawbacks of D-T nuclear fusion reactors are eliminated in these advanced fuel cycles.

Unfortunately, the reaction rates for the proton-based fuel cycles do not become appreciable until high fuel temperatures can be achieved (ion temperatures of roughly 300 keV). A detailed plot of reaction rate parameter $\langle v\sigma \rangle$ versus ion temperature for various advanced fuel cycles is provided in Figure 2.1. New effects become important such as fast (nonthermal) fusion, nuclear elastic and inelastic scattering, and the Doppler broadening of cross sections due to the energy distribution of reaction products. We should note that recent interest in the p-^{11}B reaction as an advanced fusion fuel cycle has stimulated a more careful evaluation of its fusion cross section. The resulting average reaction rate parameter is now felt to lie roughly 65% lower (at the ignition temperature of 300 keV) than the earlier data shown in Figure 2.1.[13]

Thus, while advanced fuel cycles certainly exhibit the desirable characteristics of minimum radioactivity and inexhaustible fuel supply, they do require far higher fuel temperatures (and/or fuel densities). The implications of these requirements for the design of advanced inertial confinement fusion fuels will become more apparent in the next section.

2.2. THERMONUCLEAR FUSION REACTION CRITERIA

2.2.1. THE LAWSON CRITERION

The usual Lawson criterion[14] for a thermonuclear fusion process is obtained by balancing the fusion energy release against the energy investment in bringing the fuel to thermonuclear temperatures and the energy lost through radiation (both bremsstrahlung and cyclotron radiation):

$$E_{\text{fusion}} = E_{\text{thermal}} + E_{\text{radiation}} \qquad (2.6)$$

The fusion energy released is given in terms of the fusion reaction rate and the time characterizing the reaction, the confinement time τ. To be more specific, we will consider a D-T fusion reaction, in which case the fusion energy release E_{fusion} can be written as

$$E_{\text{fusion}} = n_D n_T \langle v\sigma \rangle W\tau = \frac{n^2}{4} \langle v\sigma \rangle W\tau \qquad (2.7)$$

where $\langle v\sigma \rangle$ is the Maxwellian-averaged reaction rate parameter, W is the energy released per fusion reaction (17.6 MeV for D-T), and τ is the confinement time. Here we have assumed equimolar concentrations of D and T so that

$$n_D = n_T = \frac{n}{2}$$

where n is the ion number density. The thermal energy, assuming ideal gas behavior, is then given by

$$E_{\text{thermal}} = \tfrac{3}{2}nkT_i + \tfrac{3}{2}nkT_e = 3nkT \qquad (2.8)$$

where we have assumed $T_i = T_e$ for convenience. We will ignore the radiation energy loss for now,

$$E_{\text{radiation}} \sim 0$$

noting that if the fuel temperature is greater than 4 keV, the fusion energy release will exceed the bremsstrahlung radiation loss. In inertial confinement schemes, the thermonuclear burn typically occurs at 20 to 100 keV. Furthermore, in such schemes, magnetic field effects can be ignored to first order so that cyclotron radiation is of little concern.

If we now balance the fusion energy release against the thermal energy,

$$\frac{n^2}{4} \langle v\sigma \rangle W\tau = 3nkT$$

we can solve for a condition on the density times the time of confinement

$$n\tau > \frac{12kT}{\langle v\sigma \rangle W} \qquad (2.9)$$

When the reaction rate is evaluated at suitable temperatures (10 keV for D-T, 100 keV for D-D), this yields the usual Lawson criteria:

$$n\tau > 10^{14} \text{ s/cm}^3 \qquad \text{for D-T reactions}$$
$$n\tau > 10^{16} \text{ s/cm}^3 \qquad \text{for D-D reactions} \qquad (2.10)$$

Similar criteria can be obtained for other possible fusion reactions.

2.2.2. CRITERIA FOR EFFICIENT INERTIAL CONFINEMENT FUSION: THE ρR CRITERION

The basic concept of inertial confinement fusion is to implode the fuel to very high densities so that the time of thermonuclear fusion reactions becomes shorter than the inertial confinement time (the fuel pellet disassembly time). That is, in the ICF process the time available for the fusion reaction is determined by the disassembly of the thermonuclear fuel under its high pressure, this expansion being retarded only by the fuel's inertia. This approach removes the constraints on fuel density posed by the structural limitations of magnetic confinement fusion systems. In so doing, it also allows far more efficient thermonuclear burn (since we have found that the fusion reaction rate scales as the square of the fuel density—see Eq. 2.2). However the drawback in inertial confinement fusion is that the time scale of the process is determined by the dynamics of the burning fuel and cannot be controlled by external means.

Such considerations motivate the use of an alternative to the Lawson criterion in characterizing the efficiency of the thermonuclear fusion reaction process. The most meaningful figure of merit for ICF plasmas is the product of the fuel density ρ and the radius R, rather than the customary density-confinement time $n\tau$ product arising in the Lawson criterion.[15, 16] To understand this, suppose we estimate the fuel disassembly time as the pellet radius R divided by the speed of sound (essentially the time required for a density disturbance to propagate from the surface of the pellet into the center):

$$\text{fuel disassembly time} = \tau_d \sim \frac{R}{c_s} \tag{2.11}$$

In a similar fashion, we can estimate the thermonuclear reaction time as the inverse of the reaction rate:

$$\text{thermonuclear reaction time} = \tau_b \sim \left[\left(\frac{\rho}{m_i} \right) \langle v\sigma \rangle \right]^{-1} \tag{2.12}$$

Thus a crude estimate of the efficiency of the thermonuclear burn is given by the ratio of these times

$$\text{thermonuclear burn fraction} = f_b = \frac{\tau_d}{\tau_b}$$

$$= \left(\frac{\langle v\sigma \rangle}{m_i c_s} \right) \rho R \tag{2.13}$$

where we have introduced the *thermonuclear burn fraction* f_b, defined as the fraction of the fuel consumed in the reaction. (We will develop a slightly more accurate expression for f_b, which accounts for fuel depletion, later in this section.)

We now recall that

$$c_s = \left(\frac{kT}{m_i}\right)^{1/2} \sim T^{1/2}$$

Furthermore, for D-T fuel at efficient burn temperatures (20 to 80 keV),[17]

$$\langle v\sigma \rangle \sim T^{1/2}$$

Hence, if we substitute in numerical values for this temperature range, we find

$$\frac{\langle v\sigma \rangle}{m_i c_s} \sim \text{constant} \sim 1$$

Therefore the burn fraction f_b, which measures the efficiency of the thermo-nuclear burn, is given approximately by

$$\text{thermonuclear burn} \atop \text{efficiency} \sim f_b \sim \rho R \text{ g/cm}^2 \qquad (2.14)$$

Hence the criterion characterizing efficient inertial confinement fusion is apparently

$$\rho R > 1 \text{ g/cm}^2$$

The product of density times radius, ρR, is important for other reasons. To sustain the thermonuclear burn, some of the fusion energy must be redeposited in the fuel. For the case of D-T reactions,

$$D + T \rightarrow \alpha(3.5 \text{ MeV}) + n(14.1 \text{ MeV})$$

To capture the energy of the 3.5-MeV alpha particle, the fuel size must exceed the range of the alpha. But the range of a 3.5-MeV alpha particle in a 20-keV D-T plasma is about 0.5 g/cm². Hence if we can achieve fuel conditions such that $\rho R > 0.5$, then the alpha energy will be deposited in the fuel and efficient self-heating will occur.

A somewhat different perspective on the importance of the ρR product is useful[18]. The most important processes in fusion reactions are *binary* collisions. Examples include the fusion reactions themselves as well as the collision processes, which lead to charged particle reaction product slowing down and energy deposition (self-heating) and electron-ion energy exchange. Since binary processes depend on the square of the density, if we could somehow increase density by a factor of 10^3, then we could increase the collision rate by 10^6.

More precisely, we can scale the three important rates characterizing the burning fuel as being proportional to the square of the density ρ:

$$\left.\begin{array}{c}\text{rate of thermonuclear burn}\\\text{rate of energy deposition by charged particles}\\\text{rate of electron-ion energy exchange}\end{array}\right\}\sim\rho^2$$

But we have found that

$$\text{inertial confinement time}\sim R$$

Hence we can scale each of the major processes occurring in the ICF fuel (per unit mass) as

$$\left.\begin{array}{c}\text{thermonuclear burn efficiency}\\\text{self-heating}\\\text{burn propagation}\end{array}\right\}\sim\rho R$$

We will demonstrate later that the optimum value of ρR is about 3 g/cm^2 for D-T fuels. Hence for inertial confinement fusion in D-T fuel, we replace the usual Lawson criterion

$$n\tau > 10^{14} \text{ s/cm}^3$$

with the new goal

$$\rho R > 3 \text{ g/cm}^2 \tag{2.15}$$

2.2.3. RELATIONSHIP BETWEEN THE ρR AND $n\tau$ CRITERIA

We can easily relate the Lawson criterion and the ρR criterion for inertial confinement fusion. We will use a slightly more accurate estimate[19] of the disassembly time for a freely expanding sphere of radius R

$$\tau_d \sim \frac{R}{4c_s} \tag{2.16}$$

which takes account of the fact that in a spherical fuel pellet, half of the mass is beyond 80% of the radius. If we furthermore note that the number density $n = \rho/m_i$, we can write

$$n\tau = \frac{\rho R}{4c_s m_i} \tag{2.17}$$

If we substitute in the appropriate numbers, we find that

$$\rho R = 3 \text{ g/cm}^2 \Rightarrow n\tau = 2 \times 10^{15} \text{ s/cm}^3$$

Hence efficient thermonuclear burn demands a $n\tau$ product well in excess of that of the usual Lawson criterion (10^{14} s/cm^3). (It should be noted that magnetic confinement fusion schemes that work close to the Lawson criterion will burn only a small fraction of their thermonuclear fuel.)

2.2.4. DEPLETION EFFECTS

We can improve our estimate of the burn fraction f_b by taking account of the depletion of the fuel as the burn proceeds. Recall the rate equation for the tritium fuel density:

$$\frac{dn_T}{dt} = -n_D n_T \langle v\sigma_{DT} \rangle \tag{2.18}$$

If we take equimolar densities, $n_D = n_T = n/2$, we find

$$\frac{dn}{dt} = -\frac{n^2}{2} \langle v\sigma \rangle \tag{2.19}$$

We can now integrate this equation from time $t=0$ to the disassembly time $t=\tau_d$ to find

$$\frac{1}{n} - \frac{1}{n_0} = \tfrac{1}{2} \langle v\sigma \rangle \tau_d$$

where n_0 is the initial fuel number density.

Let us now define the burn fraction f_b as

$$f_b = \frac{n_0 - n}{n_0} = 1 - \frac{n}{n_0} \tag{2.20}$$

If we now use our estimate for the disassembly time $\tau_d = R/4c_s$ and $\rho = nm_i$, we find

$$\rho R = \left(\frac{8m_i c_s}{\langle v\sigma \rangle} \right) \frac{f_b}{1 - f_b}$$

or rearranging:

$$f_b = \frac{\rho R}{(8m_i c_s / \langle v\sigma \rangle) + \rho R} \tag{2.21}$$

We can evaluate the bracketed quantity $(8m_i c_s / \langle v\sigma \rangle)$ for D-T fuel conditions at 20 keV to find

$$\frac{8m_i c_s}{\langle v\sigma \rangle} \sim 6.3 \text{ g/cm}^2 \tag{2.22}$$

Thus our more accurate expression for the burn fraction f_b in D-T fuel that accounts for fuel depletion is

$$f_b = \frac{\rho R}{6.3 + \rho R} \tag{2.23}$$

In particular, note that $\rho R \sim 3$ g/cm^2 implies a burn fraction of $f_b = 0.30$, that is, a burn of some 30% of the fuel.

2.2.5. POSSIBLE ICF FUEL CANDIDATES

It is apparent from our earlier discussion that the D-T fuel cycle will be used in early fusion reactors because of its significantly larger fusion reaction rate at relatively low fuel temperatures (10 keV). However the copious quantities of 14-MeV neutrons produced by this reaction present difficult radiation damage problems. Furthermore the tritium inventory in D-T reactor concepts ranges from 10 to 100 kg (10^9 curies) and presents a significant radioactivity hazard. Hence some attention has been given to advanced fuel cycles that minimize fast neutron radiation damage and eliminate the need for tritium breeding. Of most interest are the D-D and p-^{11}B fusion reactions. Since these fuels are generally characterized by lower reaction rates than D-T at low temperatures, the success of utilizing them in ICF systems will depend on increased values of both ignition temperatures and ρR.[11,12]

We can estimate the ρR requirements for these fuels by returning to our expression for the burn fraction f_b given by Eq. 2.21. This equation provides us with a means to compare the attractiveness of various fusion reactions since we can evaluate their ρR requirements for efficient thermonuclear burn:

$$\rho R > \frac{m_i c_s}{\langle v\sigma \rangle}$$

We have tabulated this requirement on ρR for the three fusion reactions of most interest in Table 2.2.

More accurate estimates of the ρR and ignition requirements necessary to burn advanced fuels require the use of ICF target simulation computer codes to study the burn dynamics of specific target designs. Using the PHD-IV code, the Wisconsin group[20] found that the optimum performance for D-D-fueled targets was achieved at ρR values between 40 and 80 g/cm^2 at an ignition

Table 2.2   Candidates for Inertial Confinement Fusion Fuels

Fuel Candidate	ρR Requirement (g/cm^2)
$D + T \rightarrow {}^4He + n$	2 to 5
$D + D \begin{array}{l} \nearrow {}^3He + n \\ \searrow T + p \end{array}$	10 to 20
${}^{11}B + p \rightarrow 3\,{}^4He$	~ 500

temperature of 20 keV. The minimum ρR value necessary to burn D-D fuels was 10 g/cm^2. Such parameters implied a target yield of 100 to 200 MJ and a gain of 200 to 300. For moderate-size targets (1 mg), such large values of ρR require very large compressions (10^5). To relax this compression requirement, D-D fuel cycles will probably require more massive pellets producing higher fusion energy yields. Massive targets also require larger driver energies, probably in the 100-MJ range (and therefore lower pellet gains on the order of 10).

The minimum target conditions for burning p-^{11}B are $\rho R \sim 50$ g/cm^2 and 400 keV, while the optimum ρR value is 500 g/cm^2. The high ρR and temperature necessary for burn propagation in p-^{11}B targets are due in part to the excessive number of electrons present in the p-^{11}B fuel. It is unlikely that such very large values of ρR can be achieved by hydrodynamic processes since this would imply an implosion velocity of 3×10^8 cm/s (an order of magnitude greater than needed for D-T). As with D-D fuels, larger ρR values will require both larger compressions and larger fuel masses, thereby implying greater yields and driver energies. The driver sizes required by the p-^{11}B fuel cycle appear to rule out its use in ICF reactor applications, at least as based on present driver or target design concepts.

2.3. A SIMPLE ANALYSIS OF ICF DRIVER REQUIREMENTS

Let us now examine the requirements on a driver (laser or charged particle beam) designed to implode ICF fuel pellets to produce a net energy gain.[21] A crude sketch of such an ICF system is given in Figure 2.2. We will define

$$\begin{matrix}\text{pellet or target} \\ \text{energy gain or} \\ \text{multiplication}\end{matrix} \equiv M \equiv \frac{E_{\text{fusion}}}{E_{\text{driver}}} \qquad (2.24)$$

$$\begin{matrix}\text{driver coupling} \\ \text{efficiency}\end{matrix} \equiv \varepsilon_D \equiv \frac{E_{\text{fuel}}}{E_{\text{driver}}}$$

The fusion energy can then be calculated in terms of the fusion reaction rate and the volume of the fuel pellet (assumed to be a simple sphere of radius R) as

$$E_{\text{fusion}} = \left(\tfrac{4}{3}\pi R^3\right) n^2 \langle v\sigma \rangle W\tau\beta \qquad (2.25)$$

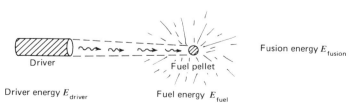

Figure 2.2. A crude diagram of an inertial confinement fusion power system.

Here we have introduced a factor β to account for reaction product self-heating and thermonuclear burn propagation, processes that we will consider in more detail later in this chapter.

We now recall that the derivation of the familiar Lawson criterion proceeded by balancing

$$E_{\text{fusion}} = E_{\text{fuel}}$$

to find $n\tau \sim 10^{14}$ s/cm^3 (for D-T reactions). Therefore we might proceed by balancing

$$E_{\text{fusion}} = ME_{\text{driver}} = \frac{M}{\varepsilon_D} E_{\text{fuel}} \qquad (2.26)$$

to find a new Lawson criterion

$$n\tau > \frac{M}{\beta\varepsilon_D} \times 10^{14} \qquad (2.26)$$

But we are after more information—the driver energy E_{driver} itself.

If we take the electron and ion temperatures to be equal, $T_e = T_i = T$, and at their initial burn temperatures, then

$$E_{\text{fuel}} = \text{thermal energy} = \left(\tfrac{4}{3}\pi R^3\right) 2\left(\tfrac{3}{2} n k T\right) \qquad (2.27)$$

We can return to substitute this into our balance condition Eq. 2.25, recalling

$$\tau \sim \tau_d \sim \frac{R}{c_s}$$

to write

$$\left(\tfrac{4}{3}\pi R^3\right) n^2 \langle v\sigma \rangle W\beta \left(\frac{R}{c_s}\right) = \frac{M}{\varepsilon_D} \left(\tfrac{4}{3}\pi R^3\right)(nkT)$$

We can now solve for R as

$$R = \left(\frac{M}{\beta\varepsilon_D}\right) \frac{kTc_s}{n\langle v\sigma \rangle W}$$

Finally, we use the definition of ε_D given by Eq. 2.24 to write

$$E_{\text{driver}} = \frac{1}{\varepsilon_D} E_{\text{fuel}} = \frac{4\pi}{\varepsilon_D \beta} \left(\frac{M}{\beta\varepsilon_D}\right)^3 \left(\frac{kTc_s}{n\langle v\sigma \rangle W}\right)^3 nkT \qquad (2.28)$$

We can write this in a more convenient form as

$$E_{driver} = \frac{M^3}{(\beta \varepsilon_D)^4} \frac{1}{\eta^2} \times 1.6 \times 10^6 J \qquad (2.29)$$

where we have introduced the compression factor

$$\eta \equiv \frac{n}{n_{liquid}} \qquad (2.30)$$

where we have chosen as a reference density the number density of liquid D-T, $n_{liquid} = 4.5 \times 10^{22}$ cm^{-3}. (The corresponding mass density $\rho_{liquid} = 0.2$ g/cm^3.)

A couple of example applications of this result are of interest. If we were to estimate the driver requirements for energy break-even, $M = 1$, assuming fuel at liquid density, $\eta = 1$, and perfect coupling but with no self-heating or burn propagation, $\beta \varepsilon_D = 1$, then the formula would yield a driver requirement of $E_{driver} = 1.6$ MJ. A reactor application with $M = 100$, $\eta = 1$, and $\beta \varepsilon_D = 1$ would required $E_{driver} = 1.6 \times 10^6$ MJ. Hence the required driver energies are quite large if we are working with fuels at liquid density.

But notice that the driver energy scales as the inverse square of the compression. If we were somehow able to compress the fuel to very high densities, say, a compression factor of 10^4, then the reactor driver energy is reduced to $E_{driver} = 16$ kJ. Although this estimate is far too optimistic, it does demonstrate the strong sensitivity of driver energy requirements on fuel compression.

As an aside, we should note that most of the driver energy will be used to compress rather than heat the fuel. We will demonstrate later that the driver coupling efficiency is typically about $\varepsilon_D \sim 5\%$, even if the driver beam is totally absorbed in the target. Hence if we are to achieve $\beta \varepsilon_D \sim 1$, we will need a multiplication due to alpha particle self-heating and thermonuclear burn propagation of $\beta \sim 20$. These latter processes will therefore play a very important role in any practical ICF scheme.

2.4.   THE SCENARIO FOR INERTIAL CONFINEMENT FUSION

We have noted that the most important processes in fusion reactions involve binary collisions (fusion reactions, alpha particle energy deposition, and electron-ion temperature equilibration) and therefore scale as the square of the compression. That is, increasing density by a factor of 10^3 increases the collision rate by 10^6. This is manifested in the dependence of thermonuclear burn efficiency, self-heating, and burn propagation on the product of fuel density times radius, ρR.

In spherical compression, the density (for constant mass) scales as $\rho \sim R^{-3}$. Hence we find that the product $\rho R \sim (R^{-3}) R \sim R^{-2}$. Thus compression by 10^3

reduces the mass required to initiate efficient thermonuclear burn by 10^6. (The typical imploded fuel masses in ICF targets are of the order of 10^{-3} g.)

But how do we achieve such densities?[22] There are several examples on an astronomical level. In the sun, compressions range as high as 10^3 with a corresponding temperature of 1 to 2 keV. This corresponds to a pressure of 10^{11} atmospheres. The fuel confinement is maintained by the enormous mass of the sun, $m_\odot \sim 10^{30}$ kg.

In white dwarf stars, compressions of 10^5 to 10^6 are found corresponding to pressures of 10^{15} atm. At these compressions the electrons become degenerate. That is, their de Broglie wavelengths become comparable to their mean separation so that the exclusion principle becomes important and results in an additional repulsive force.

So the necessary compressions (and densities) do occur on an astronomical scale. But how can we generate sufficient pressures (10^{12} atm) to compress the fuel to $\eta \sim 10^4$ in a terrestrial environment? There are several possibilities:

1. *Chemical explosives.* The pressures produced by chemical explosives are limited by the strengths of chemical bonds to roughly 10^6 atm. These pressures can be increased by another order of magnitude (10^7 atm) by using geometric convergence (e.g., a spherical implosion), but they still fall far short of the required magnitude.

2. *Light pressure.* Suppose we focus the intense beams of high-powered lasers on the fuel. Then the ponderomotive force exerted by the light on the pellet surface corresponds to a pressure of

$$p \sim \frac{I}{c} \sim \frac{10^{17} \text{ W/cm}^2}{3 \times 10^{10} \text{ cm/s}} \sim 10^8 \text{ atm}$$

   still not high enough even with very powerful lasers.

3. *Ablation pressure.* Here the idea is to use the driver energy beam (laser light or charged particle beams) to heat surface material and ablate it off into the vacuum surrounding the pellet. The back reaction to the ablating surface generates a pressure, much as that generated by a rocket exhaust. For the same reason that matter-ejecting rockets have much larger thrust than photon rockets, the pressure is multiplied to

$$p \sim \frac{I}{v} \sim \frac{10^{17}}{10^8} \sim 10^{10} \text{ atm}$$

4. *Geometric convergence.* We can multiply this pressure further by taking advantage of geometric convergence, for example, in a spherical implosion. This convergence can increase the pressure by a factor of 100.

5. *Isentropic compression.* If we compress the fuel isentropically so that it is not heated to high temperatures, then we can compress it into a Fermi

degenerate state (that is, until the electrons obey Fermi-Dirac rather than Maxwell-Boltzmann statistics). Then the pressure required for a given compression or fuel density is at a minimum and given by[22]

$$p \sim \tfrac{2}{3} n_e \varepsilon_F \left[\tfrac{3}{5} + \frac{\pi^2}{4} \left(\frac{kT}{\varepsilon_F} \right)^2 - \frac{3\pi^4}{80} \left(\frac{kT}{\varepsilon_F} \right)^4 + \cdots \right] \qquad (2.31)$$

where ε_F is the Fermi energy.

$$\varepsilon_F = \tfrac{1}{8} \left(\frac{3}{\pi} \right)^{2/3} \frac{h^2}{m_e} n^{2/3}$$

For example, at a compression of $\eta \sim 10^4$ (corresponding to an electron density of $n_e \sim 5 \times 10^{26}$), the minimum pressure is $p \sim 10^{12}$ atm when $\varepsilon_F \gg kT$. This should be compared with a required pressure of $p \sim 10^{13}$ for a temperature of $kT \sim 5$ keV (which corresponds to the ideal ICF fuel ignition temperature for D-T).

To illustrate just how this enormous compression might be achieved, let us consider the simplest scenario of the implosion of a fuel pellet to the required densities using high-powered laser beams (see Figure 2.3)[22, 23]:

1. We begin by irradiating a 1-mm sphere of liquid D-T fuel uniformly about its surface with intense laser light (which will reach a peak power intensity of 10^{17} W/cm^2).

2. The outer surface of the pellet heats, ionizes, and ablates off to surround the pellet in a cloud or "corona" of low density plasma, characterized by electron densities $n_e \sim 10^{19}$ to 10^{22} cm^{-3}.

3. The electrons in the corona continue to absorb more energy from the incident laser beams, but now the beam can only penetrate into the critical density where the plasma frequency equals the light frequency. This critical density surface occurs at 10^{21} cm^{-3} for Nd laser light at 1.06 μm and 10^{19} cm^{-3} for CO$_2$ laser light at 10.6 μm.

4. The energy deposited by the laser at the critical surface is then transported into the surface of the pellet by processes such as electron thermal conduction. This energy continues to heat the pellet surface, driving the ablation process and producing high pressures.

5. As the ablation of the surface continues, a shock front is formed that converges (implodes) inward, pushing cold D-T fuel ahead of it to higher and higher densities along the "Fermi degenerate adiabat." The various hydrodynamic and energy transport phenomena involved in this stage of the implosion process are shown in Figure 2.4. Here it is important to compress the pellet fuel isentropically (without appreciable heating) in an effort to bring it to very high density while still leaving it relatively cold.

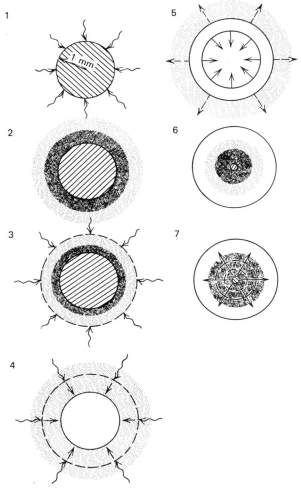

Figure 2.3.   The stages in the implosion of an inertial confinement fusion target: (1) irradiation by driver beams, (2) formation of plasma atmosphere, (3) driver beam absorption in atmosphere, (4) ablation driven imploding shocks, (5) compression of fuel core, (6) ignition, (7) burn propagation.

6. When the shock fronts converge at the center of the highly compressed pellet core, they shock heat a small region at the center of the compressed core to thermonuclear ignition temperatures (2 to 5 keV). If $\rho R > 0.5$ g/cm^2 alpha particle self-heating will occur, and the intense spark at the center of the compressed core will rapidly heat to optimum burn temperatures of 20 to 100 keV.

7. As the central spark burns, some alpha particles are deposited in adjacent cold fuel, bringing it to ignition temperatures. The tendency of the burning

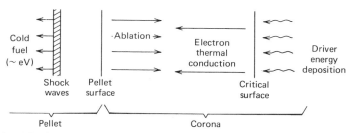

Figure 2.4. The hydrodynamic and energy transport processes involved in an ICF target implosion.

fuel spark to become more transparent to alphas as it heats up enhances this process. The adjacent fuel material layer burns, producing further self-heating in cold fuel material and producing a thermonuclear burn wave that propagates outward, consuming the dense pellet core. If $\rho R > 3$ g/cm^2, an efficient thermonuclear burn will occur, with some 30% of the fuel in the dense core being consumed.

Of course a great many physical processes, some of which are only marginally understood, are involved in this scenario. These processes will be the primary focus of Chapters 3, 4, and 5. Furthermore, this ICF target consisting of liquid DT is oversimplified. More realistic and complex designs will be discussed in Chapter 9.

2.5. TARGET GAIN REQUIREMENTS

Thus far we have considered the driver energies required for efficient thermonuclear burn in ICF targets, and have run through a brief scenario of one possible implosion scheme. Let us turn this discussion around a bit by asking just what value of pellet energy gain will be required in ICF applications.[23]

Recall that we have defined the pellet or target energy gain as

$$\text{pellet energy gain} \equiv M \equiv \frac{E_{\text{fusion}}}{E_{\text{driver}}} \tag{2.32}$$

Also recall the definition of the driver-coupling efficiency:

$$\text{coupling efficiency} \equiv \varepsilon_D \equiv \frac{E_{\text{fuel}}}{E_{\text{driver}}} \tag{2.33}$$

It is useful to introduce one further definition that takes into account the fact that the thermonuclear burn will occur only in the highly compressed core of the fuel pellet. We will define the *fuel gain* G_F by

$$\text{fuel gain} \equiv G_F \equiv \frac{E_{\text{fusion}}}{E_{\text{fuel}}} \tag{2.34}$$

The fuel gain and pellet gain are related by the driver-coupling efficiency

$$M = \varepsilon_D G_F \qquad (2.35)$$

Here we recall that ε_D is determined not only by the efficiency of the driver beam absorption process, but also by the efficiency in converting absorbed energy into the energy of the compressed pellet core through the ablation process. For example, a high gain pellet with $M = 200$ would be characterized by a coupling efficiency of roughly $\varepsilon_D \sim 0.05$ and therefore would require a very large fuel gain of $G_F = 4000$.

To determine the pellet gain requirements for a power reactor, suppose we consider the use of the fusion energy to produce electricity through a thermal cycle as shown schematically in Figure 2.5. Here we have noted that a certain fraction of the produced electrical energy must be circulated back to power the driver. We can associate an efficiency with each aspect of this process as follows:

$$\text{driver efficiency} \equiv \eta_D \equiv \frac{\text{driver energy output}}{\text{driver electrical input}}$$

$$\text{gross plant thermal efficiency} \equiv \eta_{\text{th}} \qquad (2.36)$$

$$\text{net plant thermal efficiency} \equiv \eta_P$$
$$\text{(taking into account driver power)}$$

From the definition of the pellet gain we can determine

$$M = \frac{1}{\eta_D(\eta_{\text{th}} - \eta_P)} - 1 \qquad (2.37)$$

Hence we can solve for the plant efficiency as a function of pellet gain

$$\eta_P = \eta_{\text{th}} - \frac{1}{\eta_D(M+1)} \qquad (2.38)$$

The fraction of the gross electrical power needed for the driver, that is, the

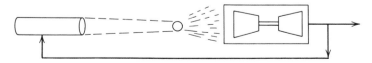

Figure 2.5. The production of electricity using an inertial confinement fusion driven thermal cycle.

recirculating power fraction F_D, is given by

$$F_D = \frac{1}{\eta_D \eta_{\text{th}}(M+1)}$$

With this background, we can now go on to define two criteria for determining pellet gain requirements:

Engineering breakeven: the fusion energy production is just sufficient to balance driver energy needs, that is, $F_D = 1$. Then the breakeven gain is

$$M^{EB} = \frac{1-\eta_D}{\eta_D} \tag{2.39}$$

Scientific breakeven: the fusion energy production is just sufficient to balance driver energy output.

$$M^{SB} = 1 \tag{2.40}$$

For typical thermal cycles, η_{th} ranges between 35 and 50%. Furthermore, most driver designs (lasers in particular) are constrained by a limit on driver efficiency of $\eta_D \sim 10\%$. Thus we find that

$$M^{EB} \sim 10 \tag{2.41}$$

For significant power production, it is apparent that we will need

$$M \gtrsim 100 \tag{2.42}$$

In fact, most power plant designs based on ICF schemes assume a pellet gain of $M \sim 150$ to 200.

For example, the Wisconsin SOLASE laser fusion reactor design[20] assumes parameters: laser efficiency $\eta_D = 0.07$, fusion energy $E_{\text{fusion}} = 150$ MJ, and laser energy output $E_{\text{driver}} = 1$ MJ. This would correspond to a pellet gain of $M = 150$.

It is apparent from Eq. 2.39 that the required pellet gain for engineering breakeven is a very sensitive function of driver efficiency. Unfortunately, most advanced laser drivers project only rather modest efficiencies in the 4 to 8% range. These would require the development of high gain targets. For example, a driver pulse energy of 1 or 2 MJ would achieve a fusion yield of 100 to 250 MJ per shot. Such low driver efficiencies also require a large circulating power ratio F_D (20 to 50%) to energize the laser. To reduce the circulating power ratio to lower levels (10%) with laser drivers would appear to require very large driver and pulse yields.

By way of contrast, light ion and heavy ion drivers should be able to achieve efficiencies in the 20 to 30% range. High gain targets ($M = 100$ to 200) would then allow a recirculating power ratio of only 10% for a driver in the 1 to 2 MJ range and a fusion energy yield of 100 to 250 MJ per shot. Such charged particle beam drivers, with their much larger efficiencies, could also be used with low gain targets ($M \sim 30$) with an increase in driver pulse energy (10 MJ) and shot yield (300 MJ).

To better understand possible ICF target gains, let us examine in more detail the fuel gain G_F for D-T fuels. Although the largest D-T reaction rate occurs in the temperature range 20 to 80 keV, let us base our estimates at the more modest ignition temperature 10 keV. The energy required for a single D-T reaction (including the thermal energy of both ions and electrons) is then 4×10 keV. Hence the energy gain from a single D-T reaction is (17.6 MeV)/(4×10 keV)$=440$. That is, a 100% burn of a uniformly heated pellet core would yield a fuel gain of only $G_F = 440$. But this is not sufficient, since we know that typical burn fractions are $f_b \sim 30$ to 50%. Furthermore, the driver-coupling efficiency of $\varepsilon_D \sim 0.05$ implies that to achieve pellet gains of $M \sim 100$, we are going to need fuel gains of $G_F \sim 2000$ or greater.

How do we design such high gain targets? It is useful to make some order-of-magnitude estimates. Suppose we have managed to compress the core of a fuel pellet to the desirable condition of $\rho R \sim 3$ g/cm^2. Since this is sufficient to capture the alpha reaction products (with range 0.5 g/cm^2), we need only heat the core to ignition temperatures of 1 to 2 keV. This corresponds to a specific ignition energy of $e_{\text{ignition}} \sim 30$ MJ/g. To achieve the necessary ρR condition requires a compression of $\eta \sim 10^4$. If we manage to compress the pellet into a degenerate state such that $kT < \varepsilon_F \sim 1$ keV, then the compression energy (p dV work) required is $e_{\text{compression}} \sim 30$ MJ/g. Hence the total energy input required is the sum of the ignition and compression energies, or 60 MJ per gram. But for a $\rho R \sim 3$ g/cm^2 fuel, the specific fusion energy release $e_{\text{fusion}} \sim 10^5$ MJ/g. Hence the maximum fuel gain of such a target would be

$$G_F = \frac{e_{\text{fusion}}}{e_{\text{core}}} \sim \frac{100,000}{60} \sim 1500$$

If we recall that roughly 95% of the driver energy goes into the ablation process required to produce the high compressions, that is, $\varepsilon_D = 0.05$, then we find a pellet gain of

$$M = \frac{E_{\text{fusion}}}{E_{\text{driver}}} = 75$$

Hence it is apparent that to achieve such high pellet gains, we must depend on the processes of self-heating and thermonuclear burn propagation to minimize the driver energy that will act as a match to light the fusion flame.

As an aside, it should be noted that target gains tend to increase with both driver energy and fusion energy yield. For example, numerical and analytic

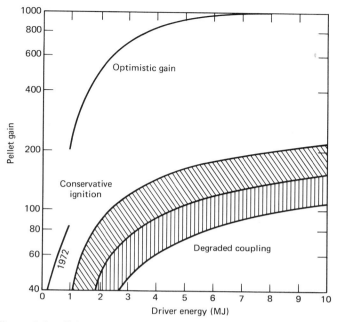

Figure 2.6. Calculated pellet gain as a function of driver energy.

studies of various target designs predict a trend (see Figure 2.6)[24, 25]

$$M = M_0 E_{\text{driver}}^{0.4} \qquad (2.43)$$

Here, depending on the model (and the degree of optimism), M_0 varies from a low of 30 to a high of 300.

2.6. INERTIAL CONFINEMENT FUSION BURN PHYSICS

The key to the practical application of inertial confinement fusion is the achievement of high target gains, that is, maximizing the ratio of fusion energy released to driver energy. Considerations in the preceding section suggested that target gains of as much as 100 would be needed to overcome driver and energy conversion inefficiencies. The achievement of such high ICF target gains will require taking advantage of several important physical processes: (1) isentropic compression of the fuel to very high densities, (2) self-heating of the compressed fuel to high temperature by trapping reaction products (alpha particles) in the reaction region, and (3) igniting only a small region in the compressed fuel and then allowing the thermonuclear burn to propagate through the remaining fuel. In this section we examine each of these important processes and their effect on target gain.

2.6.1. TARGET COMPRESSION

To assess the importance of fuel compression, it is a useful exercise to examine in some detail the heating of an uncompressed fuel target to fusion temperatures. In particular, we will consider the target to be a simple spherical droplet of D-T fuel at liquid density, $\rho_{liquid} = 0.2$ g/cm³ (particle density $n = 4.5 \times 10^{22}$ cm⁻³). The general idea is to attempt to heat the fuel pellet to thermonuclear temperatures very rapidly so that an appreciable number of D-T fusion reactions will occur before the pellet disassembles. The important physical processes in this problem include: the heating of the pellet fuel to fusion temperatures using driver energy deposited on its surface, the equilibration of electron and ion temperatures in the fuel, and the hydrodynamic expansion or disassembly of the target. We will base our discussion of these processes on a simple model developed by Brueckner and Jorna.[21,23]

The key to the physics of the problem is to recognize that the driver beam can deposit energy only in the outer layers of the target. For example, laser light will only penetrate into the critical density (10^{21} cm⁻³ for 1.06-μm light). Hence the interior of the target must be heated to fusion temperatures by some mechanisms other than laser energy deposition. (Note that if we could develop ultraviolet lasers, say with wavelengths in the 0.1 to 0.2 μm range, then the light beam could penetrate into even liquid density plasma.)

We will take the principal heating mechanism to be electron thermal conduction. The small electron mass makes these particles sufficiently mobile that the plasma becomes an excellent conductor of heat. The thermal conductivity is strongly temperature dependent, scaling as $\kappa \sim T^{5/2}$. This temperature dependence leads to a number of interesting nonlinear conduction effects which will be considered in detail in Chapter 4.[26] However, most relevant to the present discussion is the formation of a thermal wave that moves into the target to heat the fuel material.

In the actual pellet-heating process we find a large variety of physical phenomena, including the absorption of the driver energy, the transport of this energy into the surface of the pellet via electron thermal conduction, the heating and ablation of the target surface, and the penetration of both thermal and shock waves into the dense pellet fuel. However for the purposes of our

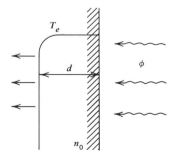

Figure 2.7. Heat wave propagation into a laser heated target.

present discussion of the heating of an uncompressed target at constant liquid density, we will consider a simple model in which the driver energy is transported into the surface of a slab (one-dimensional) target in such a way that the temperature at the surface is maintained at a constant magnitude T_e. The driver is assumed to maintain the surface of the target at this temperature.

This process will drive a thermal wave into the target which heats the dense fuel material. We can estimate the distance d_{HW} the heat wave will propagate into the target in a time t by a simple energy balance (see Figure 2.7). If n_0 is the number density of the target and ϕ is the driver power intensity (W/cm^2) at its surface, then

$$\text{energy incident on surface} = \phi t \sim n_0 k T_e d_{HW} = \text{thermal energy per unit area of target} \qquad (2.44)$$

We can balance this energy flux against the heat flux conducted into the target by using Fourier's law of heat conduction

$$q = -\kappa \frac{\partial T}{\partial x}$$

where κ is the thermal conductivity. If we balance this against the incident energy flux ϕ and estimate the temperature gradient $\partial T/\partial x \sim (T_e/d_{HW})$, we find

$$\phi \sim \kappa \frac{T_e}{d_{HW}}$$

But for a plasma we have noted that the thermal conductivity depends on temperature as

$$\kappa = \kappa_0 T_e^{5/2}$$

Hence we can solve for the distance of penetration of the heat wave as

$$d_{HW} \sim \left[\frac{\kappa_0 T_e^{5/2} t}{k n_0} \right]^{1/2} \qquad (2.45)$$

(In Chapter 4 we will note that an exact solution of the nonlinear diffusion equation describing heat conduction in a plasma will give a similar result.)

But this is only part of the story. We must now account for the fact that the pellet will disassemble through the propagation of a rarefaction wave inward from its surface at the speed of sound. The depth of penetration of this hydrodynamic disassembly wave is

$$d_H = c_s t = v_0 \left[\frac{k(T_e + T_i)}{2} \right]^{1/2} t \qquad (2.46)$$

If we compare Eqs. 2.45 and 2.46, we note that since $d_{HW} \sim t^{1/2}$ and $d_H \sim t$, the heat wave will initially propagate into the target more rapidly than the disassembly wave. The rarefaction wave will catch up with the heat wave when $d_{HW} \sim d_H$ or

$$d_{HW} = \left(\frac{\kappa_0}{2v_0 n_0} \right) \frac{T_e^{5/2}}{[k(T_e + T_i)/2]^{1/2}} \tag{2.47}$$

At this point the target density n_0 will drop, and the driver beam (e.g., the laser beam) can penetrate into the target and heat the fuel directly.

The thermal energy (per unit area) of the heated region of the target is given by

$$E_{thermal} = \tfrac{3}{2} n_0 k (T_e + T_i) d_{HW}$$

If we heat this region to fusion temperatures, the fusion energy produced is given in terms of the reaction rate as

$$E_{fusion} = \frac{n_0^2}{4} \langle v\sigma \rangle W \tau_R d_{HW}$$

where the reaction time τ_R can be identified as the disassembly time for the region

$$\tau_R \sim \tau_d \sim \frac{d_{HW}}{c_s} = \frac{2\kappa_0 T_e^{5/2}}{v_0^2 n_0 k^2 (T_e + T_i)}$$

At what temperature do we evaluate $\langle v\sigma \rangle$? The heat wave will heat the electrons first. The ion temperature will then equilibrate with that of the electrons according to the rate equation

$$\frac{dT_i}{dt} = \frac{1}{\tau_{ei}} (T_e - T_i), \qquad \tau_{ei} = \frac{3m_e m_i}{8(2\pi)^{1/2} ne^4 \ln \Lambda} \left(\frac{T_e}{m_e} \right)^{3/2}$$

We can again use simple estimates of the derivative term $dT_i/dt \sim T_i/\tau_R$ to find

$$T_i \sim \frac{\tau_R}{\tau_{ei}} (T_e - T_i)$$

If we now use our expression for the disassembly time for τ_R, we can solve for

$$T_i \sim 0.315 T_e \tag{2.48}$$

This simple analysis suggests that the ions will only partially equilibrate with the electrons.[27,28] The ion temperature in the reaction zone will always be much lower than the electron temperature. (This feature is confirmed by computer simulations.)

We can now combine these equations to estimate the thermal energy of the target fuel as

$$E_{\text{thermal}} = \tfrac{3}{2} \frac{\kappa_0}{v_0} T_e^{5/2} \left(\frac{T_e + T_i}{2k} \right)^{1/2}$$

Thus we can determine

$$\frac{E_{\text{fusion}}}{E_{\text{thermal}}} = \frac{1}{6} \langle v\sigma \rangle W \left[\frac{\kappa_0 T_e^{5/2}}{v_0^2 k^3 (T_e + T_i)^2} \right] \tag{2.49}$$

Brueckner and Jorna have evaluated this expression for several typical cases. For temperatures $T_e = 10$ keV, the corresponding ion temperature is $T_i = 3.15$ keV and the ratio of fusion to thermal energy is 0.0755. For electron temperatures $T_e = 30$ keV, the ion temperature is $T_i = 9.45$ keV and the fusion to thermal energy ratio is 3.54. If we assume the driver beam is incident on a target area of πd_{HW}^2, then the driver energy required for the first case is 505 kJ, while that required for the second case (which exceeds breakeven) is 1,120 MJ.

Hence this simple model suggests that the driver requirements for breakeven are of the order of 10^3 MJ because of the balance among the time required to conduct heat into the target, the time required for the electron and ion temperatures to equilibrate, and the time available before the pellet disassembles. It is apparent from this analysis that the direct heating of an uncompressed pellet looks quite out of the question. (And we still have not accounted for incomplete driver energy absorption, energy lost to surface ablation, or temperature and density gradients in the heated layer.)

Hence fuel compression will be essential for appreciable gain. We have already seen the importance of this feature in our earlier expression Eq. 2.29 for the driver energy requirement (keeping in mind the limitations of this result).

$$E_{\text{driver}} \sim \frac{M^3}{(\beta\varepsilon)^4} \frac{1}{\eta^2} \times 1.6 \text{ MJ}$$

Here we found that the required driver energy scales as the inverse square of the compression. Thus a premium is placed on achieving high fuel densities.

2.6.2. SELF-HEATING

A second important goal in inertial confinement fusion is to achieve fuel densities sufficient to trap an appreciable fraction of the high energy reaction products. If this can be accomplished, then the reaction products will deposit their energy in the fusion reaction zone, heating it to higher temperatures and thereby increasing the fusion reaction rate. The requirement for efficient thermonuclear burning is that the range of the charged particle reaction products be only a small fraction of the dense fuel core. Then, once burning starts, the reaction production energy deposition will quickly "bootstrap" heat the core to high temperatures (20 to 80 keV) where the fusion reaction rate is the greatest.[29-31]

The reaction products of the D-T reaction are 3.5-MeV alpha particles and 14.1-MeV neutrons. The ratio of the alpha particle range to the compressed fuel radius R is given in terms of the electron temperature as[27]

$$\frac{\lambda_\alpha}{R} \sim \frac{1.9}{\left(1+122/T_e^{5/4}\right)} \frac{1}{(\rho R)} \qquad (T_e \text{ in keV}) \qquad (2.50)$$

The ratio of the range of 14.1-MeV neutrons to the fuel radius R is

$$\frac{\lambda_n}{R} \sim \frac{4.6}{\rho R} \qquad (2.51)$$

At electron temperatures of 10 keV the neutron range is some 20 times that of the alpha range, so to first order we will assume that only the alpha particle self-heating needs to be treated. (We will examine the case of neutron self-heating in Chapter 4.)

Of the alpha particle energy deposited in the fuel, a fraction

$$f_i = \left(1+\frac{32}{T_e}\right)^{-1}$$

is deposited in the ions. In particular, for a temperature $T_e = 10$ keV, $f_i = 0.24$, so that roughly 25% of the alpha particle energy will contribute to bootstrapping the fuel temperature to higher values (and higher reaction rates).

Brueckner and Jorna[21] have developed a simple model to estimate the effect of alpha self-heating on fuel ignition and burn. If W_{dep} is the alpha particle energy deposited in the fuel region, then the rate of energy deposition can be equated to the increase in thermal energy of the fuel (assuming for the moment temperature equilibration $T_e = T_i$)

$$\frac{dE}{dt}\bigg|_{\text{self-heating}} = \frac{4\pi R^3}{3} \frac{n^2}{4} \langle v\sigma \rangle W_{dep}$$

$$= \frac{dE_{\text{th}}}{dt} = \frac{4\pi R^3}{3} 3nk \frac{dT}{dt} \qquad (2.52)$$

The time available for heating and fusion is again determined by the disassembly time. If the temperature were uniform, we could use $\tau_d \sim R/c_s$. However in this case we must take into account the temperature dependence of the speed of sound, $c_s = v_0(kT)^{1/2}$. If we change variables from time t to $R(t)$, where dR/dt is identified as the sound velocity c_s, then we can integrate Eq. 2.52 as

$$\frac{1}{12} \int_0^R n \, dr = \int_{T_0}^{T_1} \frac{v_0 T^{1/2} \, dT}{\langle v\sigma \rangle W_{\text{dep}}}$$ (2.53)

where T_0 is the initial temperature and T_1 is the temperature after a time corresponding to the disassembly time. The thermal energy of the pellet just prior to ignition is

$$E_{\text{thermal}}(0) = \frac{4\pi R^3}{3} 3nkT_0$$

We can integrate the fusion reaction rate to find the total fusion energy released as

$$E_{\text{fusion}}(t) = \frac{4\pi R^3}{3} \frac{W}{4} \int_0^t n^2 \langle v\sigma \rangle \, dt'$$

Therefore we can identify

$$\frac{E_{\text{fusion}}}{E_{\text{thermal}}(0)} = \frac{W}{12kT_0} \int_0^t n \langle v\sigma \rangle \, dt' = \frac{W}{T_0} \int_{T_0}^{T_1} \frac{dT}{W_{\text{dep}}(T)}$$ (2.54)

To complete this calculation, we must estimate the alpha particle energy deposition W_{dep}. Brueckner and Jorna use a simple interpolation

$$W_{\text{dep}} = W_\alpha \left(\frac{R}{R + \lambda_\alpha} \right)$$ (2.55)

where $W_\alpha = 3.5$ MeV and λ_α is the range of the alpha particle. They have calculated the integrals in Eq. 2.54 to determine the ratio $E_{\text{fusion}}/E_{\text{thermal}}(0)$ for various driver energies. Appreciable self-heating can lead to a very significant decrease in required driver energies. For example, a self-heating ratio of $E_{\text{fusion}}/E_{\text{thermal}}(0) = 100$ would reduce the required driver energy by a factor of 400 over that for an uncompressed pellet, since the driver only has to provide the relatively small energy for ignition.

The optimum energy for ignition depends on several factors. For example, bremsstrahlung radiation is a serious loss mechanism for low temperatures. Below 4 keV, the bremsstrahlung energy loss rate is some four times that of the alpha particle heating rate. The condition for pellet transparency to bremsstrahlung radiation is given by[23]

$$n^2 R < 6.43 \times 10^{47} (kT)^{7/2} \text{ cm}^{-5}$$ (2.56)

If we use the disassembly time $\tau_d = R/c_s$, we find this implies a condition on compressions of

$$\eta = \frac{n}{n_L} < 7.75 \times 10^4 \left[\frac{E_{\text{fusion}}}{E_{\text{thermal}}(0)} \right] \tag{2.57}$$

Hence for compressions of 10^3 to 10^4 and values of $E_{\text{fusion}}/E_{\text{thermal}}(0)$ of 100 to 1000, the fuel becomes opaque to bremsstrahlung radiation, and ignition can occur for temperatures T_0 less than 4 keV. For large self-heating ratios, the optimum ignition temperature is about 2 keV.

2.6.3. THERMONUCLEAR BURN PROPAGATION

Fuel compression and alpha particle self-heating are insufficient in themselves to achieve the very high gains required of ICF targets. In addition, the fuel target must be designed to take advantage of the process of thermonuclear burn propagation. Here the general idea is to use shock compression to induce central ignition of the fuel, that is, to ignite a central "spark" surrounded by cold fuel below the ignition temperature. As the spark burns it becomes transparent to the reaction products, and they stream out and rapidly heat adjacent fuel material to ignition temperatures. In this way one can ignite a spherically expanding burn wave which propagates outward, leading to complete ignition of the compressed pellet core. The propagation of such an expanding thermonuclear burn wave through a spherical target is shown in the computer simulation results plotted in Figure 2.8.[20]

Energy transfer from the burning central region to the adjacent cold fuel can occur through three mechanisms[21]: (1) hydrodynamic energy transfer (due to rapid pressure buildup in the burning region), (2) electron thermal conduction from the hot burning region to the cold fuel material, and (3) energy deposition by escaping reaction products (alphas, neutrons, and X rays). Usually the propagation of the burn front is highly supersonic, so that hydrodynamic energy transfer (1) is not a dominant effect.

Since hydrodynamic effects are only of secondary importance in the propagation of the burn wave, Brueckner and Jorna modeled this phenomenon by ignoring hydrodynamic motion and determined the rate of advance of the burn wave based on energy conservation. First they noted that if the central spark region is characterized by a density n_0 and a radius r, one can write the rate of energy production in the uniform central region

$$\frac{dE_{\text{fusion}}}{dt} = \frac{4\pi r^3}{3} \langle v\sigma \rangle W_\alpha \frac{n_0^2}{4}$$

(where the neutron energy deposition is ignored). The rate of change of the internal energy in the expanding region can be calculated as

$$\frac{d}{dt}\left(\tfrac{4}{3}\pi r^3 n_0 k T_0 \right) = \tfrac{4}{3}\pi r^3 n_0 k \dot{T}_0 + 4\pi n_0 r^2 k T_0 \dot{r}$$

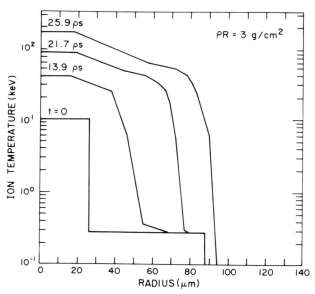

Figure 2.8. The propagation of a thermonuclear burn wave outwards through a fuel core heated to 10 keV. (Courtesy of the University of Wisconsin Fusion Engineering Program.)

Equating these two energy expressions will yield an equation for the radius of the burning region r:

$$\dot{r} = \frac{n_0 \langle v\sigma \rangle W_\alpha}{12kT_0} r - \frac{1}{3} r \frac{\dot{T}_0}{T_0} \tag{2.58}$$

The temperature of the burning region, T_0, will increase until the alpha particle range λ_α becomes large enough to allow alphas to escape into the surrounding fuel. That is, the burn region temperature T_0 will adjust itself so that $\lambda_\alpha \sim r$,

$$r \sim \lambda_\alpha = \lambda_0 \frac{T_0^{3/2}}{n_0} \qquad (\text{for } T_0 > 40 \text{ keV})$$

Hence one can write

$$\frac{\dot{T}_0}{T_0} \sim \frac{2}{3} \frac{\dot{r}}{r}$$

But we recall that the speed of sound is given by $c_s = v_0 T^{1/2}$. Hence we can calculate

$$\frac{\text{burning front speed}}{\text{speed of sound}} = \frac{\dot{r}}{c_s} \sim 3 \langle v\sigma \rangle W_\alpha \frac{\lambda_0}{44 v_0}$$

For temperatures $kT_0 > 15$ keV, $\dot{r}/c_s > 2$, that is, the propagation of the burn front is supersonic in the cold fuel material.

If we return to the self-heating equation 2.54, we can estimate the required driver energy for ignition. We will take $kT_1 = 4$ keV and $kT_0 = 2$ keV. Then the initial thermal energy required for central ignition is found to be

$$E_{\text{thermal}} = \tfrac{4}{3}\pi n_0 r^3 kT_0 = 7.99 \times 10^6 \eta^2 \text{ kJ}$$

The minimum energy of the cold fuel is its degeneracy energy (that is, the energy due to the repulsion of the degenerate electrons):

$$E_{\text{degeneracy}} = \tfrac{4}{3}\pi R^3 n_0 \varepsilon_{\text{deg}} = 80.5 R^3 \eta^{5/3} \text{ kJ}$$

where we have taken the individual degeneracy energy per electron as $\varepsilon_{\text{deg}} = 2.68\eta^{2/3}$ eV per electron. After ignition, the fuel will burn at 40 to 150 keV. If fuel depletion is accounted for by solving

$$\frac{dn}{dt} = -\frac{n^2}{2}\langle v\sigma \rangle$$

for

$$\frac{n(t)}{n_0} = \frac{1}{1 + \tfrac{1}{2}\langle v\sigma \rangle n_0 t}$$

we find a fusion energy yield of

$$E_{\text{fusion}} = \tfrac{4}{3}\pi R^3 (n_0 - n) W_{\text{fusion}}/2$$

$$= \tfrac{4}{3}\pi R^3 \frac{n_0^2}{4} W_{\text{fusion}} \left[\frac{\langle v\sigma \rangle t}{1 + \tfrac{1}{2}\langle v\sigma \rangle n_0 t} \right] \tag{2.59}$$

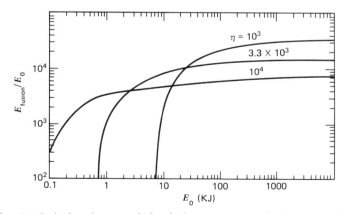

Figure 2.9. Analytical estimates of the fusion energy production in cold D-T fuel ignited by a spherically propagating burn wave. (After Brueckner and Jorna, Ref. 21.)

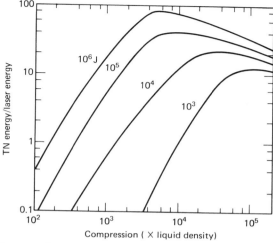

Figure 2.10. Estimates of target gain versus compression based on early versions of the LASNEX code. (After Emmett et al., Ref. 19.)

If we take t to be the disassembly time $\tau_d = R/c_s$ and furthermore assume that burning occurs at 80 keV, we find

$$E_{\text{fusion}} = \frac{1.75 \times 10^7 \eta^2 R^4}{1 + 0.00648 R\eta} \text{ kJ}$$

Brueckner and Jorna[21] used this expression to calculate the fuel gain $G_f = E_{\text{fusion}}/E_0$ for various ignition energies E_0 and compressions η. These results are plotted in Figure 2.9. Similar estimates based on LASNEX computer calculations at Livermore[32] are given in Figure 2.10. These results suggest that very large fuel gains are possible if burn propagation occurs.

2.6.4. SOME FURTHER COMMENTS

Hence the key ingredients in achieving high ICF target gain are compression, self-heating, and thermonuclear burn propagation. The trick is to bring the fuel to very high density, but in such a way that only the central region is heated to ignition temperatures. Then self-heating and burn propagation will ignite the remainder of the fuel material.

High compressions will require a high implosion velocity of the material surrounding the fuel region. To be more precise, if the ignition temperature of the fuel is taken as 3 keV, then the thermal velocity of ions in the fuel is $v_{\text{th}} = (kT_i/m_i)^{1/2} = 3.3 \times 10^7$ cm/s. If we take this as a rough measure of the required implosion velocity, then we find a rough estimate of the required implosion energy[31] for a 1-mg target as

$$E_{\text{imp}} = \frac{1}{2} m u_{\text{imp}}^2 \sim \frac{1}{2} m v_{\text{th}}^2 = 45 \text{ kJ}$$

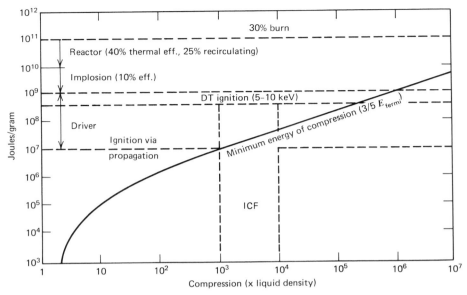

Figure 2.11. The energy densities required to achieve fuel ignition for various compressions. (After Nuckolls et al., Ref. 22.)

A second important consideration in achieving high gain is to compress the fuel in such a way that one avoids heating the fuel material—that is, isentropically. As we will show in the next chapter, the minimum work required for the compression of fuel can be estimated by using the ideal gas relationship for adiabatic compression, $pV^\gamma = $ constant, to find

$$W_{1 \to 2} = \frac{nk}{1-\gamma} T_1 \left[\left(\frac{V_1}{V_2} \right)^{\gamma-1} - 1 \right]$$

For example, the work required to compress a 1-mg target through a volume change of 1000, assuming the fuel starts at a temperature of $T_1 = 1$ eV, is only 6 kJ.

We have summarized these discussions in Figure 2.11, which indicates the energy densities required to achieve fuel ignition for various compressions.

REFERENCES

1. R. F. Post, "Controlled Fusion Research: An Application of the Physics of High Temperature Plasmas," *Rev. Mod. Phys.* **28**, 338 (1956).

2. D. J. Rose and M. Clark, *Plasmas and Controlled Fusion* (New York, MIT Press–Wiley, 1961).

3. W. Stacey, *Magnetic Fusion Physics* (New York, Wiley-Interscience, 1981).

4. J. Rand McNally, Jr., K. E. Rothe, and R. D. Sharp, "Fusion Reactivity Graphs and Tables for Charged Particle Reactions," Oak Ridge National Laboratory Report ORNL TM-6914, August, 1979.

5. J. L. Tuck, "Thermonuclear Reaction Rates," Los Alamos Scientific Laboratory Report LAMS-1640 (March, 1954).

6. W. B. Thompson, "Thermonuclear Reaction Rates," Harwell Laboratory Report AERE-T/M-138 (May, 1956).

7. W. B. Thompson, "Thermonuclear Reaction Rates," *Proc. Phys. Soc.* **70B**, 1 (1957).

8. N. Jarmie and J. D. Seagrave, eds., "Charged Particle Cross Sections," Los Alamos Scientific Laboratory Report LA-2014 (March, 1956).

9. S. L. Greene, Jr., "Maxwell Averaged Cross Sections for Some Thermonuclear Reactions on Light Isotopes," Lawrence Livermore Laboratory Report UCRL-70522 (May, 1967).

10. J. Rand McNally, Jr., and R. D. Sharp, "Advance Fuels for Inertial Confinement Fusion," *Nucl. Fusion* **16**, 868 (1976).

11. G. W. Shuy, "Advanced Fusion Fuel Cycles and Fusion Reaction Kinetics," University of Wisconsin Fusion Project Report UWFDM-335 (December, 1979).

12. R. W. Conn and G. W. Shuy, "Alternate Fusion Fuel Cycle Research," presented at the 8th Int. Conf. on Plasma Phys. and Controlled Nuclear Fusion Research, Brussels, 1980.

13. M. Gordinier and R. Conn, University of Wisconsin Fusion Project, October, 1976.

14. J. D. Lawson, "Some Criteria for a Power Producing Thermonuclear Reactor," *Proc. Phys. Soc. London* **B70**, 6 (1957).

15. K. Boyer, "Laser Fusion," *Aero. and Astro.* 28 (July, 1973).

16. R. E. Kidder, "Lectures on Inertial Confinement Fusion: The Inside Story," presented at the AUA-ANL Faculty Workshop on Inertial Confinement Fusion, Argonne National Laboratory (1978).

17. R. Grande, "Laser Driven Fusion," Les Hoches Lectures on Strongly Coupled Plasmas (1980).

18. E. Teller, "A Future ICE (Thermonuclear, That Is!)," *IEEE Spectrum* 60 (January, 1973).

19. J. L. Emmett, J. Nuckolls, and L. Wood, "Fusion Power by Laser Implosion," *Sci. Am.* **231**, 24 (June, 1974).

20. G. A. Moses, R. W. Conn, and S. I. Abdel-Khalik, "The SOLASE Conceptual Laser Fusion Reactor Study," *Proc. Third Topical Meeting on the Technology of Controlled Nuclear Fusion*, Santa Fe, NM (1978).

21. K. Brueckner and S. Jorna, "Laser Driven Fusion," *Rev. Mod. Phys.* **46**, 325 (1974).

22. J. Nuckolls, J. L. Emmett, and L. Wood, "Laser Induced Thermonuclear Fusion," *Physics Today* 46 (August, 1976).

23. K. A. Brueckner, "Introduction to Laser Driven Fusion," *Energy*, Vol. 3, *Nuclear Energy and Energy Policies*, Chap. 23, ed. by S. S. Penner (Reading, Mass., Addison-Wesley, 1976), p. 349.

24. K. A. Brueckner, "An Assessment of Drivers and Reactors for Inertial Confinement Fusion," K. A. Brueckner Associates, prepared for the Electric Power Research Institute, Report EPRI-AP-1371 (1980).

25. *Laser Program Annual Reports*, Lawrence Livermore Laboratory Report UCRL-50021 (1976).

26. Ya. B. Zel'dovich and Yu. P. Raizer, *Physics of Shock Waves and High-Temperature Hydrodynamic Phenomena* (New York, Academic, 1966).

27. G. S. Fraley, E. J. Linnebur, R. J. Mason, and R. L. Morse, "Thermonuclear Burn Characteristics of Compressed Deuterium-Tritium Microspheres," *Phys. Fluids* **17**, 474 (1974).

28. H. Brysk, "Electron-Ion Equilibration in a Partially Degenerate Plasma," *Plasma Phys.* **16**, 927 (1974).

29. D. B. Henderson, "Burn Characteristics of Marginal Deuterium-Tritium Microspheres," *Phys. Rev. Lett.* **33**, (1974).

30. G. D. Beynon and G. Constantine, "A Study of Fusion-Neutron Heating in Laser-Compressed Deuterium-Tritium Spheres," *J. Phys.* **G3**, 81 (1977).

31. R. J. Mason and R. L. Morse, "Hydrodynamics and Burn of Optimally Imploded Deuterium-Tritium Spheres," *Phys. Fluids* **18**, 814 (1975).

32. *Laser Program Annual Reports*, Lawrence Livermore Laboratory Report UCRL-50021 (1977).

THREE

The Physics of Hydrodynamic Compression

The primary objective in inertial confinement fusion is to compress the fuel pellet in such a way that only a central region of the compressed fuel mass is brought to ignition temperatures, leaving the rest of the compressed pellet as cold as possible. In this way one can ignite a central spark that propagates through the compressed fuel mass as a thermonuclear burn wave. This approach minimizes driver energy requirements by using the driver only to compress the fuel to densities sufficient for ignition and efficient thermonuclear burn.

The essential processes involved in this scheme are shown in Figure 3.1.[1, 2] The driver energy is deposited in the outer layers of the fuel pellet (perhaps even in the outer layers of the plasma corona surrounding the pellet). This energy is then transported inward to the ablation surface by mechanisms such as electron thermal conduction where it heats the surface material, ablating it outward. This ablation process produces large pressures that drive imploding shock waves into the pellet, compressing the fuel to very high densities in such a manner as to avoid premature fuel heating (isentropic compression). In this way the minimum amount of work is required to achieve fuel conditions necessary for efficient thermonuclear burn. At peak compression, the central region of the compressed fuel mass is brought to ignition temperature by the converging shock waves. A thermonuclear burn wave is produced that propagates outward, consuming the cold compressed fuel material and releasing thermonuclear energy.

To achieve this energy-efficient thermonuclear burn, we place two requirements on the hydrodynamic implosion process: first, we must isentropically compress the fuel core to ρR values greater than 1 g/cm^2, second, we must implode the fuel in such a way that the shock strength at the center is great

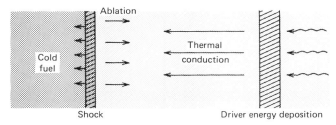

Figure 3.1. The physical processes involved in ICF target implosion.

enough to heat and ignite a central spark or hot-spot of fuel when the shock waves reflect at the origin.

We examine each of the important physical processes involved in this approach to inertial confinement fusion, working backward from the compression of the fuel material to optimum inertial confinement fusion burn conditions ($\rho R > 1$), to the generation of pressures by ablation, to energy transport, and finally to the deposition of driver energy at the outer surface of the pellet. More specifically, our sequence of considerations is as follows:

1. Isentropic compression by convergent shock waves.
2. Equations of state for highly compressed matter.
3. Ablation-generated pressures.
4. Energy transport in plasmas.
5. Driver energy deposition.

We begin with a discussion of the physics of the hydrodynamic compression of matter.

3.1. PLASMA HYDRODYNAMICS

We begin by briefly reviewing the hydrodynamics of a single-species fluid such as a gas. We will characterize the state of the fluid by three hydrodynamic variables:

$$\text{mass density: } \rho(\mathbf{r}, t) = mn(\mathbf{r}, t)$$

$$\text{local fluid velocity: } \mathbf{u}(\mathbf{r}, t)$$

$$\text{temperature: } T(\mathbf{r}, t)$$

We can implement the laws of conservation of mass, momentum, and energy to write [3-5]:

$$\frac{\partial \rho}{\partial t} + \nabla \cdot \rho \mathbf{u} = 0 \qquad \text{(mass)}$$

$$\rho \left(\frac{\partial}{\partial t} + \mathbf{u} \cdot \nabla \right) \mathbf{u} - \frac{\rho}{m} \mathbf{F} = -\nabla \cdot \mathbf{P} \qquad \text{(momentum)} \qquad (3.1)$$

$$\rho c_v \left(\frac{\partial}{\partial t} + \mathbf{u} \cdot \nabla \right) T = -\nabla \cdot \mathbf{q} - \mathbf{P} : \Lambda \qquad \text{(energy)}$$

where

$$\mathbf{P} \equiv \text{pressure tensor}$$

$$\Lambda_{ij} \equiv \frac{m}{2} \left(\frac{\partial u_i}{\partial x_j} + \frac{\partial u_j}{\partial x_i} \right)$$

$$\mathbf{q} \equiv \text{heat flux density}$$

These equations can be derived in a number of ways, including control volume arguments or by taking moments of the Boltzmann equation.[6,7] Although they are exact, they are incomplete as they stand since the pressure tensor and heat flux are as yet unspecified. To close the equations, it is customary to introduce the usual approximate transport laws:

Stokes law of viscosity:

$$\mathbf{P} = p\mathbf{I} - 2\frac{\mu}{m} \left(\Lambda - \frac{m}{3}\mathbf{I}\nabla \cdot \mathbf{u} \right) \tag{3.2}$$

where p is the local hydrostatic pressure, $p = \rho(k/m)T$, and μ is the shear viscosity.

Fourier's law of thermal conduction:

$$\mathbf{q} = -\kappa \nabla T \tag{3.3}$$

where κ is the thermal conductivity.

The *hydrodynamics equations* characterizing the fluid can be written as:

$$\frac{\partial \rho}{\partial t} + \nabla \cdot \rho\mathbf{u} = 0$$

$$\rho\left(\frac{\partial}{\partial t} + \mathbf{u} \cdot \nabla \right)\mathbf{u} - \frac{\rho}{m}\mathbf{F} = -\nabla p + \tfrac{1}{3}\nabla\mu\nabla \cdot \mathbf{u} + \mu\nabla^2\mathbf{u} \tag{3.4}$$

$$\rho c_v\left(\frac{\partial}{\partial t} + \mathbf{u} \cdot \nabla \right)T = -p(\nabla \cdot \mathbf{u})T + \nabla \cdot \kappa\nabla T$$

These are sometimes referred to as the *Navier-Stokes equations*.

Thus far we have confined our attention to a single-species gas. But a plasma must often be modeled as a two-component fluid, accounting for the dynamics of both the ions and electrons.[8] In this case we would choose six hydrodynamic variables:

$$\rho_e, \rho_i, u_e, u_i, T_e, T_i$$

Therefore we might expect to need six hydrodynamics equations. Furthermore, we now have the added complication of the electric and magnetic fields associated with the long-range Coulomb interaction of the electrons and ions. Hence to these six hydrodynamics equations we must also add Maxwell's equations, which self-consistently describe electromagnetic forces on the fluids. Such an extended set of hydrodynamics equations is rather complex.[9]

Fortunately for most inertial confinement fusion applications, we can simplify somewhat this set of equations for a plasma by first recognizing that over the length scales of most concern, there is no charge separation.[10] More precisely, we can ignore charge separation if the ratio of the Debye length to the electron mean free path is much less than one:

$$\frac{\lambda_D}{\lambda_{ee}} \ll 1$$

Here we have defined[11]

$$\lambda_D \equiv \text{Debye length} = \left(\frac{kT_e}{4\pi n_e e^2}\right)^{1/2}$$

$$\lambda_{ee} \equiv \text{electron mean free path} = \frac{(3kT_e)^2}{8(0.714)\pi n_e e^2 \ln \Lambda}$$

In inertial confinement fusion applications, this ratio typically ranges from 10^{-3} (for $n_e \sim 10^{23}$ and $kT_e \sim 1$ keV) to 0.03 (for $n_e \sim 10^{27}$ and $kT_e \sim 1$ keV).

The fact that the Debye length is much shorter than the mean free path (and hence hydrodynamic length scales in ICF plasmas) allows us to eliminate the force term $\rho \mathbf{F}/m$ and avoid the need to solve Maxwell's equations along with the hydrodynamics equations. (It should be noted that this approximation can frequently be relaxed by computing the transport coefficients that appear in the hydrodynamics equations in such a way that accounts for electric or magnetic field effects.)[12]

The short Debye length also allows us to assume that electron fluid and ion fluid charge densities and velocities are equal,

$$n_e \sim Z n_i \qquad u_e \sim u_i$$

that is, the plasma behaves as a "single fluid" rather than as a mixture of two fluids. However in most cases the time scales of interest are much shorter than the electron-ion temperature equilibration times (although usually larger than the electron or ion self-equilibration times):

$$\tau_{ee} < \tau_{ii} \ll \tau \ll \tau_{ei}$$

Hence we must characterize each of the components of the plasma fluid by a different temperature, $T_e \neq T_i$.

Therefore the model most frequently used to describe the hydrodynamics of an inertial confinement fusion plasma is a single-fluid, two-temperature description in which

$$n = n_i(1+Z)$$

$$\rho = n(m_i + Zm_e)$$

$$u_i = u_e = u$$

$$p = p_i + p_e$$

$$T_i \neq T_e$$

Several other remarks are in order before we write down the full set of the hydrodynamics equations for an inertial confinement fusion plasma. First, we note that because of the large mass difference between the electrons and the ions, $m_i \gg m_e$, the ions are responsible for momentum transport (and hence viscosity) while the electrons are responsible for energy transport (and hence thermal conduction) in the plasma.[9,13] In mathematical terms,

$$\mu_i \frac{\partial u_i}{\partial x} \gg \mu_e \frac{\partial u_e}{\partial x}$$

$$\kappa_e \frac{\partial T_e}{\partial x} \gg \kappa_i \frac{\partial T_i}{\partial x} \qquad (3.5)$$

Furthermore, the very high thermal conductivity in a plasma leads to a Prandtl number, $\mathrm{Pr} = \mu c_p / \kappa \sim 0.065$, which is very small. That is, inertial confinement fusion plasmas are nearly inviscid.

The single-fluid, two-temperature hydrodynamics equations used to describe an inertial confinement fusion plasma can now be written as

$$\frac{\partial \rho}{\partial t} + \nabla \cdot \rho \mathbf{u} = 0$$

$$\rho \left(\frac{\partial}{\partial t} + \mathbf{u} \cdot \nabla \right) \mathbf{u} - \frac{\rho}{m} \mathbf{F} = -\nabla p + \tfrac{1}{3} \nabla \mu_i \nabla \cdot \mathbf{u} + \mu_i \nabla^2 \mathbf{u}$$

$$\rho c_{ve} \left(\frac{\partial T_e}{\partial t} + \mathbf{u} \cdot \nabla \right) T_e = \nabla \cdot \kappa_e \nabla T_e - p_e (\nabla \cdot \mathbf{u}) - \omega_{ei}(T_e - T_i) + S_e$$

$$\rho c_{vi} \left(\frac{\partial T_i}{\partial t} + \mathbf{u} \cdot \nabla \right) T_i = \nabla \cdot \kappa_i \nabla T_i - p_i (\nabla \cdot \mathbf{u}) + \omega_{ei}(T_e - T_i) + S_i \qquad (3.6)$$

It is this set of equations that is normally solved in the target design computer codes discussed in Chapters 6 and 9. The two-temperature nature of these equations is most important in those regions of the pellet where the driver energy is deposited and in the electron thermal conduction region—that is, in the corona or ablation regions. Here the electron and ion temperatures are likely to be only weakly coupled.

Fortunately, in the cold, compressed plasma core of primary concern in this chapter, the electron-ion equilibration time is still very short, being comparable to 10^{-12} s. So for the purposes of the following discussion, it is appropriate to use the single-species fluid model of the plasma.

It is with this restricted viewpoint that we now consider in some detail the processes involved in imploding fuel pellets to very high densities.

3.2. SHOCK WAVES

The compression of the fuel in an ICF target is driven by the ablation of the target surface using driver energy deposition. The velocity of the material ablating off of the surface is determined essentially by the local speed of sound in the high temperature ablation region. But the speed of sound in the cold fuel region ahead of the ablation front is quite low. Hence the inward motion of the pellet surface due to ablation pressure is supersonic with respect to the cold fuel material, and shock waves form. Such shock waves play an important role in the ICF pellet implosion process. In this section we study the general physics of shock wave propagation.[4,14-19]

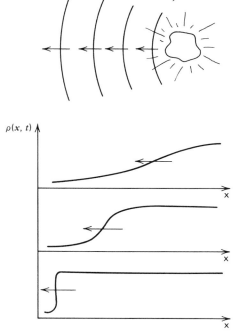

Figure 3.2. Formation of a shock wave.

When a large disturbance is suddenly introduced into a gas, say by the rupture of a diaphragm maintaining a pressure differential in the gas or by a rapid local deposition of energy, this disturbance will propagate into the adjacent gas with roughly the local speed of sound c_s. But the speed of sound is proportional to the square root of the gas density, $c_s \sim \rho^{1/2}$. Hence regions of the disturbance with higher densities will tend to propagate faster than those of lower density, thereby causing the density perturbation to steepen into a sharp wave front or *shock wave* propagating faster than the speed of sound in the ambient gas ahead of the wave. (See Figure 3.2).

Mathematically, we can define a shock wave as any abrupt disturbance that propagates through the gas, causing a change of state. The Euler equations for an ideal fluid predict that such shock waves will propagate as a discontinuity in ρ, u, and T. But dissipative phenomena such as viscosity and thermal conduction will yield a finite shock wave thickness (although the shock thickness is frequently on the order of the molecular mean free path).

To be more specific, we will consider the propagation of a plane (one-dimensional) shock wave propagating from right to left in a medium with a speed D. We will furthermore assume that the fluid ahead of and behind the shock wave is in steady state, described by the state variables indicated in Figure 3.3.

It is customary to take the flow velocity ahead of the shock to be zero. That is, the ambient gas ahead of the shock wave is assumed to be at rest: $u_0 = 0$.

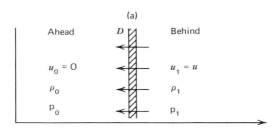

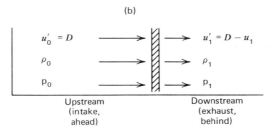

Figure 3.3. Coordinate frames for shock wave analysis. (*a*) Fixed frame of reference. (*b*) Frame of reference moving with shock wave.

The gas behind the shock is then set into motion with a velocity u_1. We are usually given the density and pressure in the ambient gas ahead of the shock, ρ_0, p_0, along with some measure of the shock wave strength such as its speed D or the driving pressure behind it, p_1. Our goal then is to determine the gas properties after the shock wave has passed such as the density ρ_1 and the flow velocity u_1. Here it should be noted that one commonly introduces the *Mach number* characterizing the shock wave, defined as the ratio of the shock speed and the speed of sound in the ambient gas ahead of the shock:

$$M = \frac{D}{c_{s0}} = \text{Mach number}$$

It should be apparent that, by definition, the Mach number characterizing the shock wave is greater than one; the shock wave propagates supersonically into the gas ahead.

To analyze the shock wave, it is convenient to shift to a coordinate frame moving along with the shock (see Figure 3.3). In this frame the gas appears to decelerate from a speed $u_0' = D$ to a slower speed $u_1' = D - u_1$. For that reason, one refers to the gas ahead of and behind the shock wave as the "upstream" and "downstream" shock regions or the "intake" and "exhaust" regions, respectively. For convenience, we drop the primes from the notation for the velocities u_0 and u_1 in the coordinate frame moving with the shock wave.

We can easily determine the downstream variables by using the conservation equations 3.1, written here in one-dimensional form (and setting viscosity and thermal conductivity equal to zero, for the present at least):

$$\frac{\partial \rho}{\partial t} + \frac{\partial}{\partial x}(\rho u) = 0$$

$$\frac{\partial}{\partial t}(\rho u) + \frac{\partial}{\partial x}(p + \rho u^2) = 0 \tag{3.7}$$

$$\frac{\partial}{\partial t}\left(\rho e + \frac{\rho u^2}{2}\right) + \frac{\partial}{\partial x}\left[\rho u\left(e + \frac{u^2}{2} + \frac{p}{\rho}\right)\right] = 0$$

Here we have found it convenient to introduce two new quantities:

$$\text{specific internal energy} = e = c_v kT$$

$$\text{specific enthalpy} = h = e + \frac{p}{\rho}$$

For steady-state flow we can ignore the time derivatives and integrate the conservation equations across the shock to find

$$\text{mass conservation:} \; \rho_0 u_0 = \rho_1 u_1$$

$$\text{momentum conservation:} \; p_0 + \rho_0 u_0^2 = p_1 + \rho_1 u_1^2$$

$$\text{energy conservation:} \; h_0 + \frac{u_0^2}{2} = h_1 + \frac{u_1^2}{2} \tag{3.8}$$

These equations are known as the *Rankine-Hugoniot relations*.[4,14,16,20] They are quite general and exact (and can be derived from a number of different perspectives, including simple physical arguments).

The Rankine-Hugoniot relations represent three equations for six unknowns: (ρ_0, u_0, p_0) and (ρ_1, u_1, p_1), since the specific enthalpy h is given in terms of the density and pressure by an equation of state

$$h_0 = h_0(\rho_0, p_0)$$

$$h_1 = h_1(\rho_1, p_1) \tag{3.9}$$

We are presumably given the density and pressure in the undisturbed gas ahead of the shock, ρ_0 and p_0. Furthermore we are frequently given the "strength" of the shock in terms of the shock speed $D = u_0 - u_1$ or the driving pressure p_1. (We will usually assume the latter situation since it most closely approximates the situation of interest in ICF applications.) Therefore we have three equations in three unknowns:

$$\rho_0, p_0, p_1 \Rightarrow \rho_1, u_1, u_0 \tag{3.10}$$

To proceed further, we must assume some form of equation of state characterizing the gas. For the moment we will leave this arbitrary and develop a slightly different perspective of the shock propagation. We begin by solving the Rankine-Hugoniot relations for the upstream and downstream velocities in terms of the specific volumes $V_0 = 1/\rho_0$ and $V_1 = 1/\rho_1$:

$$u_0^2 = V_0^2 \left(\frac{p_1 - p_0}{V_0 - V_1} \right)$$

$$u_1^2 = V_1^2 \left(\frac{p_1 - p_0}{V_0 - V_1} \right)$$

but

$$\tfrac{1}{2}\left(u_0^2 - u_1^2 \right) = \tfrac{1}{2}\left(p_1 - p_0 \right)\left(V_0 + V_1 \right)$$

so that

$$h_1 - h_0 = \tfrac{1}{2}\left(p_1 - p_0 \right)\left(V_0 + V_1 \right)$$

When combined with equations of state, this yields the pressure behind the shock as a function of the pressure ahead of the shock and the specific volumes:

$$p_1 = H(V_1, p_0, V_0) \tag{3.11}$$

This function relating p_1 to $V_1 = 1/\rho_1$ is known as the *shock Hugoniot*.[4] It is

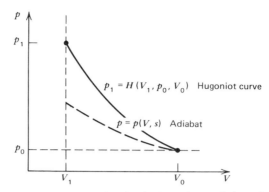

Figure 3.4. A comparison of the shock Hugoniot and the adiabat for a gas.

most convenient to represent this function as a curve on the usual p-V diagram familiar from thermodynamics (see Figure 3.4).

It should be noted at the outset that the shock Hugoniot differs significantly from the p-V relation characterizing the reversible, adiabatic (or isentropic) compression of a gas which behaves as $pV^\gamma =$ constant. We recall that the area under the p-V curve represents the work required to compress the fluid. Hence the fact that the shock Hugoniot lies above the *adiabat* or *isentrope*[4,21,22] for the gas implies that more work is required to compress a material by the passage of a shock than would be required by an isentropic compression (that follows the adiabatic curve). We will note later that this is evidence of the fact that the propagation of a strong shock wave is not isentropic. Irreversible processes such as viscosity and thermal conduction increase the internal energy (i.e., temperature) of the shocked medium beyond the minimum energy necessary to merely compress it. We will also demonstrate that the stronger the shock wave, the more the shock Hugoniot will depart from the adiabat (the more work required for compression by the shock—the more energy dissipated).

Sometimes the shock Hugoniot is written in a slightly different form

$$p - p_K = \frac{\gamma}{V}(e - e_K)$$

where p_K and e_K are the pressure and specific internal energy as functions of volume at 0 K. γ is a parameter dependent only on volume and is known as the *Grüneisen ratio*. This particular form of the Hugoniot, known as the Mie-Grüneisen equation of state, is very useful for studying the propagation of shock waves in solids.

To make this discussion more explicit, let us consider the equation of state for an ideal gas[4,14,16] (for the details involved in the derivation of this equation of state, see Section 3.7). The ideal gas model is characterized by

$$e = c_v T = \left(\frac{1}{\gamma - 1} \right) pV$$

$$h = c_p T = \left(\frac{\gamma}{\gamma - 1} \right) pV \qquad (3.12)$$

Hence we can solve for

$$u_0^2 = \frac{V_0}{2} \left[(\gamma - 1) p_0 + (\gamma + 1) p_1 \right]$$

$$u_1^2 = \frac{V_0}{2} \left[\frac{[(\gamma + 1) p_0 + (\gamma - 1) p_1]^2}{(\gamma - 1) p_0 + (\gamma + 1) p_1} \right]$$

In this way we can calculate the relationship between the upstream and downstream variables:

$$\frac{p_1}{p_0} = \left[\frac{(\gamma + 1)\rho_1 - (\gamma - 1)\rho_0}{(\gamma + 1)\rho_0 - (\gamma + 1)\rho_1} \right]$$

$$\frac{\rho_1}{\rho_0} = \left[\frac{(\gamma + 1)p_1 + (\gamma - 1)p_0}{(\gamma - 1)p_1 + (\gamma + 1)p_0} \right] \qquad (3.13)$$

$$\frac{T_1}{T_0} = 1 + \left[\frac{2\gamma}{(\gamma + 1)^2} \right] \left[\frac{\gamma M_0^2 + 1}{M_0^2} \right] (M_0^2 - 1)$$

Notice that if we substitute the specific volume into the pressure-density relation, we find an explicit form for the shock Hugoniot in an ideal gas

$$p_1 = p_0 \left[\frac{(\gamma + 1)V_0 - (\gamma - 1)V_1}{(\gamma + 1)V_1 - (\gamma - 1)V_0} \right] = H(V_1, p_0, V_0) \qquad (3.14)$$

We can also calculate the upstream and downstream Mach numbers:

$$M_0 = \left[\frac{(\gamma - 1) + (\gamma + 1)p_1/p_0}{2\gamma} \right]^{1/2} \rightarrow \left(\frac{\gamma + 1}{2\gamma} \right)^{1/2} \left(\frac{p_1}{p_0} \right)^{1/2}$$

$$M_1 = \left[\frac{(\gamma - 1) + (\gamma + 1)p_0/p_1}{2\gamma} \right]^{1/2} \rightarrow \left(\frac{\gamma - 1}{2\gamma} \right)^{1/2} \rightarrow 0.45 \qquad (3.15)$$

As we might expect, $M_0 > 1$ and $M_1 < 1$ imply that the motion of the shock is supersonic into the material ahead of it and subsonic with respect to the gas behind the shock.

The limiting form of the relations 3.13 between upstream and downstream variables for very strong shock waves ($p_1/p_0 \to \infty$) is particularly interesting:

$$\frac{p_1}{p_0} \to 1 + \frac{2\gamma}{\gamma+1}\left(M_0^2 - 1\right) \to \infty$$

$$\frac{\rho_1}{\rho_0} \to \left[\frac{(\gamma+1)M_0^2}{(\gamma-1)M_0^2 + 2}\right] \to \frac{\gamma+1}{\gamma-1} \qquad (3.16)$$

$$\frac{T_1}{T_0} \to \frac{(\gamma-1)}{(\gamma+1)}\frac{p_1}{p_0} \to \infty$$

This dependence is shown in Figure 3.5. In particular, we note that while the temperature and pressure rise across the shock will increase indefinitely with the strength of the shock, the compression or density change approaches an asymptotically limiting value of

$$\frac{\rho_1}{\rho_0} \to \frac{\gamma+1}{\gamma-1} \qquad (3.17)$$

For an ideal monatomic gas, $\gamma = \frac{5}{3}$. Hence we find the important result that the maximum compression that can be achieved by a single plane shock wave in a monatomic gas is 4.

We can also calculate the entropy change across the shock. If we define the specific entropy s as

$$s = c_v \ln p\rho^{-\gamma}$$

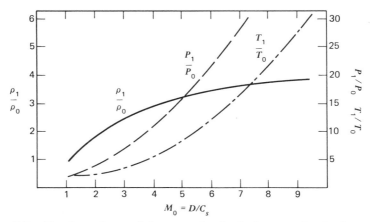

Figure 3.5. The dependence of the state of a shocked gas on the shock strength.

then the entropy change is

$$\Delta s \equiv s_1 - s_0 = c_v \ln \left\{ \frac{p_1}{p_0} \left[\frac{(\gamma-1)p_1 + (\gamma+1)p_0}{(\gamma+1)p_1 + (\gamma-1)p_0} \right]^\gamma \right\} \qquad (3.18)$$

In particular, we note that the entropy increases as the logarithm of the shock strength, p_1/p_0:

$$\Delta s \sim c_v \ln C \frac{p_1}{p_0} \to \infty$$

(It should be noted that while we have not specifically included the effects of viscosity and heat conduction in our analysis based on the Rankine-Hugoniot relations, such dissipative behavior is still properly described by calculating the change in the state of the gas as we have done. A specific inclusion of viscous and thermal conduction effects would only affect the shock wave structure, not the states of the gas ahead of and behind the shock.) We should also note that in the limit of weak shock waves,

$$\Delta s \to 0 \qquad \text{as} \qquad \frac{p_1}{p_0} \to 1$$

that is, the propagating disturbance tends to the limiting case of an isentropic acoustic (sound) wave.

With this background, let us return to our diagram of the shock Hugoniot (Figure 3.4) and address the question of how we might use shock waves to isentropically compress thermonuclear fuel to high density while leaving it relatively cold. Since weak shock waves approach an isentropic sound wave, we might attempt to use a series of many weak shocks to approach isentropic compression. This approach is shown in detail in the p-V diagram of Figure 3.6, in which a series of multiple shocks are used to approximate the adiabat and isentropically compress the fuel to a much higher density for the same final pressure (and therefore requiring far less p-V work than would be required by a single shock wave).

From a somewhat different point of view, we recall that the maximum compression achievable by a single plane shock wave in a monatomic gas is 4. Therefore, by subjecting the shocked gas to a second shock wave, we can increase the density by a factor of $4 \times 4 = 16$. We can continue on in this fashion to multiple-shock the fuel to higher and higher density.[4]

But as we noted earlier, the compression of a gas using a strong shock is highly nonisentropic and therefore rather inefficient. Our analysis suggests that blasting a fuel pellet with arbitrarily large pulses of energy, thereby driving strong shock waves through it, will not serve to efficiently compress the fuel. However these shocks will increase the temperature of the fuel to arbitrarily large values—provided enough driver energy is available. Hence it appears

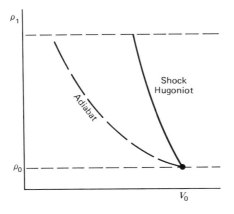

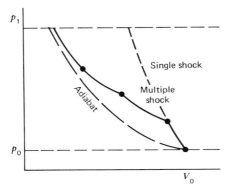

Figure 3.6. Multiple shock compression.

that simple shock compression might be used in part to meet the ignition temperature conditions for efficient thermonuclear burn, even though the required compressions of 1000 to 10,000 times liquid density cannot be achieved in this manner.

Although the general features of shock waves are illustrated by our study of plane shock wave propagation, there are some important differences that arise in the propagation of shock waves in convergent geometries. Hence we now turn our attention to a study of the implosion of a spherically convergent shock wave, which is the phenomenon of most interest in inertial confinement fusion applications.

3.3. SPHERICALLY CONVERGENT SHOCK WAVES

A key idea in the use of shock waves to compress inertial confinement fusion fuel pellets is to use the convergence properties of a spherical implosion to multiply the driving pressure and hence the compression. The earliest analysis of this phenomenon was performed by Guderley,[23] who considered a self-similar solution of the Euler equations for a spherically convergent shock wave in an ideal gas. Before discussing Guderley's solution, it is useful to provide a brief introduction to the concept of self-similar solutions in hydrodynamics.

In fluid dynamics one frequently finds that the fluid variables $\rho(x,t)$, $u(x,t)$, and $T(x,t)$ become a function of a combination of space and time, say

$$\rho(x,t), u(x,t), T(x,t) \rightarrow \text{fcn}(xt^{\alpha})$$

This corresponds in essence to a frozen picture of the flow. That is, all distributions with respect to x change with time without changing form; they remain "similar" to themselves.

The type of flow in which the distributions of flow variables remain similar to themselves with time and vary only as a result of changes of scale is called *self-similar*.[24,25] The most common case is $\alpha = -1$, that is

$$\rho(x,t), u(x,t), T(x,t) \rightarrow \text{fcn}\left(\frac{x}{t}\right)$$

The reason for this behavior lies in the fact that the Euler equations contain no characteristic length or time scales. (Indeed the only length and time scales in a gas are the mean free path and the collision time, which are related to viscosity and thermal conductivity.) The only dimensional parameter is the speed of sound, c_s. Hence the flow can depend only on the combination x/t.

The mathematical importance of self-similar flow is that it reduces the usual partial differential equations describing hydrodynamics to ordinary differential equations. These ordinary differential equations can then be studied (and perhaps even integrated) to determine the hydrodynamic behavior.

Guderley obtained a self-similar solution of the Euler equations describing a spherically convergent geometry in terms of reduced variables related to the radius-time diagram of the shock front shown in Figure 3.7. He found a solution for the shock radius $r_s(t)$ of the form

$$r_s(t) = S\left(1 - \frac{t}{t_c}\right)^{\alpha} \tag{3.19}$$

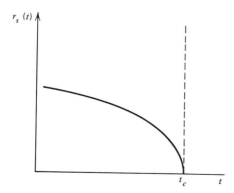

Figure 3.7. The radius-time diagram for an imploding spherical shock front.

where the time t_c corresponds to convergence of the shock at the center. Here, $\alpha = 0.717$ for a monatomic gas ($\gamma = \frac{5}{3}$) and S is a measure of shock strength.

Of more interest, however, was the state of the gas behind the converging shock wave. Guderley found that the passage of a converging shock gives a density increase or compression of 4 (for a gas with $\gamma = \frac{5}{3}$), just as for a plane shock wave. However, this is followed in the spherically convergent case by an adiabatic compression to a total compression of about 15. The shock wave is then reflected at the center, and upon returning gives a further shock compression to 33. In summary, then, Guderley found that the maximum compression from a single convergent shock wave was 33.

Therefore one approach to achieving the ultra–high density conditions necessary for efficient thermonuclear burn would be to deposit energy uniformly on the surface of the fuel pellet in such a way as to produce a strong, convergent shock wave. Brueckner and Jorna[2] have analyzed this situation and predict a driver energy requirement of

$$E_{\text{driver}} = 1.6 \frac{M^3}{\varepsilon_D^4} \eta^2 \qquad \text{MJ} \qquad (3.20)$$

where we recall that η is the desired compression, M is the fusion energy multiplication or gain, and ε_D is the driver-coupling efficiency. Surprisingly enough, this result indicates that the compression of the fuel by a strong convergent shock wave does not appreciably reduce the required driver energy. A more detailed analysis indicates that the effect of compression is offset by an inefficient temperature distribution produced by the passage of the strong shock which reduces fusion energy production in most of the fuel. The compressed pellet core which is strongly heated is too small, and the compression time is too short to produce an appreciable fusion energy yield.[2] More detailed computer calculations indicate that the required driver energy for breakeven using single shock compression is roughly 500 MJ. This is far too large for any conceivable driver.

3.4. ISENTROPIC COMPRESSION

The key to achieving the very high fuel densities necessary for efficient thermonuclear burn is to compress the pellet isentropically in such a way that heating of the dense fuel core is minimized so that the minimum compression energy is required. This can be accomplished by producing a sequence of shock waves that approach the adiabatic curve of compression in the p-V diagram of the fuel. Such isentropic compression requires a gradually rising pressure on the surface of the pellet which generates a sequence of shock waves of increasing strength which are adjusted in time so that successive shocks do not overtake each other before arriving at the center of the pellet.

More precisely, the compression and temperature history of the fuel after the passage of the first shock follow (approximately) an adiabat until the shock reaches the center, where its kinetic energy is converted into internal energy (temperature) and a reflected shock forms. The final temperature is determined by the initial shock strength. To avoid excessive preheating, we want to reach only the minimum temperature required for ignition. Then the achievable compression is limited only by the degeneracy pressure of the electrons or by the ignition of the fuel before maximum compression has been achieved.

Perhaps the simplest model of this process is obtained by assuming we can achieve perfectly isentropic compression by compressing the fuel along an adiabat. In this case, thermodynamics can tell us the energy required for compression. When a gas undergoes an isentropic compression, we first recall that

$$pV^\gamma = \text{constant}$$

where $\gamma = c_p/c_v$. From this relationship, we immediately find that the states before and after compression are related by

$$p_1 V_1^\gamma = p_2 V_2^\gamma$$

or

$$\frac{p_2}{p_1} = \left(\frac{V_1}{V_2}\right)^\gamma$$

hence

$$\frac{T_2}{T_1} = \left(\frac{V_1}{V_2}\right)^{\gamma-1} = \left(\frac{p_2}{p_1}\right)^{(\gamma-1)/\gamma}$$

If we model the compression process using a piston analogy, we find that the work done on the fuel during the isentropic compression is

$$W_{1\to 2} = \int_1^2 p\, dV = \text{constant} \int_1^2 V^{-\gamma} dV$$

$$= \frac{p_2 V_2 - p_1 V_1}{1-\gamma} = \frac{nk}{1-\gamma} T_1 \left[\left(\frac{V_1}{V_2}\right)^{\gamma-1} - 1\right] \qquad (3.21)$$

For example, suppose we take a pellet mass of 1 mg, an initial temperature of $T_1 = 1$ eV, and a volume compression ratio of $V_1/V_2 = 1000$. Then for $\gamma = \frac{5}{3}$, we find the required work is $W = 5.9$ kJ. That is, the amount of work required to compress a milligram of $D-T$ fuel to 1000 times its liquid density can be as low as 6 kJ.

Of course the actual driver energy required will be considerably larger, not only because it is impossible to achieve a perfectly isentropic compression but also because most of the driver energy (95%) is expended in producing the driving pressure (via ablation). One can reduce the compression requirements by using spherical shell targets. The shell stores kinetic energy during the implosion and then produces the necessary temperatures and pressures in the fuel by energy transfer upon convergence. One can also use a shell of $D-T$ fuel inside a tamper shell of massive material. We study several of these more sophisticated target designs in some detail in Chapter 9.

Kidder[26-28] has developed a theory of isentropic compression of shells using self-similarity concepts. Such a theory indicates that it is possible to compress an ideal gas to arbitrarily large compressions with a pressure-time profile of the form

$$\frac{p(\tau)}{p_0} = F(\tau) = \frac{1}{(1-\tau^2)^{5/2}} \tag{3.22}$$

where

$$\tau \equiv t/t_c$$

$$t_c \equiv \frac{R_{20}^2 - R_{10}^2}{3c_{20}^2}$$

These parameters are defined in terms of the shell geometry in Figure 3.8.

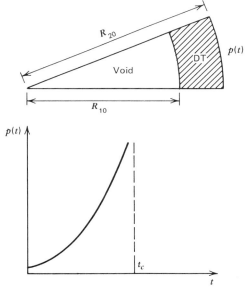

Figure 3.8. Parameters characterizing the isentropic implosion of a spherical shell.

In practice, we want to bring the fuel to ignition temperatures during the latter stages of compression. To this end, we might vary the pressure profile just prior to shock convergence to deviate from isentropic compression. Kidder proposes a pressure-time profile of

$$
\frac{p(\tau)}{p_0} =
\begin{cases}
F(\tau), & 0 \leqslant \tau \leqslant \tau_a \leqslant 1 \\
F(\tau_a)\exp\left[5\tau_a\dfrac{(\tau-\tau_a)}{(1-\tau_a^2)}\right], & \tau > \tau_a
\end{cases}
\tag{3.23}
$$

where $\tau_a = t_a/t_c$ and is usually about 0.9. His analysis indicates that roughly one sixth of the original pellet mass can be compressed by a factor of 10,000.

3.5. SHOCK WAVE PROPAGATION IN PLASMAS

Thus far we have analyzed the propagation of a shock wave in a single-component gas. However a thermonuclear plasma is in fact a two-component gas with a dramatic difference in the mass of each species (electrons versus ions). This leads to a complex shock structure.[9, 13, 29]

To be more precise, we have shown in Figure 3.9 the shock wave structures for a plane shock wave in an ideal gas (described as a sharp discontinuity by the Euler equations), a shock in a real gas (in which viscosity and thermal conductivity broaden out the shock, providing a structure or shock thickness several mean free paths in thickness), and a shock wave in a plasma. We can understand the more complex structure of the shock wave in a plasma if we recall that thermal conduction or energy transport is due to electrons, while viscosity or momentum transport is due to the ions. If we assume a single-fluid, two-temperature model, then the shock structure of the density profile is determined essentially by the ions and therefore a thickness of the order of the ion mean free path (See Figure 3.10).

However, the new feature is the role played by electron thermal conduction. In the shock waves typical of inertial confinement fusion plasmas, the driver energy is deposited in the electrons, resulting in an increase of electron temperature behind the shock. But the very large thermal conductivity of the electrons transports this thermal energy in a thermal conduction wave ahead of the shock. This electron conduction wave leads the ion shock structure by the mean free path for electron-ion collisions.[10] This thermal energy in the electrons ahead of the shock is then transferred to the ions by electron-ion collisons, resulting in the preheating of ions ahead of the shock wave. The ion temperature then rises through the shock wave from viscous heating and may overshoot the electron temperature behind the shock. Eventually, far behind the shock, the electron and ion temperatures will equilibrate.

The presence of a preheating "foot" ahead of the shock due to electron thermal conduction is a very important phenomenon since it reduces the strength and therefore the compression of the plasma shock wave. Another

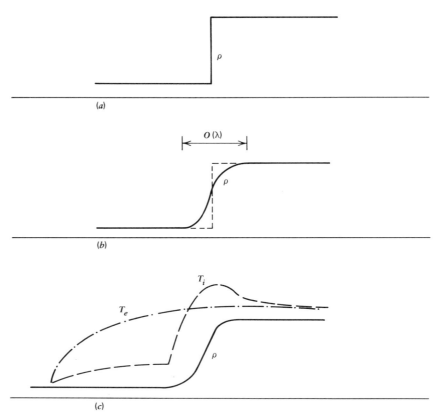

Figure 3.9. Comparison of shock wave propagation in gases and plasmas. (*a*) shock wave in an ideal gas. (*b*) Shock wave in a real gas. (*c*) Shock wave in a plasma.

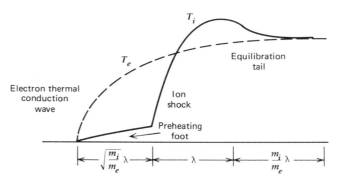

Figure 3.10. The structure of a shock wave in a plasma.

interesting feature of plasma shock waves is the presence of two Mach numbers, one characterizing ion flow (which is essentially the Mach number of the shock)

$$M^2 = \frac{\frac{1}{2}m_i u_i^2 + \frac{1}{2}m_e u_e^2}{\gamma_i kT_i + \gamma_e kT_e} \sim \frac{m_i u_i^2}{2\gamma_i kT_i}$$

and another, M_e, characterizing electron flow

$$M_e = \frac{u}{c_{se}}$$

Since $M \sim (m_i/m_e)^{1/2} M_e$, the electron flow will remain subsonic in the shock even for very large Mach numbers, $M < 30$.

3.6. HYDRODYNAMIC STABILITY OF THE IMPLOSION PROCESS

In the inertial confinement fusion implosion process, the general approach is to use ablation-generated pressures to compress the fuel to high density. In effect, we are attempting to accelerate a dense fluid (the cold fuel) by pushing against it with a lighter fluid (the ablating surface material). This is an unstable hydrodynamic configuration and may lead to the classical Rayleigh-Taylor instability.[30-33]

Perhaps an analogy will make this more apparent. Suppose we try to float a layer of water on top of a layer of lighter fluid such as oil. If we are very careful, we can prepare two such fluid layers in a container. However the slightest disturbance will trigger surface oscillations that will grow until globs of oil begin to pass through the water to the surface under buoyancy forces (gravity).

The pellet implosion scheme may also be subject to a thermal phenomenon known as the Benard instability.[2] This instability arises when a heated fluid is pushing against a colder fluid. A familiar example would be the convection cells that appear in cloud formation processes in the atmosphere. The high temperature ablation front pushing against the cold fuel in an ICF implosion might also be subject to such instabilities.

The early analysis of such hydrodynamic instabilities proceeded by linearizing the hydrodynamics equations characterizing the fuel motion and then examining the stability of normal mode expansions of disturbances as described by these linearized equations.[34] These studies suggested that the Benard instability would not be present, while the Rayleigh-Taylor instability would be mitigated by the ablation process. In particular, it was felt that the ablation process would convect instabilities away from the ablation surface before they could grow to large amplitudes. Such analysis was supported by early implosion experiments on glass microballoons.

However, more recent studies have recognized that such microballoon targets behave in an "exploding pusher" mode rather than an isentropic ablation mode. In these implosions the shell target is heated isothermally very rapidly. The shell explodes, both inward and outward, thereby driving the fuel ahead of it to high density. This nonisentropic implosion process is now felt to be less subject to Rayleigh-Taylor instabilities than true ablatively driven implosions.

Targets designed for ablative implosions typically consist of shells of varying composition to improve driver beam absorption, energy transport, and implosion. In the pellet implosion process, the acceleration force can cause the boundary between the heavier tamper shell and the lighter fuel to become Rayleigh-Taylor unstable. This is most important when the fuel and tamper begin to decelerate as the fuel reaches its final stages of compression just prior to ignition. At this point the large inertial force of the heavy tamper material can result in jets of high Z material streaming into the fuel. This can destroy the ignition process much in the same way that high Z impurities are detrimental to magnetic fusion plasmas. Even if ignition does occur, this mixing of impurities with the fuel can degrade the efficiency of the thermonuclear burn (i.e., fractional burnup). This, in turn, affects the yield and gain of the target in an adverse fashion.

It is felt by some that the Rayleigh-Taylor instability problem will seriously limit the allowable aspect ratio (radius to thickness, $R/\Delta R$) of the target shells.[35-38] Very thin-walled targets tend to be more susceptible to the instabilities. The linear growth of Rayleigh-Taylor instabilities can be modeled as

$$\Delta = \Delta_0 e^{\gamma t}$$

where the growth rate parameter γ is given by

$$\gamma = (\alpha k a)^{1/2}$$

$$\alpha = \frac{\rho_1 - \rho_2}{\rho_1 + \rho_2}$$

$$k = \frac{2\pi}{\lambda}$$

$$a = \text{acceleration}$$

Hence the instability grows most rapidly for large density differences at shell interfaces, for large accelerations, and for short wavelength disturbances. However very short wavelength disturbances quickly grow out of the linear instability regime and cannot be considered using this model. In fact it is found that the most serious wavelengths are those that are about equal to the shell thickness, for these do not saturate before becoming disruptive and, according to this model, these modes have the most rapid growth rate.

However, care must be taken when using such simple linear models of instability growth. Without the inclusion of dissipative effects such as thermal conduction and viscosity, they can be misleading.

Fluid instabilities are an extremely important aspect of the implosion process and cannot be ignored in ICF target design. Their presence (or absence) remains a crucial unanswered question in inertial confinement fusion research.

3.7 EQUATIONS OF STATE

The analysis of the ICF pellet implosion process requires some knowledge of the behavior of matter under extreme conditions of density, temperature, and pressure. Of most interest is the equation of state for highly compressed matter, commonly written in the form

$$p = p(\rho, T)$$

In simple models of the implosion process, the fuel is sometimes modeled as an ideal gas. However, the detailed understanding and design of ICF targets requires a more accurate description of the equation of state, including both collision processes and quantum effects.

We are commonly used to thinking of the properties of matter in terms of the familiar (although complex) states of solids, liquids, and gases. However, at high temperatures ionization and radiation processes become important. Furthermore, at the very high densities characterizing imploded ICF fuels, the electron de Broglie wavelengths become comparable to interparticle spacing, and quantum effects become very important. We can roughly classify the various states of matter on a density-temperature diagram as shown in Figure 3.11.[6,39] The regions of most interest in ICF applications are shaded.

We begin our discussion of equation of state models by reviewing perfect gas models, including the particularly simple case of an ideal gas with only translational degrees of freedom. Not only do such models serve as a point of reference for more complex models, but they are also occasionally used in the analyses of the ICF implosion process. We will then turn our attention to a brief discussion of more complex models in which particle interactions are included such as the Thomas-Fermi-Dirac model and tabulated equations of state.

3.7.1 PERFECT GAS MODELS

By definition, the interaction among particles is ignored in perfect gas models. In many cases of practical importance this model is quite satisfactory, particularly if the contributions to the internal energy of the gas from atomic excitation and ionization are included.[6,22] The equation of state for a perfect

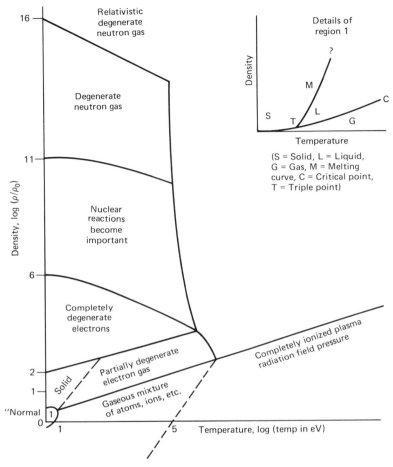

Figure 3.11. The states of matter as classified on a density-temperature diagram. (After Brush, Ref. 39.)

gas can be written as

$$p = nkT = \frac{NkT}{V} = \rho RT$$

where R is the gas constant per unit mass.

The internal energy of a perfect gas can be attributed to a variety of contributions, including translational motion, molecular rotation and vibration, atomic excitation and ionization, and so on. However, the interaction among various particles (atoms, molecules, electrons, and ions) is ignored.

The Ideal Gas Model. At sufficiently low temperatures, or for fully ionized gases, one can ignore internal degrees of freedom and consider only the contribution to the internal energy of the gas from translational motion. If we note that the neglect of particle interactions in the perfect gas model implies that $(\partial e/\partial p)_T=0$ and $(\partial e/\partial V)_T=0$ then it is apparent that the internal energy is a function of temperature only. That is, the perfect gas model ignores the dependence of internal energy on pressure that would arise from particle interactions.

If we now recall the first law of thermodynamics

$$dQ=de+p\,dV$$

and the definition of the specific heat at constant volume, $c_v=(\partial e/\partial T)_V$, we can use the ideal gas assumption, $e=e(T)$ only, to write

$$c_v=\left(\frac{\partial e}{\partial T}\right)_V=\frac{de}{dT}$$

We can substitute this into the first law to find

$$dQ=\frac{de}{dT}dT+p\,dV=c_v\,dT+p\,dV$$

But we recall the equation of state, $pV=NRT$, so that

$$p\,dV+V\,dp=NR\,dT$$

Hence the first law becomes

$$dQ=(c_v+NR)\,dT-V\,dp$$

or

$$\frac{dQ}{dT}=c_v+NR-V\frac{dp}{dT}$$

At constant pressure,

$$c_p=\frac{dQ}{dT}\bigg|_p=c_v+NR=\text{fcn}\,(T)\qquad\text{only}$$

Thus we can find also

$$dQ=c_p\,dT-V\,dp$$

Let us apply these results to describe the adiabatic, reversible (isentropic) compression of an ideal gas. We begin with our two alternative forms of the first law:

$$dQ = c_v\, dT + p\, dV$$

$$dQ = c_p\, dT - V\, dp$$

In an adiabatic process, by definition $dQ=0$, so that

$$V\, dp = c_p\, dT$$

$$p\, dV = -c_v\, dT$$

We can take the ratio of these two expressions to find

$$\frac{dp}{p} = -\frac{c_p}{c_v}\frac{dV}{V} = -\gamma\frac{dV}{V}$$

where we have defined $\gamma = c_p/c_v$. If we integrate this equation, we find the familiar relationship for an adiabatic gas

$$pV^\gamma = \text{constant} \qquad (3.24)$$

This is just the equation for the adiabat on the p-V diagram describing the isentropic compression of an ideal gas. Alternative forms of this result are:

$$p\rho^{-\gamma} = \text{constant}$$

$$T\rho^{1-\gamma} = \text{constant}$$

In an ideal gas,

$$e = \frac{1}{\gamma-1}\frac{p}{\rho} = c_v T$$

$$h = \frac{\gamma}{\gamma-1}\frac{p}{\rho} = c_p T$$

If the gas is monatomic (3 degrees of freedom, $\frac{1}{2}kT$ per degree), then $\gamma = \frac{5}{3}$.

It should be apparent that if we want to maximize the final compressed fuel density ρ_F for a given final pressure p_F, we wish as low an adiabat as possible with a minimum initial p_0 for a given ρ_0 since

$$\frac{\rho_F}{\rho_0} = \left(\frac{p_F}{p_0}\right)^{1/\gamma} \qquad (3.25)$$

that is, we must avoid preheating of the fuel.

We must be careful in applying these results to ICF implosions, however, since, as the density increases at fixed temperature the fuel departs further and further from ideal gas behavior.

Ionization and Electronic Excitation.   It is essential to include the effects of atomic excitation and ionization in equation of state models. Typically, the ionization of atoms begins at values of kT much lower than the ionization potential. In most materials ionization will begin at $kT \sim 7$ to 15 eV (although for the alkali metals this is lowered to several eV).[4] The internal energy of the ionized gas will include contributions from the thermal energy of the particles (ions and electrons) and the potential energy represented by ionization.

At very high temperatures, the energy and pressure due to thermal radiation may become comparable to the hydrodynamic energy and pressure of the gas. If the radiation field is in thermal equilibrium with the gas, one can simply add the radiation energy and pressure to those of the gas. In Chapter 5 we will discuss in more detail how the radiation field can be coupled into the hydrodynamic description of the gas (plasma).

It is customary to assume that the materials comprising ICF targets are rapidly ionized and can be treated as plasmas. Hence it is important to examine the equation of state for such Coulomb gases.

3.7.2.   THE FERMI DEGENERATE ELECTRON GAS

If the density of the fuel becomes high enough, while the fuel temperature remains relatively low, the de Broglie wavelength of the electrons will become comparable to the interparticle spacing, and the exclusion principle will become important. The electrons become a "degenerate electron gas" obeying Fermi-Dirac statistics rather than the Maxwell-Boltzmann distribution:[4,6,21]

$$n(\varepsilon) \sim \frac{1}{\exp\left[(\varepsilon-\mu)/kT\right]+1}$$

As the electrons are compressed, their density is limited by the number of available quantum states. Once a level of states has been filled, no more electrons may be added at this energy. Additional electrons must then be added at higher energy.

If n_e is the electron density, then the maximum momentum state filled in the degenerate gas is given by

$$n_e = \int_0^{p_F} \frac{8\pi p^2}{h^3} \, dp$$

The corresponding maximum energy ε_F is known as the Fermi energy and given by

$$\varepsilon_F = \frac{p_F^2}{2m_e} = \frac{1}{8}\left(\frac{3}{\pi}\right)^{2/3} \frac{h^2}{m_e} n_e^{2/3} = 2.19 \times 10^{-15} n_e^{2/3} \, (eV) \qquad (3.26)$$

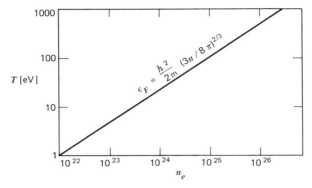

Figure 3.12. Identification of regions of Fermi degeneracy as functions of density and temperature.

The corresponding pressure exerted by the degenerate gas is given by

$$p_F = \frac{1}{20}\left(\frac{3}{\pi}\right)^{2/3}\frac{h^2}{m_e}n_e^{5/3} = 3.3\rho\left(\frac{g}{cm^3}\right)^{5/3} \qquad \text{(Mbar)} \qquad (3.27)$$

(Note that the pressure scales with density just as it did for the ideal gas.) The usual condition for Fermi degeneracy can be expressed by comparing the thermal energy kT to the Fermi energy ε_F. For example, at liquid-state density, $n_e = 4 \times 10^{22}$, and the Fermi energy is $\varepsilon_F = 5$ eV. Hence if the thermal energy of the electrons is below 5 eV, the electrons will behave as a degenerate gas. This, of course, is the situation for electrons in a metal.

By way of contrast, in the highly compressed core of an ICF fuel pellet $n_e \sim 10^{26}$ and $\varepsilon_F \sim 500$ eV. Hence the fuel will be in a degenerate state until it ignites and heats to appreciably higher temperatures. We have plotted the Fermi energy as a function of density in Figure 3.12.

More detailed theories of the electron gas yield the pressure-density relationship as:

slightly degenerate gas:

$$p = nkT\left[1 + n\left(\frac{h^2}{2\pi mkT}\right)^{3/2}2^{-7/2} + \cdots\right] \qquad (3.28)$$

almost completely degenerate gas:

$$p = \left(\frac{3}{4\pi}\right)^{2/3}\frac{h^2}{5m}n^{5/3}\left[1 + \frac{5\pi^2 m^2}{3h^4\beta^2}n^{-4/3} + \cdots\right] \qquad (3.29)$$

3.7.3 THOMAS-FERMI AND THOMAS-FERMI-DIRAC MODELS[4,40-42]

The degenerate electron gas model ignores the Coulomb forces between ions and electrons. For dense fuels, we must account for the Coulomb interaction energy since usually $Ze^2/r_{ei} > \varepsilon_F$, where r_{ei} is the electron-ion separation distance which scales as $\rho^{-1/3}$. For example, in D-T fuel at density $n \sim 4.5 \times 10^{22}$, $T = 5$ eV, and $Z = 1$ the Fermi energy and Coulomb energy are comparable, and therefore we must consider Coulomb effects.

The inclusion of the Coulomb interaction is usually accomplished by using the Thomas-Fermi model, which treats the electron energy as the sum of kinetic and Coulomb terms

$$E_e = \frac{p_e^2}{2m_e} - e\phi(\mathbf{r})$$

Poisson's equation is then solved to determine the self-consistent electron density $n_e(\mathbf{r})$ and potential $\phi(\mathbf{r})$. In essence, the Thomas-Fermi theory models the nuclei as a classical gas, moving freely within a background of Fermi degenerate electron gas. The electrons are not bound in quantum states but instead are influenced by the potential of the nearest nucleus in such a way that the average electron density depends on the local potential energy.

This model can be improved to account for exchange effects (that is, the effective interaction of electrons with parallel or antiparallel spins through the Pauli exclusion principle). This more complex theory is known as the Thomas-Fermi-Dirac model.

3.7.4. OTHER EQUATION OF STATE MODELS

One can use limited experimental data obtained from shock wave experiments on solids to infer equation of state behavior. However of perhaps more use are direct computer simulations of the microscopic behavior of dense matter. One can identify two classes of simulations: those based on random sampling or Monte Carlo methods, and those based on particle dynamics simulations. In Monte Carlo methods,[43] one randomly samples the particle phase space to construct ensemble averages characterizing the equilibrium behavior of a dense system. In essence, one samples from an ensemble of particle configurations weighted by a canonical ensemble distribution. In microscopic particle simulation methods,[44] one integrates the equations of motion characterizing a number of particles (500 or so) as they interact. Time averages of these motions can then be performed to determine macroscopic properties.

As data are accumulated from experiment, theory, or computer simulation they are evaluated and placed in tabulated equation of state data bases. Such tabulated equations of state represent the most accurate data for dense ICF fuels.[45]

3.8. ABLATION-DRIVEN COMPRESSION

The implosion of ICF targets is driven by the forces produced by surface ablation. The incident driver beam energy is absorbed in the outer layers of the target or in the plasma cloud surrounding the target. This energy is then transported into the pellet surface, where it generates high temperatures leading to surface ablation. This thermal wave front at the ablation surface acts like a leaky piston, compressing the cold fuel ahead of it to high density while the hot material at the ablation front expands and ablates away from the pellet surface.

In this section we will ignore the detailed mechanisms of driver beam energy deposition and transport and focus instead on the ablation-driven compression process. Of particular interest is the partitioning of the deposited driver energy among four processes: the thermal and kinetic energy of the ablation layer, and the thermal and kinetic energy of the dense fuel being compressed ahead of the ablation layer.

We will examine a simple model of the spherical, ablatively driven implosion process. This will be followed by experimental results that determine the ablation efficiency for laser irradiated targets. Finally, a few words will be said about different classes of ablative acceleration.

3.8.1. A SIMPLE MODEL OF A SPHERICAL ABLATIVE IMPLOSION

Mayer, Steele, and Larsen[46] have developed a simple model of the ablative implosion of a high aspect ratio (shell radius to shell thickness) spherical shell that examines the efficiency of energy transfer from ablation pressure to shell implosion kinetic energy. The implosion of the shell is driven by two forces: that due to the reactive force of the ablating material (the "rocket effect"), and that due to the ablation pressure caused by the deposition and transport of driver energy as heat into the ablation surface. To model the implosion Mayer et al. assume that the high aspect ratio shell can be adequately represented by an infinitely thin mass shell of mass M_s at radius $R(t)$. The ablation material density ρ_a and ablation velocity v_a (relative to the moving shell surface) are taken as constant. The shell is assumed to contain an adiabatic fuel ($p_F V^\gamma =$ constant) that eventually compresses to a pressure sufficient to reverse the inward motion of the imploding shell.

Newton's law for this model can be written as

$$\frac{d}{dt}\left(M_s \dot{R} \right) = \left(\dot{R} + v_a \right) \dot{M}_s + 4\pi R^2 \left(p_F - p_a \right)$$

The shell loses mass because of ablation at a rate

$$\dot{M}_s = -4\pi R^2 \rho_a v_a$$

To analyze this model, Mayer et al.[46] introduce a characteristic implosion time

$$\tau = \left(\frac{M_0}{4\pi R_0 \rho_a} \right)^{1/2} \frac{1}{c_a}$$

so that Newton's law can be written in a dimensionless form as

$$\eta \frac{d^2 y}{dt^2} = -(M^2 + 1 - \beta y^{-3\gamma}) y^2$$

where

$$\eta(t) = \frac{M_s(t)}{M_0} = 1 - \int_0^t M\alpha y^2 \, dt'$$

$$y(t) = \frac{R(t)}{R_0}$$

$$t' = \frac{t}{\tau}$$

$$M_0 = 4\pi R_0^2 \delta \rho_{\text{shell}} \qquad \text{(initial shell mass)}$$

$$M = \frac{v_a}{c_a} = \text{Mach number}, \qquad c_a = \left(\frac{p_a}{\rho_a} \right)^{1/2}$$

$$\beta = p_0/p_a, \qquad \alpha = \left(\frac{R_0}{\delta} \frac{\rho_a}{\rho_{\text{solid}}} \right)^{1/2}, \qquad \delta = \Delta R$$

This equation has been numerically integrated for initial conditions: $\eta(0) = 1$, $y(0) = 1$, $\dot{y}(0) = 0$ for various choices of the parameters M, α, and β. Before examining these solutions, it is useful to compute the various energies associated with this model. If we multiply by $\dot{R}$ and integrate, we find

$$E_{KE}^{\text{shell}} = \tfrac{3}{2} E_0 \left(\frac{dy}{dt'} \right)^2 \eta = \tfrac{1}{2} M_s \dot{R}^2 \qquad \text{shell kinetic energy}$$

$$E_{KE}^{\text{ablation}} = \tfrac{1}{2} \int_0^t \dot{M}_s (\dot{R} + v_a)^2 \, dt = \tfrac{3}{2} E_0 \int_0^t \dot{\eta} \left(\dot{y} + \frac{M_s}{\alpha} \right) dt' \qquad \text{blow off kinetic energy}$$

$$E_{\text{mech}} = -\int_0^t 4\pi R^2 p_a \dot{R} \, dt = -3E_0 \int_0^t y^2 \dot{y} \, dt' \qquad \begin{matrix} \text{mechanical energy due to} \\ \text{ablation pressure} \end{matrix}$$

$$E_{\text{exh}} = \tfrac{1}{2} \int_0^t \dot{M}_s v_a^2 \, dt = \tfrac{3}{2} E_0 \frac{M^2}{\alpha^2} \int_0^t \dot{\eta} \, dt' \qquad \text{``rocket exhaust'' energy}$$

Here,

$$E_0 = \tfrac{4}{3}\pi R_0^3 p_a$$

is the energy needed to fill the original shell volume at the ablation pressure. Notice the energy balance:

$$\text{input energy} \to E_{KE}^{\text{shell}} + E_{KE}^{\text{ablation}} = E_{\text{mech}} + E_{\text{exh}} \to \text{compression}$$

Detailed examination of the model reveals that the transfer of energy from the "mechanical" input (ablation pressure) to the shell kinetic energy is quite efficient. That is, the fraction of energy turned into pressure at the ablation surface is effectively utilized in the implosion process. The energy transfer efficiency can be defined as

$$\varepsilon = \frac{E_{KE}^{\text{shell}}}{E_{\text{mech}} + E_{\text{exh}}}$$

A useful measure of the relative contributions of the rocket exhaust force and the ablation pressure force driving the shell is provided by the Mach number M. For the smaller values of M expected in shell implosions, the ablation pressure force dominates.

In summary, then, this simple model indicates that it is the ablation pressure that most strongly influences the pellet implosion, and the energy delivered to the shell from the mechanical ablation pressure is the most efficient energy transfer mechanism.

ABLATIVE ACCELERATION: EXPERIMENT AND THEORY

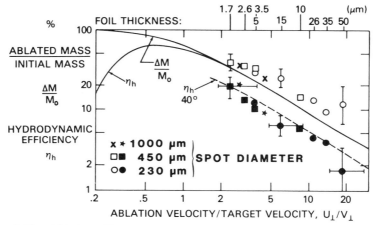

Figure 3.13. Ablation efficiency of laser irradiated thin foils. (After Ripin, et. al., Ref. 48.)

3.8.2. THE EFFICIENCY OF ABLATION-DRIVEN IMPLOSIONS

We wish to estimate the efficiency of coupling thermal energy deposited by the driver into an ablatively driven implosion process. We recognize here that the essential process of interest involves the conduction of heat into the ablation front where it produces pressures that drive the ablation and implosion process. Any analysis of this phenomenon, short of a complete hydrodynamics computer simulation, is susceptible to significant inaccuracy due to over-simplification. For instance, an isothermal blowoff model estimates the maximum implosion efficiency to be about 10%. Rather than resorting to such a simple analysis a priori we start with experimental results and use these as a guide to an appropriate simple theory. In Figure 3.13 are presented results of ablatively accelerated planar thin foil targets by a 1.05-μm Nd glass laser at NRL.[47,48] Thin foil targets are used to simulate spherical shell targets at large radii. The hydrodynamic efficiency is plotted as a function of the ratio of ablation velocity and target velocity. We see that efficiencies of as high as 20% are achievable. Hence the simple isothermal blowoff approximation is in error by a factor of 2. To describe the data, Ripin et al. used a simple rocket model. During the acceleration phase the target of mass M and velocity v is accelerated by the steady state blowoff of the ablated plasma at constant velocity u in the target reference frame. Hence the momentum conservation relationship

$$M\frac{dv}{dt} = -u\frac{dM}{dt}$$

is integrated to yield the well known rocket equation

$$\frac{v}{u} = \ln\left(\frac{M_0}{M}\right).$$

The hydrodynamic efficiency η_h is defined as the kinetic energy of the accelerated target divided by the absorbed laser energy

$$\eta_h = \frac{\frac{1}{2}Mv^2}{E_a}.$$

Since the absorbed laser energy must be balanced by the energy of ablation and the acceleration of the target we can obtain an expression for the hydrodynamic efficiency as

$$\eta_h = (v/u)^2[\exp(v/u) - 1]^{-1}.$$

For small fractional mass loss, this reduces to

$$\eta_h \simeq v/u \simeq \Delta m/M_0.$$

This simple rocket model, plotted along with the experimental data on Figure 3.13 shows surprisingly good agreement.

3.8.3. DETERMINATION OF ABLATION-GENERATED PRESSURES[2,47−49]

The essential physics of the ablation driven compression process are shown in Figure 3.14. The incident driver energy is absorbed in the outer regions of the pellet corona. This energy is conducted into the surface of the pellet, where it generates the high temperatures leading to surface ablation. These temperatures and the surface ablation produce large pressures that drive shock waves into the pellet to compress the fuel.

We can identify three classes of pressure produced by the driver: (1) Ablation pressure due to the flow or ablation of heated plasma from the pellet surface. The ablation pressure p_A is largest where ρT_e is largest. (2) Superthermal particle preheat pressure due to energetic electrons produced in the driver energy deposition region that then stream in to the pellet surface and deposit their energy. (3) Light pressure. If a laser driver is used, the incident light can generate a ponderomotive force or pressure at the critical surface. The magnitude of this light pressure is

$$p_L = \frac{I}{c} \sim 3 \times 10^{-16} I \left(\frac{W}{cm^2} \right) \text{ Mbar}$$

Although the light pressure can affect the blowoff plasma density profile, it cannot directly drive the pellet compression.

The ablation front can be analyzed in a manner very similar to that used to study a shock wave. That is, we can model the front as a discontinuity in the plasma properties.[4] If we move to a coordinate frame fixed at the ablation front (see Figure 3.15), we can again apply the Rankine-Hugoniot relations:

$$\rho_0 u_0 = \rho_1 u_1$$

$$p_0 + \rho_0 u_0^2 = p_1 + \rho_1 u_1^2$$

$$h_0 + \tfrac{1}{2} u_0^2 + \frac{W}{\rho_0 u_0} = h_1 + \tfrac{1}{2} u_1^2 + \frac{W}{\rho_1 u_1} \qquad (3.30)$$

Here we have inserted a new term, W, into the energy equation to represent the heat source due to electron thermal conduction in from the energy deposition region.

There are two characteristic propagation velocities characterizing the ablation process:

heat wave velocity: $v_{HW} \sim W/\rho_0 h_1$ (which can be obtained by equating the absorbed power to the enthalpy flux as we did in Section 2.6.1)

shock wave velocity: $v_{SW} \sim (p_1/\rho_0)^{1/2}$

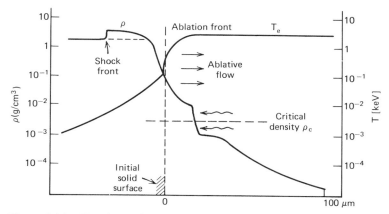

Figure 3.14. Density and temperature profiles in a laser heated plasma.

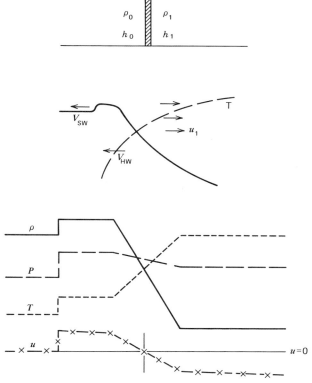

Figure 3.15. Profiles across the ablation front.

These propagation velocities can be used to distinguish among three different cases:

1. If the shock speed v_{SW} is greater than the heat wave speed v_{HW}, then the shock is driven into the dense pellet medium by the ablation as shown in Figure 3.15. The driving pressure is then $p_1 + \rho_1 u_1^2$. The case in which $v_{SW} > v_{HW}$ is known as *subsonic deflagration*.[4]

2. In the opposite situation, the heat front velocity v_{HW} is greater than the shock wave velocity v_{SW}, and the heat wave moves into the pellet material ahead of the shock, preheating it and reducing the compression achieved by the following shock. This is known as a *supersonic thermal conduction wave*. We will consider it in more detail in the next chapter.

3. The limiting case of $v_{HW} = v_{SW}$ is known as the *Chapman-Jouget deflagration*.

If we assume an equation of state for the material, for example, an ideal gas with $h = (\gamma/\gamma - 1)(p/\rho)$, the Rankine-Hugoniot relations provide us with four equations in five unknowns, $u_0, u_1, \rho_1, h_1,$ and p_1 for a given $\rho_0, h_0,$ and W. We need one more equation relating the driver energy and the ablation parameters. This last equation is the heating law representing the interaction physics between the driver and the plasma. It allows us to complete the set of equations and solve for the jump relations across the ablation front, just as for the shock wave problem. Such an analysis[49] indicates that the pressure at the Chapman-Jouget point (where $v_{HW} = v_{SW}$) scales as

$$p_1 \sim 2\rho_1^{1/3}\left(\frac{I}{4}\right)^{2/3} \qquad \text{Mbar}$$

where I is the incident driver intensity in TW/cm^2. In the particular case of Nd laser drivers, we can set ρ_1 approximately equal to the critical density to find that the ablation pressure scales as $p_1 \sim (0.6) I^{2/3}$ Mbar. The pressure ahead at the deflagration front $p_0 \sim 2p_1$.

3.8.4. ELECTRON AND ION BEAM DRIVEN ABLATION

The analysis in this section has been confined to laser-driven ablation, in which the incident laser beam is shielded from the ablation surface by the plasma blowoff cloud surrounding the target. In this process, the incident beam energy is absorbed in the plasma corona in the vicinity of the critical density, and this energy is then transported into the ablation surface.

This situation is modified for electron and ion beam drivers. Here, the incident beam particles penetrate much more deeply into the target. There is nothing analogous to a critical surface beyond which the beam cannot penetrate. Hence the analysis of the ablation process driven by charged particle beams is somewhat different. We further discuss the detailed energy deposition mechanisms for charged particle beams in Chapter 5.

REFERENCES

1. J. Nuckolls, L. Wood, A. Thiessen, and G. Zimmerman, "Laser Compression of Matter to Super-High Densities: Thermonuclear (CTR) Applications," *Nature* **239**, 139 (1972).

2. K. A. Brueckner and S. Jorna, "Laser Driver Fusion," *Rev. Mod. Phys.* **46**, 325 (1974).

3. L. D. Landau and E. M. Lifshitz, *Fluid Mechanics* (New York, Pergamon, 1959).

4. Ya. Zel'dovich and Y. P. Raizer, *Physics of Shock Waves and High Temperature Hydrodynamic Phenomena* (New York, Academic, 1966).

5. H. Lamb, *Hydrodynamics* (New York, Dover, 1945).

6. K. Huang, *Statistical Mechanics* (New York, Wiley, 1963).

7. W. G. Vincenti and C. H. Kruger, *Introduction to Physical Gas Dynamics* (New York, Wiley, 1965).

8. N. A. Krall and A. W. Trivelpiece, *Principles of Plasma Physics* (New York, McGraw-Hill, 1973).

9. M. J. Jaffrin and R. F. Probstein, *Phys. Fluids* **7**, 1658 (1964).

10. M. S. Chu, "Structure of a Plasma Shock Wave," *Phys. Fluids* **15**, 413 (1972).

11. L. Spitzer, *Physics of Fully Ionized Gases.*, 2nd ed. (New York, Wiley, 1962).

12. S. Braginskii, *Review of Plasma Physics*, Vol. 1 (New York, Consultants Bureau, 1965), p. 205.

13. J. D. Jukes, "The Structure of a Shock Wave in a Fully Ionized Gas," *J. Fluid Mech.* **3**, 175 (1957).

14. W. G. Vincenti and C. H. Kruger, *Introduction to Physical Gas Dynamics* (New York, Wiley, 1965).

15. R. A. Gross, in *The Physics of High Energy Density*, ed. by P. Caldirola and H. Knoepfel (New York, Academic, 1971), p. 245.

16. H. W. Liepmann and A. Roshko, *Elements of Gas Dynamics* (New York, Wiley, 1957).

17. R. Courant and K. O. Friedrichs, *Supersonic Flow and Shock Waves* (New York, Wiley-Interscience, 1957).

18. G. B. Whitham, *Linear and Nonlinear Waves* (New York, Wiley-Interscience, 1974).

19. J. von Neumann, "Theory of Shock Waves," in *Collected Works*, Vol. 6 (1943) pp. 178–202.

20. H. Hugoniot, *J. l'Ecole Polytech.* **58**, 1 (1899).

21. L. D. Landau and E. M. Lifshitz, *Statistical Physics* (Reading, Mass., Addison-Wesley, 1958).

22. M. W. Zemansky, *Heat and Thermodynamics* (New York, McGraw-Hill, 1943).

23. G. Guderley, "Starke Kugelige und Zylindriche Verdichtungsstösse in der Nahe des Kugelmittelpunktes bzw. der Zylinderachse," *Luftfahrforschung* **19**, 302 (1942).

24. L. D. Landau and E. M. Lifshitz, *Fluid Mechanics* (New York, Pergamon, 1959).

25. L. I. Sedov, *Similarity and Dimensional Methods in Mechanics* (New York, Academic, 1959).

26. R. E. Kidder, "Theory of Homogeneous Isentropic Compression and Its Application to Laser Fusion," in *Laser Interaction and Related Plasma Phenomena*, ed. by H. Hora and H. Schwarz 1973).

27. R. E. Kidder, "Laser Driven Compression of Hollow Shells: Power Requirements and Stability Limitations," *Nucl. Fusion* **16**, 3 (1976).

28. R. E. Kidder, "Energy Gain of Laser-Compressed Pellets: A Simple Model Calculation," *Nucl. Fusion* **16**, 405 (1976).

29. R. A. Gross and C. K. Chu, "Plasma Shock Waves," *Adv. Plasma Phys.* **2**, 139 (1969).

30. G. Taylor, "The Instability of Liquid Surfaces When Accelerated in a Direction Perpendicular to Their Planes," *Proc. Roy. Soc.* **A201**, 192 (1950).

31. Lord Rayleigh, "Investigation of the Character of the Equilibrium of an Incompressible Heavy Fluid of Variable Density," *Scientific Papers* (Cambridge, 1900), pp. 200–207.

32. R. Lelevier, G. Lasher, and F. Bjorkland, "Effect of a Density Gradient on the Taylor Instability," Lawrence Livermore Laboratory Report UCRL-4459 (1955).

33. G. Wolf, "The Dynamic Stabilization of the Rayleigh-Taylor Instability and the Corresponding Dynamic Equilibrium," *Z. Phys.* **227**, 291 (1969).

34. K. A. Brueckner, S. Jorna, and R. Janda, "Hydrodynamic Stability of a Laser-Driven Plasma," *Phys. Fluids* **17**, 1554 (1974).

35. J. D. Lindl and W. C. Mead, "Behavior of Fluid Instabilities in Laser Fusion Pellets: Results of 2-D Calculations," *Phys. Rev. Lett.* **34**, 1273 (1975).

36. J. Lindl, R. O. Bangerter, J. H. Nuckolls, W. C. Mead, and J. J. Thomson, "Effects of Density Gradient Modification on Fluid Instability in Thermonuclear Micro-Implosions," Lawrence Livermore Laboratory Report UCRL-78470 (1976).

37. J. Boris, "Dynamic Stabilization of the Imploding Shell Rayleigh-Taylor Instability," *Comments on Plasma Physics and Controlled Fusion* **31**, (1977).

38. Yu. Afans'sev, N. G. Basov, E. G. Gamalii, O. N. Krokhin, and V. B. Rozanov, "Symmetry and Stability of Laser-Driven Compression of Thermonuclear Targets," *JETP Lett.* **23**, 566 (1976).

39. S. G. Brush, "On the Equation of State at High Temperatures and Densities," translated from H. Steinwedel, J. Hans, and D. Jensen, *Z. Physik* **125**, 394 (1949).

40. R. Latter, "Temperature Behavior of the Thomas-Fermi Statistical Model for Atoms," *Phys. Rev.* **99**, 1854 (1955).

41. E. E. Saltpeter and H. S. Zapolsky, "Theoretical High Pressure Equations of State Including Correlation Energy," *Phys. Rev.* **158**, 876 (1967).

42. R. P. Feynman, N. Metropolis, and E. Teller, "Equations of State Based on the Generalized Fermi-Thomas Theory," *Phys. Rev.* **75**, 1561 (1949).

43. S. G. Brush, H. L. Sahlin, and E. Teller, "Monte Carlo Study of a One-Component Plasma," *J. Chem. Phys.* **45**, 2102 (1966).

44. H. P. Hansen, E. L. Pollock, I. R. McDonald, and P. Vieillefosse, "Statistical Mechanics of Dense Ionized Matter. III. Dynamical Properties of the Classical One-Component Plasma," *Phys. Rev. A* **11**, 1025 (1975).

45. The SESAME Equation of State Library, Los Alamos Scientific Laboratory, 1979.

46. F. J. Mayer, J. T. Steele, and J. T. Larsen, "A Simple Spherical Ablative-Implosion Model," KMS Fusion, Inc. Report U856 (1980).

47. R. Decoste, S. Bodner, B. Ripin, E. McLean, S. Obenshain, and C. Armstrong, "Ablative Acceleration of Laser-Irradiated Thin-Foil Targets," Phys. Rev. Lett. **42**, 1673 (1979).

48. B. Ripin, R. Descoste, S. Obenshain, S. Bodner, E. McLean, F. Young, R. Whitlock, C. Armstrong, J. Green, J. Stamper, S. Gold, D. Nagel, R. Lehmberg, and J. McMahon, "Laser Plasma Interaction and Ablative Acceleration of Thin Foils at $10^{12}-10^{15}$ w/cm^2," Phys. Fluids **23**, 1012 (1980).

49. J. L. Bobin, F. Delobeau, G. De Giovanni, C. Fauquignon, and F. Floux, "Temperature in Laser-Created Deuterium Plasmas," *Nucl. Fusion* **9**, 115 (1969).

50. J. L. Bobin, D. Colombant, and G. Toton, "Fusion by Laser-Driven Flame Propagation in Solid DT-Targets," *Nucl. Fusion* **12**, 445 (1972).

51. J. Orens, "Accurate Analytic Approximations and Numerical Solutions for the Structure of Quasi-Static Laser Driven Ablation Layers," Naval Research Laboratory Report NRL-4167 (1980).

FOUR

Energy Transport in ICF Plasmas

In inertial confinement fusion, the energy deposited by laser or charged particle beams is used to implode the fuel in tiny pellets to the high density conditions necessary for ignition and efficient thermonuclear burn. In Chapter 2 we considered the various processes occurring in an inertially confined, thermonuclear fusion reaction. In Chapter 3 we turned our attention to the hydrodynamic processes involved in the pellet implosion, such as shock wave propagation, isentropic compression, and ablation-generated pressures. In this chapter we consider the various mechanisms by which energy is transported from the driver energy deposition region into the ablation surface of the target.

In most inertial confinement fusion schemes, the incident driver energy is shielded from the surface of the target by the plasma cloud or corona of blowoff material. For example, an incident laser beam will be unable to propagate to densities higher than the critical density at which the plasma frequency characterizing the blowoff plasma equals the frequency of the incident light. Since this density is usually quite low (10^{21} electrons/cm^3 for 1.06 μm light and 10^{19} electrons/cm^3 for 10.6 μm light), most driver energy will be absorbed in regions of the plasma corona far from the ablation surface. Light and heavy ion beams will also tend to be shielded from the target surface by the blowoff plasma surrounding the target.

Hence the mechanisms for transporting the driver energy deposited in the outer regions of the plasma corona into the ablation surface are of considerable importance in inertial confinement fusion (see Figure 4.1). We will consider three such energy transport mechanisms in this chapter: classical electron thermal conduction, hot (superthermal) electron transport, and radiation transport.

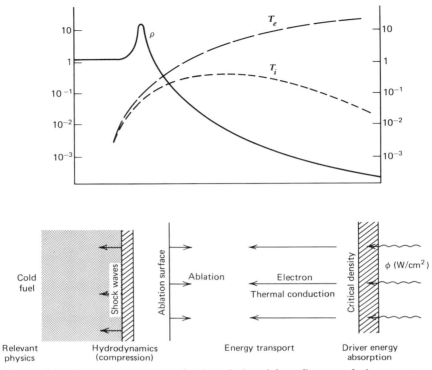

Figure 4.1.   Energy transport mechanisms in inertial confinement fusion targets.

4.1. ELECTRON THERMAL CONDUCTION

An important energy transport mechanism in high temperature plasmas is electron thermal conduction. The small mass of the electrons coupled with the high temperature of the pellet plasma corona or blowoff cloud make its thermal conductivity quite high. Although the process of conventional thermal conduction in plasmas is well understood, there are additional phenomena present in ICF plasmas which complicate the conduction process considerably. The strong temperature dependence of the electron thermal conductivity makes this a highly nonlinear process. The driver energy absorption produces a number of very high energy electrons (so-called superthermal electrons, since they are not characterized by the usual thermal distribution assumed in the plasma hydrodynamic model). These electrons can stream into the dense pellet fuel, preheating it before the ablation-driven shock waves can compress it. The presence of both hot and cold electrons can lead to plasma instabilities which produce a turbulent state in the plasma corona, tending to inhibit the thermal conduction process. Density and temperature gradients can produce large magnetic fields which also inhibit the conduction process. Therefore it should be apparent that the thermal conduction energy transport process is rather complex in ICF plasmas and must be considered in some detail.

4.1.1. ELECTRON THERMAL CONDUCTIVITY

In the classical theory of electron thermal conduction, one assumes that the heat flux is given by Fourier's law[1] of thermal conduction

$$\mathbf{q} = -\kappa \nabla T \tag{4.1}$$

where κ is the thermal conductivity. The heat conduction process is dominated by the fast moving electrons; the contribution of the much slower ions can be ignored to first order. The thermal conductivity characterizing noninteracting electrons diffusing through a background of fixed ions (the Lorentz gas model) has been calculated by Spitzer[2] as

$$\kappa = \frac{5n_e k^2 T_e}{m_e \nu_{ei}} = 20 \left(\frac{2}{\pi} \right)^{3/2} \frac{(kT_e)^{5/2} k}{m_e^{1/2} e^2 Z \ln \Lambda} \tag{4.2}$$

The kinetic theory of gases indicates that heat flow is possible only with a skewed or distorted particle distribution function. This implies that the flow of hot electrons carrying the thermal energy must be compensated by a return drift of cold electrons (See Figure 4.2). An electric field is established by this motion, and this field will contribute to the heat flux. We can represent this contribution by writing

$$\mathbf{q} = -\kappa \nabla T - \beta \mathbf{E}$$

where β is the Peltier coefficient characterizing the thermoelectric contribution

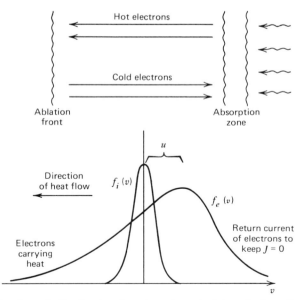

Figure 4.2. The skewed distribution function characterizing electron thermal conduction.

to the heat flux. We can relate the electric field to the temperature gradient by demanding that the net current density be zero:

$$\mathbf{j} = \frac{1}{\eta}\mathbf{E} + \alpha\nabla T = 0$$

where α is the Seebeck coefficient and η is the resistivity. We can solve this for

$$\mathbf{E} = -\alpha\eta\nabla T$$

to rewrite the heat flux as

$$\mathbf{q} = -(\kappa - \beta\alpha\eta)\nabla T = -\kappa_{eff}\nabla T$$

where we have defined an effective thermal conductivity κ_{eff} which takes account of the induced electric field:

$$\kappa_{eff} = \kappa - \beta\alpha\eta = \left(1 - \frac{\beta\alpha\eta}{\kappa}\right)\kappa = \delta(T, Z)\kappa$$

The scaling parameter $\delta(T, Z)$ is a function of temperature and charge and is tabulated by Spitzer. For conditions of interest in the pellet corona, we can approximate

$$\delta = \frac{0.095(Z + 0.24)}{1 + 0.24Z}$$

The general form of the thermal conductivity can therefore be written as

$$\kappa_e = \delta_e(T, Z)20\left(\frac{2}{\pi}\right)^{3/2}\frac{k(kT_e)^{5/2}}{m_e^{1/2}e^4 Z\ln\Lambda_{ei}}$$

$$\kappa_i = \delta_i(T, Z)20\left(\frac{2}{\pi}\right)^{3/2}\frac{k(kT_i)^{5/2}}{m_i^{1/2}e^4 Z^4\ln\Lambda_{ii}} \tag{4.3}$$

For point of future reference, we will also write the electron-ion temperature equilibration frequency as

$$\omega_{ei} = \frac{8(2\pi)^{1/2}n_e Z^2 e^4\ln\Lambda_{ei}}{3m_e m_i k^{3/2}}\left[\frac{T_e}{m_e} + \frac{T_i}{m_i}\right]^{-3/2} \tag{4.4}$$

We should note in particular that all of these coefficients depend strongly on temperature, making thermal processes in plasmas highly nonlinear.

4.1.2. THE CONVENTIONAL THEORY OF HEAT CONDUCTION IN PLASMAS[3]

When energy is deposited locally in a fluid, it gives rise to local perturbations in fluid properties such as density, pressure, and temperature. These disturbances will then propagate away from the source, transporting energy to other regions of the fluid. In most fluids the two principal energy transport modes are hydrodynamic motion (sound waves or shock waves) and thermal conduction. In certain types of high temperature phenonema, radiative transfer can also become an important energy transport mechanism.

The velocities characterizing energy transport via thermal conduction are usually much smaller than those characterizing hydrodynamic disturbances such as sound waves. Therefore in most fluids, pressure disturbances will propagate out more rapidly and equilibrate before temperature disturbances. In these cases, we can effectively decouple thermal conduction from hydrodynamic motion and consider the energy transport to be governed by the equation of thermal conduction

$$\rho c_p \frac{\partial T}{\partial t} = - \nabla \cdot \mathbf{q} + W$$

where the heat flux vector $\mathbf{q}(\mathbf{r}, t)$, is given by Fouriers's law

$$\mathbf{q} = -\kappa \nabla T$$

In conventional heat conduction problems, one usually assumes that the thermal conductivity is a constant so that we can write

$$\rho c_p \frac{\partial T}{\partial t} = \kappa \nabla^2 T + W \tag{4.5}$$

If we divide through by ρc_p, then we find the usual diffusion equation

$$\frac{\partial T}{\partial t} = D_T \nabla^2 T + Q$$

where $D_T = (\kappa / \rho c_p)$ is the thermal diffusivity and $Q = W / \rho c_p$ is a normalized source term. Although the thermal diffusivity is the diffusion coefficient for energy transport rather than particle transport, in gases we can estimate

$$D_T \sim D \sim \frac{\lambda}{3} v_{\text{th}}$$

where λ is the mean free path and v_{th} is the thermal velocity of the particles. (As an example, in air at standard temperature and pressure (STP) conditions, $D_T \sim 0.205$ cm^2/s, while in water, $D_T \sim 1.5 \times 10^{-3}$ cm^2/s.)

When the thermal conductivity is a strong function of temperature, as it is in a plasma, the equation of heat conduction becomes highly nonlinear:

$$\rho c_p \frac{\partial T}{\partial t} = \nabla \cdot \kappa(T) \nabla T + W \tag{4.6}$$

A variety of new phenomena arise such as the presence of supersonic thermal conduction waves. The situation in which the thermal conductivity becomes strongly temperature dependent arises in several phenomena. We have noted that $\kappa \sim T^{5/2}$ in a plasma. Furthermore, in radiative transfer problems one can sometimes define an effective thermal conductivity that scales as $\kappa \sim T^3$.

The general theory of such nonlinear heat conduction has been considered in detail by Zel'dovich and Raizer,[3] who consider a general form for the thermal diffusivity of

$$D_T = \frac{\kappa}{\rho c_p} = a T^n \tag{4.7}$$

We will briefly summarize the results of their analysis with particular applications to heat conduction in plasmas.

Linear Heat Conduction.   We begin by considering the classical problem of a pulsed heat source plane at the origin of an infinite medium for the case in which the thermal conductivity is constant:

$$\frac{\partial T}{\partial t} = D_T \frac{\partial^2 T}{\partial x^2} + Q\delta(x)\delta(t) \tag{4.8}$$

This yields the classical spreading Gaussian shape solutions of the form

$$T(x,t) = \frac{Q}{(4\pi D_T t)^{1/2}} \exp\left(-x^2/4D_T t\right)$$

shown in Figure 4.3. The area under the Gaussian curves is constant and given by

$$\int_{-\infty}^{+\infty} dx\, T(x,t) = Q$$

This, of course, is the usual Green's function solution to the time-dependent diffusion equation. As such, these solutions are not waves since they exhibit an infinite propagation speed. That is, for any time $t>0$, there will be some response in the temperature $T(x,t)$, no matter how far one is from the source plane at $x=0$. There is no true wave front.

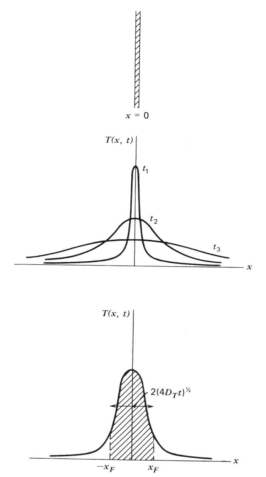

Figure 4.3. Thermal diffusion of a pulsed source as described by linear heat conduction.

We can nevertheless define a psuedo-wave-front speed by noting that most energy is localized in a zone out to a distance

$$x_F \sim (4D_T t)^{1/2} \sim (D_T t)^{1/2}$$

If we characterize the thermal conduction "wave" by a wave front at position $x_F(t)$, we can then calculate a speed of propagation

$$v_{HW} = \frac{dx_F}{dt} \sim \left(\frac{D_T}{t}\right)^{1/2} \sim \frac{D_T}{x_F} \sim \frac{\lambda}{x_F} v_{th}$$

This is an important result, for it implies that after the heat wave has propagated a distance x_F greater than a mean free path (mfp) λ, its speed of propagation will be less than the thermal velocity v_{th}. Since v_{th} is of the same order as the speed of sound c_s, this result implies that as soon as the heat wave has propagated further than a mfp, it will be moving much more slowly than the speed of sound:

$$v_{HW} \sim \frac{\lambda}{x_F} v_{th} \ll v_{th} \sim c_s$$

(Actually, the diffusion equation is only valid for distances greater than a mfp from the source plane in any event.)

Hence we conclude that the diffusion "wave" characterizing a linear heat conduction process always travels much more slowly than does a hydrodynamic disturbance (sound wave or shock wave) and will therefore decouple from hydrodynamic motion.

Nonlinear Heat Conduction. The situation changes dramatically when the thermal conductivity becomes a function of temperature. Suppose, for example, that $\kappa \sim T^n$. If we recall that the heat flux is given by

$$q = -\kappa \frac{\partial T}{\partial x} \sim -T^n \frac{\partial T}{\partial x}$$

and assume that ahead of the heat disturbance, the temperature is essentially zero, then we find that there can be no heat flux ahead of the heat disturbance: $T \sim 0 \Rightarrow \kappa \sim 0 \Rightarrow q \sim 0$. That is, we find a sharp wave front for the heat disturbance—a heat wave (see Figure 4.4). (In the case of linear heat conduction, the conductivity does not vanish for vanishing temperatures, and hence the heat flux is always nonzero for any x.)

We can estimate the shape of the heat wave front by assuming a wave behavior[3]

$$T(x, t) = T(x - v_{HW} t)$$

where v_{HW} is the velocity of the heat wave. If we substitute this trial solution into the nonlinear diffusion equation written in the form

$$\frac{\partial T}{\partial t} = \frac{\partial}{\partial x} a T^n \frac{\partial T}{\partial x}$$

we find

$$-v \frac{dT}{dx} = a \frac{d}{dx} T^n \frac{dT}{dx}$$

We can integrate this equation twice with respect to x, using the boundary

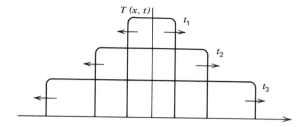

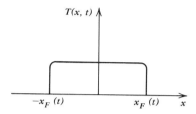

Figure 4.4. Heat wave propagation from a pulsed source as described by nonlinear heat conduction.

condition that $T=0$ at some wave front $x=x_F(t)$, to find a wave shape

$$T(x,t)=\left[\frac{nv_{HW}}{a}|x_F(t)-x|\right]^{1/n} \qquad (4.9)$$

Here $x_F(t)$ and $v_{HW}=dx_F/dt$ are as yet undetermined. Although a precise determination of these quantities requires a detailed solution of the nonlinear thermal conduction equation, we can use dimensional analysis to estimate

$$x_F(t)\sim(aQ^n t)^{1/(n+2)}=(aQ^n)^{1/(n+2)}t^{1/(n+2)}$$

$$v_{HW}(t)=\frac{dx_F}{dt}\sim(aQ^n)^{1/(n+2)}t^{1/(n+2)-1}\sim\frac{x_F}{t}\sim\frac{aQ^n}{x_F^{n+1}}$$

This suggests that the heat wave slows down as it travels away from the source. The stronger the dependence of the thermal conductivity on temperature (the larger the exponent n), the more rapidly the wave slows down.

Zel'dovich and Raizer[3] have constructed an exact self-similar solution to the nonlinear heat conduction equation for a pulsed source at the origin. For the case of a plasma in which $\kappa\sim T^{5/2}$, this solution takes the form

$$T(x,t)=T_C(t)\left[1-\frac{x^2}{x_F^2(t)}\right]^{2/5}$$

where

$$T_C(t) = \frac{Q}{2x_F(t)} \frac{2\pi^{1/2}\Gamma(0.4)}{9\Gamma(0.9)}$$

$$x_F(t) = \xi_0(aQ^{5/2}t)^{2/9}$$

$$\xi_0 = \left[\frac{(\frac{9}{2})^{7/2}2^{-3/2}}{(\frac{5}{2})\pi^{5/4}}\right]^{2/9} \left[\frac{\Gamma(\frac{9}{10})}{\Gamma(\frac{2}{5})}\right]^{5/9}$$

The heat conduction wave velocity can be calculated as

$$v_{HW}(t) = \frac{dx_F}{dt} = \xi_0(aQ^{5/2})^{2/9}t^{-7/9} \tag{4.10}$$

The structure of the heat conduction wave is compared against that for linear heat conduction in Figures 4.3 and 4.4.

A problem of more direct interest to inertial confinement fusion is that of a half-space subjected to a constant heat flux S_0 on its boundary. That is, one considers the nonlinear heat conduction equation

$$\frac{\partial T}{\partial t} = a\frac{\partial}{\partial x}T^{5/2}\frac{\partial T}{\partial x}$$

subject to the boundary condition

$$S_0 = -\kappa\frac{\partial T}{\partial x}\bigg|_0 = -c_v\rho aT^{5/2}\frac{\partial T}{\partial x}\bigg|_0$$

Zel'dovich and Raizer[3] have also obtained a self-similar solution to this problem. For our purposes, however, it is sufficient to use order of magnitude estimates. The average temperature in the wave must be given by balancing

$$S_0 \sim c_v\rho a\frac{T^{5/2+1}}{x_F}$$

But if we note $T/t \sim S_0/x_F$, we can find

$$x_F(t) \sim \left(\frac{S_0^2}{c_v\rho a}\right)^{2/9}t^{7/2}$$

$$v_{HW}(t) \sim \frac{7}{9}(c_v\rho aS_o^{5/2})^{2/9}\frac{1}{t^{2/9}} \tag{4.11}$$

We have compared the structure of the heat conduction wave for a constant

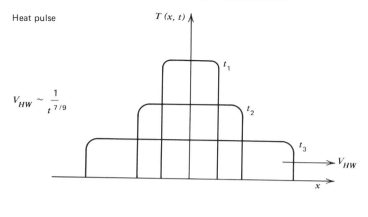

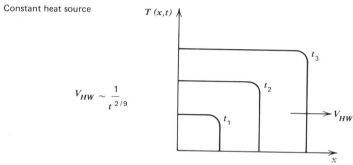

Figure 4.5. Comparison of nonlinear thermal wave propagation from pulsed and steady-state sources.

heat flux on a boundary with that for a pulsed source in Figure 4.5. Although the speed of the heat wave decreases more slowly for the constant heat source problem, once again we find that in the early stage of the thermal wave the propagation speed is large. In fact, if the heat source is large enough, the heat wave will propagate supersonically with

$$v_{HW} > v_{SW} \gtrsim c_s$$

In these cases, the fluid simply does not have enough time to get moving before the heat wave moves into it, heating it to high temperatures. Eventually, however, the heat wave slows down to the speed of sound or below (subsonic propagation).

It is an interesting exercise to determine that heat source or temperature time dependence on the boundary that will match the speeds of the heat wave and the shock wave—that is, satisfy the Chapman-Jouget condition. Let us suppose that the temperature on the boundary is programmed to increase as

$$T_0(t) = Ct^q$$

Then the distance that heat is carried into the medium in a given time scales as

$$x_F \sim (D_T t)^{1/2} \sim T^{n/2} t^{1/2} \sim t^{(nq+1)/2}$$

Hence the speed of the thermal wave behaves as

$$v_{HW} = \frac{dx_F}{dt} \sim \frac{x_F}{t} \sim t^{(nq-1)/2}$$

The shock wave travels into the medium with a speed

$$v_{SW} \sim c_s \sim T^{1/2} \sim t^{q/2}$$

Therefore the speeds of the thermal wave and shock wave will scale similarly with time if

$$\frac{nq-1}{2} = \frac{q}{2}$$

or

$$q = \frac{1}{n-1}$$

For the case of heat conduction in a plasma, $n = \frac{5}{2}$, so that the required temperature dependence is

$$T_0(t) = Ct^{2/3}$$

If we assume that all of the incident driver energy is absorbed at $x = 0$,

$$\rho c_v \frac{dT}{dt}\bigg|_{x=0} = Q(t)$$

then we can infer a driver pulse profile of

$$Q(t) \sim \frac{d}{dt}(t^{2/3}) \sim \frac{1}{t^{1/3}}$$

The implications of this analysis are important: if the rate of energy deposition is too rapid, the thermal wave will propagate supersonically into the pellet before it can be compressed by the following shock wave. If we recognize that a 1-keV electron moves with a thermal speed of $v_{th} \sim 2.3 \times 10^9$ cm/s = 23μm/ps, then it is apparent that these thermal waves can propagate very fast indeed. We should also recall that these nonlinear thermal conduction waves play a very important role in the shock wave structure in a plasma, even when $v_{HW} < v_{SW}$ (see Section 3.6) as we have indicated in Figure 3.10.

4.1.2. CORONA-CORE DECOUPLING

Kidder and Zink[4] have analyzed the coupling of thermal energy between the pellet core and the plasma corona in which the driver energy deposition occurs under the assumption of classical collision processes. The absorption and transport of energy into the pellet core via thermal conduction is governed by several significant time scales:

driver pulse rise time	$\longrightarrow$	rate at which energy is dumped into outer plasma corona or pellet
thermal conduction time	$\longrightarrow$	rate at which energy is conducted into ablation surface
electron-ion and electron-electron collision time (at the ablation front)	$\longrightarrow$	rate at which electrons can transfer energy into the pellet core

We can develop a simple model of how these times affect the transfer of energy from the energy deposition region in the corona to the pellet core. If we characterize the corona and core by bulk temperatures T_{corona} and T_{core}, respectively, then we can write the balance equations

$$\frac{dT}{dt}_{corona} = Q(t) - \frac{\lambda_0}{T_{corona}^{3/2}}(T_{corona} - T_{core})$$

$$\frac{dT}{dt}_{core} = \frac{\lambda_0}{T_{corona}^{3/2}}(T_{corona} - T_{core})$$

Here, $Q(t)$ represents the effective heat source seen by the corona:

$$Q(t) = \left(\begin{array}{c} \text{driver energy} \\ \text{deposition rate} \end{array}\right) \times \left(\begin{array}{c} \text{probability of heat} \\ \text{transfer into core} \end{array}\right)$$

If we assume that the driver energy deposition is governed by classical collision processes (e.g., laser light absorption via inverse bremsstrahlung), then we can scale

$$Q(t) = \frac{Q_0(t)}{T_{corona}^{3/2}} P(T_{corona})$$

One expects that the heat transfer probability P is a decreasing function of the corona temperature.

When these equations are solved, they reveal that if the driver energy increases too rapidly, then the core and corona temperatures will decouple. That is, if the corona is heated too rapidly, it will tend to decouple thermally from the pellet core, and further energy deposition merely heats up the corona to very high temperatures without affecting the core ("burning the fuzz off of the peach"). This effect becomes more pronounced for longer wavelength light in the case of laser drivers since the critical surface where most energy absorption occurs is at lower densities and therefore characterized by smaller collision frequencies.

4.1.3. THERMAL FLUX LIMITERS

Recall that Fourier's law gives the heat flux as

$$\mathbf{q}_e = -\kappa_e \nabla T_e$$

But in a plasma we have noted that the thermal conductivity scales with temperature as $\kappa_e \sim T_e^{5/2}$. Hence for large temperatures, the thermal conductivity becomes very large and Fourier's law will predict too large a heat flux.

Actually, in these instances, Fourier's law breaks down because the electron mfp becomes larger than the temperature gradient, that is

$$\frac{T}{|\nabla T|} \sim mfp \sim 3.81 \times 10^{12} \frac{[T_e(keV)]^3}{n_e(cm^{-3})}$$

A brief comparison of electron mean free paths for typical densities and temperatures is given in Table 4.1. We can compare these estimates against the temperature gradient scales predicted by computer code simulations of target dynamics in the corona region:

$$n_e \sim 10^{22} \ cm^{-3}, \qquad T_e \sim 5 \ keV: \qquad mfp \sim 25 \ \mu m \qquad |T/\nabla T| \sim 10 \ \mu m$$

$$n_e \sim 10^{21} \ cm^{-3}, \qquad T_e \sim 5 \ keV: \qquad mfp \sim 250 \ \mu m \qquad |T/\nabla T| \sim 100 \ \mu m$$

Hence Fourier's law is clearly invalid for these cases.[5-10] This situation is

Table 4.1.   Electron-Ion Mean Free Path (in Microns) as a Function of Electron Density and Temperature

	$n_e(cm^{-3})$			
$T_e(eV)$	10^{19}	10^{21}	10^{23}	10^{25}
10^2	10	1	—	—
10^3	1000	10	0.1	—
10^4	10^5	1000	10	0.1

particularly serious in computer hydrodynamic simulations since the unrealistically high thermal fluxes will lead to nonphysical predictions (e.g., thermal waves propagating faster than the speed of light).

In practice, it is customary to artificially limit the thermal flux to a maximum value which corresponds to free particle flow.[5-8] That is, one chooses a thermal flux which interpolates between the Fourier's law and free streaming limits:

$$q = \left[\frac{1}{q_{FL}} + \frac{1}{q_{\text{streaming}}} \right]^{-1} \tag{4.12}$$

To calculate the free streaming limit, we can simply estimate the flux of particles across a plane surface (assuming an isotropic particle distribution), and then multiply this by the mean kinetic energy per particle

$$q_{\text{streaming}} = \left(\overline{\tfrac{1}{2}mv^2} \right) \tfrac{1}{4} n_e \bar{v} = \left(\tfrac{3}{2}kT_e \right) \tfrac{1}{4} \left(\frac{kT_e}{m} \right)^{1/2} n_e$$

In many computer codes, one simply uses a flux limiting form for the heat flux:

$$q_e = \left[\frac{1}{\kappa_e |\nabla T_e|} + \left(\overline{\tfrac{1}{2}mv^2} \, n_e \bar{v} \right)^{-1} \right]^{-1} \tag{4.13}$$

(This interpolation form has been used for many years in gas dynamics where it is referred to as Sherman's universal relation. It can be derived more rigorously from kinetic theory arguments for linear heat flow problems.)

4.1.4. THERMAL CONDUCTION INHIBITION

Thus far we have discussed only one mechanism of energy transport from the region of energy deposition into the ablation surface, electron thermal conduction. However, one of the most important conclusions drawn from the laser-target interaction experiments performed during the 1970s was that other processes such as fast electron transport, spontaneous magnetic field generation,[11-13] and plasma turbulence may strongly affect the transport of energy in the plasma corona surrounding the target.

For example, a variety of experimental measurements (including X-ray diagnostics) have established that suprathermal or high energy electrons are produced by the laser light absorption process at the critical surface.[14-16] For large light intensities (10^{15} W/cm^2), an appreciable fraction of the absorbed energy appears to go into the production of these fast electrons. If we characterize the hot electrons by an effective temperature T_H, then the measurement of X-ray spectra and fast ions suggests that the hot electron temperature is from 10 to 20 times that of the background electron temperature (1 to

10 keV). The presence of such suprathermal electrons is of particular importance, since their long mfp allows them to penetrate into the target and preheat the fuel core. We return to consider this important phenomenon in some detail in the next section.

A second important phenomenon is an inhibition of electron thermal conduction. More precisely, it is found that the actual value of the thermal flux in the corona region is almost an order of magnitude smaller than that predicted by classical physics (Fouriers's law).[15] It has been suggested that this effect might be explained by either the generation of ion turbulence in the corona region or by the presence of strong magnetic fields produced by density and temperature gradients. In this section we discuss this thermal conduction inhibition process.

The evidence of this inhibition was provided first by a number of foil-irradiation experiments performed at Los Alamos.[14,15] Thin foils were illuminated from one side with intense laser light. The foils were thin enough that classical thermal conduction was expected to be sufficiently strong to transport energy quickly through the foil and yield a symmetric pattern of fast ion blowoff on either side. In fact, however, a strongly asymmetric ion blowoff pattern was observed that could be explained by reducing the heat flux by roughly a factor of 30.

A variety of subsequent experiments have supported the presence of some thermal conduction inhibition mechanism in laser driven targets. For example, researchers at NRL[17] measured the X-ray line spectrum given off by a slab target composed of layers of Al and SiO_2 to determine the penetration depth of the heat conduction wave. They inferred a reduction or inhibition factor of 15 to 25 of the classical thermal conduction. Streak camera measurements of the radial implosion velocity of glass microballoons also indicated a thermal flux inhibition of about this magnitude.[18] Other experiments on a variety of targets using light at differing intensities and pulse lengths confirm the presence of this phenomenon.[19]

To summarize existing experimental data, there appears to be a strong inhibition of thermal conduction in the corona region, estimated to be of magnitude

$$q \sim \frac{1}{20}(-\kappa \nabla T) \quad \text{or} \quad \tfrac{1}{30}(n_e k T_e) v_e$$

These experiments have found strong inhibition in both long and short pulse experiments, in both high and low Z targets, both laterally and axially with respect to the incident laser beam. It has also been determined that suprathermal electrons do not appear to be strongly influenced by this inhibition process.[20,21]

What might be causing the inhibition? Several mechanisms have been proposed that include the presence of strong magnetic fields caused by density and temperature gradients in the corona region, ion turbulence due to a two-stream instability caused by the return current of cold electrons, and even an inadequate modeling of the suprathermal electron transport mechanism.

The large thermal and density gradients induced in the plasma corona can generate currents and hence spontaneous magnetic fields of some strength.[11-13] If we ignore the Hall effect and thermoelectric terms, we can write

$$\frac{\partial \mathbf{B}}{\partial t} = \nabla \times \left[\mathbf{v} \times \mathbf{B} + \frac{c}{en_e} \nabla (n_e kT_e) \right] \tag{4.14}$$

If the electron density gradient and blowoff velocity are parallel to the incident laser beam (see Figure 4.6), we find that in the steady state

$$B \sim \frac{c}{e} \frac{kT_e}{v} \frac{1}{L}$$

where L is the scale length of the electron temperature gradient perpendicular to the radial direction, $L \sim T_e / |\nabla T_e|$. We can take as a rough estimate

$$B = \frac{10^4 T_e(\text{eV})}{L(\mu m)}$$

For example, at $T_e = 1$ keV, $L = 20$ μm, we find a spontaneous field of $B = 500$ kG.

To determine the effect[22] of this field on thermal conduction, we can use the form given by Braginskii for the transport coefficient across a magnetic field:

$$\kappa_e = \frac{\kappa_e(B=0)}{1 + (\Omega_{ce}/\nu_{ei})^2} \tag{4.15}$$

where Ω_{ce} is the electron cyclotron frequency. We can estimate

$$\frac{\Omega_{ce}}{\nu_{ei}} \sim 20 \frac{B(\text{MG})}{Z} \frac{T_e^{3/2}(\text{keV})}{n_e(10^{21})}$$

For example, for $B \sim 1$ MG and $Z = 30$, a density of $n_e \sim 10^{22}$ and temperature of $T_e \sim 1$ keV would yield a frequency ratio of $\Omega_{ee}/\nu_{ei} \sim 0.7$. By way of contrast, in the underdense corona where $n_e \sim 10^{21}$ and $T_e \sim 10$ keV, this ratio becomes

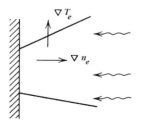

Figure 4.6. Electron density and temperature gradients occurring during laser beam irradiation of slab targets.

$\Omega_{ce}/\nu_{ei} \sim 210$. Hence it is apparent that intense magnetic fields could inhibit heat flow in the underdense corona region. But it seems unlikely that magnetic fields can cause the strong inhibition observed in the overdense corona region between the critical surface and the ablation surface.[21]

An alternative mechanism which might explain the inhibition in the overdense region is the possible presence of ion acoustic density fluctuations (turbulence) which could scatter electrons, thereby inhibiting electron thermal conduction.[23-26] We noted earlier that there is a counter flow of cold electrons to balance the hot electron flow. There is also a flow of ions due to the plasma blowoff from the ablation surface. Since the ion distribution is at rest in a frame of reference moving with the plasma blowoff or flow, while the electron distribution is skewed (to yield a thermal flux), there is a displacement of the maxima of each distribution. (See Figure 4.2.) This is a condition suitable for the presence of a two-stream instability (when $T_e \gg T_i$). The instability results in the formation of ion turbulence. The electrons would then scatter off of the turbulent ion fluctuations, thereby effectively increasing the electron-ion collision frequency ν_{ei}. Calculations at Los Alamos suggest that this process can be modeled by limiting the thermal flux characterizing free particle flow by factors of 10 to 30.[23]

Although there have been several additional studies predicting the presence of ion turbulence in the corona region, there is no general agreement on the size of its effect on thermal conduction. The influence of turbulence on the transport process can be crudely modeled by replacing the usual collision frequency by an effective value due to the scattering of particles from turbulent fluctuations. If we recall that

$$\kappa_e \sim \frac{5 n_e k^2 T_e}{m_e \nu_{ei}}$$

then this model would suggest that in the turbulent region, κ_e could be reduced by turbulence by a factor $\nu_{ei}/\nu_{\text{eff}}$

$$\kappa_e^{\text{turb}} \sim \left(\frac{\nu_{ei}}{\nu_{\text{eff}}} \right) \kappa_e^{\text{class}} \qquad (4.16)$$

Some estimates have suggested that ν_{eff} could range as large as the ion plasma frequency ω_{pi}, in which case the inhibition would be quite strong. However, it is still not clear how effective ion turbulence is in inhibiting heat conduction in the corona region.

Another suggested process is the presence of a Weibel-like instability due to the anisotropy in the electron distribution function caused by the heat flow. This anisotropy could cause small scale magnetic perturbations which would grow exponentially.[27] The electrons would then be deflected by many small encounters with the magnetic field fluctuations to result in an effective collision frequency

$$\nu_{\text{eff}} \sim \frac{v_e}{L} \left(\frac{L\omega_{pe}}{c} \right)^{1/3}$$

where L is the density scale length. While the Weibel instability effect would not be as strong ($\nu_{\text{eff}} < \nu_{ei}$) as the maximum predicted for ion turbulence, it seems to persist over wider variations in temperature and heat flux.

There are also several possible classical effects which might reduce the heat flux. For example, the corona core decoupling mechanism mentioned in a previous section can give rise to hot electrons with mfp's paths much larger than the target radius.[4] These hot electrons can bounce about the corona many times before hitting the core and depositing their energy. This effect might look like an inhibited transport process. The electric fields produced by the counter-flowing hot and cold electron streams could also inhibit the conduction process.[28,29] It might also be that the inhibition process arises from the failure to correctly calculate the thermal conductivity when the electron mfp becomes appreciable[30] (say, $\lambda_{\text{mfp}}/L \sim \frac{1}{30}$).

However, although the particular mechanism which inhibits the thermal transport has not been precisely identified, experimental evidence points to the presence of this effect in laser driven targets. If we recall that most mechanisms that could give rise to the inhibition can be modeled in terms of an effective collision frequency, ν_{eff}, then for the inhibition process to play a significant role, we must have $\nu_{\text{eff}} > \nu_{ei}$. Since ν_{ei} at the critical density scales as the inverse square of the wavelength of the incident light, $\nu_{ei} \sim \lambda_L^{-2}$, we might expect thermal flux inhibition to become more serious at longer wavelengths (e.g., 10.6 μm), while being mitigated to some extent at shorter wavelengths (0.2 μm.[25-27]

4.2. SUPRATHERMAL ELECTRON TRANSPORT

The implosion of ICF targets to efficient thermonuclear burn conditions depends sensitively on the transport mechanisms that couple driver energy deposition to the ablation-generated pressures that produce the implosion. Most driver energy is deposited initially as electron heating in the energy deposition region. We have already noted that there is strong evidence that the process of electron thermal conduction into the ablation surface is sharply reduced below the values predicted by classical theory.

Of perhaps even more concern is the mounting evidence that laser drivers deposit a significant fraction of their energy in very high energy or suprathermal electrons. That is, the various mechanisms giving rise to laser light absorption in plasmas (e.g., resonance absorption or parametric process) can produce high energy tails on the electron distribution function. Numerous experiments and computer simulations have confirmed the presence of these fast electrons.

Their presence is of particular importance, since they can stream into the core of the target, ahead of the ablation front, preheating the fuel and resulting in significantly reduced compressions. In addition, suprathermal electrons can give rise to a lower ablation pressure for a given driver power, since the energy coupling between hot and cold electrons decreases as the energy of the hot electrons increases.

Evidence of the production of suprathermal electrons in laser irradiated plasmas comes from several sources. A primary source has been the analysis of X-ray emission from targets.[11, 31-33] As we indicate in Chapter 9, the continuum or bremsstrahlung emission from these targets can be used to infer their temperatures. Detailed X-ray measurements have revealed that the X-ray emission cannot be characterized by one temperature.[34] Instead, there is a low temperature distribution with a high energy tail. Fits to these data suggest that the temperature of the hot or suprathermal component is some 10 to 20 times that of the thermal electron component.

It is also possible to measure the fast ions that are accelerated in the space charge field established by the freely expanding electrons leaving the low density corona.[35-38] Some recent experiments have attempted to measure the hot electron temperatures directly.

Several theories have been advanced to explain the production of suprathermal electrons as a result of the laser-plasma interaction process.[39, 40] Perhaps the most likely explanation involves a wavebreaking process induced by resonance absorption. As we discuss in more detail in the next chapter, resonance absorption is a process in which light incident obliquely to a plasma density gradient can excite and drive electron plasma oscillations. The energy coupled into the plasma oscillations can then dissipate through damping mechanisms (e.g., Landau damping) to appear as kinetic energy of heated electrons. If the incident light intensity is strong enough, the electron plasma waves are driven sufficiently strongly that electrons can be accelerated to high velocity through one wave period. At this point, wave breaking occurs in which electrons are accelerated out of the thermal distribution and to very high velocities—that is, suprathermal electrons are produced. A detailed analysis of the wavebreaking process indicates that these appear as very fast electrons moving outwards, away from the denser regions of the target. After a few of these electrons escape the target, a space charge field develops that reflects these hot electrons back in toward the target core, while accelerating ions in the plasma blowoff. Theoretical calculations[32] suggest that the suprathermal electrons can be characterized by an effective temperature

$$T_{\text{hot}}\,(\text{keV}) \sim 8.5\left[T_{\text{cold}}\,(\text{keV})\right]^{1/4}\left[I_L(10^{15}\,\text{W/cm}^2)\lambda_L(1.06\,\mu\text{m})\right]^{0.39}$$

$$(4.17)$$

The transport of these suprathermal electrons is a particularly complex process because of their exceptionally long mfp. If we recall that the mfp of electrons scales as

$$\lambda \sim \frac{m_e^2}{4\pi e^4 \ln \Lambda} \frac{v^4}{n_e}$$

then we can estimate the mfp of 100 keV electrons as 10^7 μm for $n_e \sim 10^{21}$, 10^5 μm for 10^{23}, 10^3 μm for 10^{25}, and 10 μm for 10^{27}. Hence, except for the very dense compressed core of the target, these suprathermal electrons see a relatively transparent plasma. They tend to bounce off the space charge potential on the outer regions of the corona, being heated to higher and higher temperatures until they finally strike the core. At this point they can penetrate ahead of the shock wave–ablation region, causing preheating of the pellet fuel.[41] (See Figure 4.7)

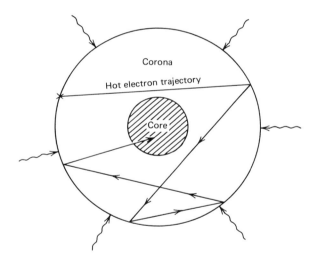

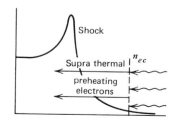

Figure 4.7. Suprathermal electron transport in ICF targets.

An accurate description of this phenomenon would require a kinetic theory analysis based upon a transport or kinetic equation for the electron distribution function. We consider such models based upon the Fokker-Planck equation[42] in Chapter 6. In most hydrodynamic computer simulation models, however, the suprathermal electrons are handled using multigroup diffusion theory.[8, 9] To account for long mfp, a flux-limiting procedure is used.

Other phenomena complicate the description of suprathermal electron transport. For example, a spatially localized hot electron component can give rise to a two-stream instability, thereby losing energy to Langmuir turbulence as the electrons stream inward toward the pellet core.[39] The streaming can also induce a background return current of cold electrons that could excite ion-acoustic turbulence.[40] Magnetic fields could also affect the suprathermal electron transport process.[26]

It is now apparent that suprathermal electron transport plays a very significant role in laser-driven ICF targets. This process not only complicates an accurate description of the energy transport process from the energy deposition region into the ablation surface, but it can also give rise to significant fuel preheating and lead to reduced ablation pressures. In fact, the concern about suprathermal electron production has been a major reason for the upsurge in interest in ion-beam drivers that do not appear to suffer from this problem.

4.3. RADIATION TRANSPORT

Radiative processes and radiation transport play an important role in inertial confinement fusion. The hot, dense plasmas produced in an ICF target emit and reabsorb radiation at soft X-ray wavelengths. This radiation represents a significant energy transfer mechanism and can strongly affect the implosion and thermonuclear burn dynamics of the target. For example, the burning thermonuclear fuel will lose energy via radiation, and this effect must be accurately described in order to predict ignition conditions. In many target designs the fuel is surrounded by a high-Z tamper layer to contain this radiation. In high gain target designs, the radiation produced by the burning fuel (at roughly 100-keV temperatures) is reabsorbed by the surrounding tamper and ablated target material (at temperatures of several keV).

Radiative transfer also plays an important role in the implosion process. This effect must be accurately described if the energy transfer to the ablation surface is to be properly programmed to achieve an optimum shock convergence. Even a small change in these energy transfer rates from the design optimum can significantly degrade the efficiency of the implosion process.

X-ray emission from the target is an important diagnostic signature in ICF experiments. Measures of the spatial and temporal distribution of the radiation emitted from ICF targets is a primary source of information about the implosion process. Furthermore, a detailed understanding of the X-ray production from ICF targets is essential to the design of suitable blast chambers in ICF reactor concepts.

The study of radiative processes in ICF targets is complicated by the fact that the models appropriate for describing the production, transport, and absorption vary significantly from region to region of the target. For example, while local thermodynamic equilibrium (LTE) models are usually sufficient for describing radiative transfer within the compressed inner layers of the tamper and fuel, in the expanding ablation material, coronal models or sometimes even fully non-LTE/rate equation models must be used. In this section we review the models used to describe radiation production and radiative transfer in ICF targets.

4.3.1. THE RADIATION FIELD

In a broad sense radiative transfer encompasses all phenomena involving the propagation of electromagnetic radiation and its interaction with matter. For example, radiative transfer problems arise in astrophysics, meteorology, photometry, high speed gas dynamics (radiation hydrodynamics), and plasma physics. In analyzing radiative transfer it is customary to emphasize the particle aspect of electromagnetic radiation by considering the radiation field to be composed of a "photon gas" and then applying traditional methods of kinetic theory.[43-49] For example, we can introduce a photon phase space density or distribution similar to those used in the kinetic theory of gases:

$$n(\mathbf{r}, \hat{\Omega}, \nu, t)\, d\nu\, d^3r\, d\hat{\Omega} = \begin{array}{l} \text{expected number of photons of} \\ \text{frequency } \nu \text{ in } d\nu, \text{ in volume } d^3r \\ \text{about } \mathbf{r}, \text{ traveling in direction} \\ \hat{\Omega} \text{ in } d\hat{\Omega}, \text{ at time } t \end{array}$$

In radiative transfer studies it is more convenient to work with the radiation specific intensity or spectral intensity function $I_\nu(\mathbf{r}, \hat{\Omega}, t)$, defined by

$$I_\nu(\mathbf{r}, \hat{\Omega}, t) = h\nu c n(\mathbf{r}, \hat{\Omega}, \nu, t)$$

If we recall that photons move with a speed c (ignoring refraction effects) and are characterized by an energy $E = h\nu$, we can identify $I_\nu(\mathbf{r}, \hat{\Omega}, t)\, d\nu\, d\hat{\Omega}$ as the radiant energy in the spectral interval $d\nu$ passing through a unit area per unit time with a direction $\hat{\Omega}$ in $d\hat{\Omega}$. That is, the spectral intensity $I_\nu(\mathbf{r}, \hat{\Omega}, t)$ can be identified as just the photon energy angular flux.

The spectral radiant energy density is then defined as

$$u_\nu(\mathbf{r}, t) = \frac{1}{c} \int_{4\pi} I_\nu(\mathbf{r}, \hat{\Omega}, t)\, d\hat{\Omega}$$

while the spectral energy flux vector is given by

$$S_\nu(\mathbf{r}, t) = \int_{4\pi} \hat{\Omega} I_\nu(\mathbf{r}, \hat{\Omega}, t)\, d\hat{\Omega}$$

Using these concepts, we can derive an equation of radiative transfer for the spectral intensity by balancing the change in I_ν due to photon streaming or transport against the change in I_ν due to sources and sinks:

$$\frac{1}{c}\frac{\partial I_\nu}{\partial t}+\hat{\mathbf{\Omega}}\cdot\nabla I_\nu=\text{change in } I_\nu \text{ due to sources and sinks}$$

$$=j_\nu(\mathbf{r},\hat{\mathbf{\Omega}},t)-k_\nu(\mathbf{r},t)I_\nu(\mathbf{r},\hat{\mathbf{\Omega}},t) \qquad (4.17)$$

where we have denoted the photon emission and absorption terms by j_ν and $k_\nu I_\nu$ (and explicitly noted that the absorption rate is linearly proportional to the spectral intensity I_ν). These terms are more commonly expressed in terms of the photon mass emission coefficient ε_ν defined by

$$j_\nu=\rho\varepsilon_\nu=\frac{\text{rate of radiant energy emitted per unit}}{\text{phase space volume}}$$

and mass attenuation coefficient

$$k_\nu I_\nu=\rho\kappa_\nu I_\nu=\frac{\text{rate of radiant energy absorption per}}{\text{unit phase space volume}}$$

where $\rho(\mathbf{r},t)$ is the mass density of the host material. Photon scattering processes are customarily included in the definitions of ε_ν and κ_ν (since a scattering event corresponds to the absorption followed by the reemission of a photon).

The equation of radiative transfer can then be written as

$$\frac{1}{c}\frac{\partial I_\nu}{\partial t}+\hat{\mathbf{\Omega}}\cdot\nabla I_\nu=\rho(\mathbf{r},t)\left[-\kappa_\nu(\mathbf{r},t)I_\nu(\mathbf{r},\hat{\mathbf{\Omega}},t)+\varepsilon_\nu(\mathbf{r},t)\right] \qquad (4.18)$$

In writing this equation, we have also neglected polarization and dispersion (dependence of the refraction index on ν) and collective effects (correlations), and we have assumed an isotropic medium, permitting us to regard photon interactions as independent, successive isolated events. Coherent phenomena such as the reflection of light are also omitted from this description.

The radiation source corresponding to spontaneous emission processes is isotropic in nature. Hence we can define the spontaneous emission coefficient J_ν as

$$j_\nu\,d\hat{\mathbf{\Omega}}=\frac{1}{4\pi}J_\nu\,d\hat{\mathbf{\Omega}}$$

The optical characteristics of a material are determined by an attenuation coefficient μ_ν which consists of the absorption and scattering coefficients

$$\mu_\nu=\rho\kappa_{\nu a}+\rho\kappa_{\nu s}$$

Attenuation of photons is described by the familiar form

$$I_\nu(x) = I_\nu(0)\exp\left[-\int_0^x \mu_\nu(x')\,dx'\right]$$

The optical thickness of a layer x with respect to the photon frequency ν is given by the dimensionless quantity

$$\tau_\nu = \int_0^x \mu_\nu(x')\,dx'$$

Materials opaque to radiation at frequency ν are characterized by $\tau_\nu \gg 1$, while materials transparent to radiation at this frequency have $\tau_\nu \ll 1$.

4.3.2. RADIATIVE PROCESSES

To calculate the absorption and emission coefficients κ_ν and ε_ν appearing in the radiative transfer equation, we must consider the possible interaction mechanisms for a photon propagating through a material. Such processes are associated with transitions between the energy levels of the atoms (or ions) comprising the host material. The change in the internal energy will be equal to the radiant energy absorbed or emitted.

A variety of different processes may be involved, but of most interest to ICF plasmas are bound-bound, bound-free, and free-free electronic transitions involving photons at X-ray wavelengths.[50, 51] When an atom or ion passes from a higher to a lower energy state, the process is accompanied by the emission of a photon with an energy (or frequency) corresponding to the difference between the energy levels. Conversely, a bound electron can absorb a photon and move to a higher energy level. For sufficiently large photon energies (or multiple photon interactions), photoionization becomes important. Photorecombination can also be an important emission process. Free-free electron processes include bremsstrahlung in which an electron emits a photon as it decelerates in the field of a ion, and inverse bremsstrahlung in which an electron in the field of an ion absorbs an incident photon.

Both capture and scattering processes contribute to the absorption coefficient κ_ν. The emission coefficient ε_ν represents the effective photon source term. Every capture process that appears in κ_ν has an inverse that contributes to ε_ν. These inverse processes may be stimulated by a preceding absorption or they may occur spontaneously. In an isotropic medium, the stimulated emission propagates in the same direction (in phase) as the incident radiation; therefore it is customary to subtract out this component by defining

$$\kappa_\nu' = \text{capture} - \text{stimulated emission} + \text{scattering}$$

$$\varepsilon_\nu' = \text{spontaneous emission} + \text{scattering}$$

The detailed calculation of the absorption and emission coefficients for bound-bound and bound-free transitions are complex and involve the detailed atomic structure of the atom or ion species involved. Bound-bound transitions (line radiation) are characterized by very large absorption coefficients that are very strongly peaked functions of frequency (corresponding to resonances at the energy level spacing, $h\nu = E_i - E_j$). Absorption coefficients for free-bound transitions behave as ν^{-3} between the bound electronic states and take large jumps at energies corresponding to the ionization potential of the bound electrons.

The determination of photon interaction rates depends directly on the state populations of the various atomic energy levels. Hence any consideration of photon transport must involve the rate equations for these population densities, N_i, which take the form[52]

$$\frac{\partial N_i}{\partial t} + \nabla \cdot (N_i \mathbf{u}) = \sum_{j=1}^{n} W_{ij} N_j, \qquad i = 1, \ldots, n \qquad (4.19)$$

The rate coefficients W_{ij} in these equations involve a variety of processes in addition to photon interactions. Of most importance are electron impact processes such as excitation and deexcitation by free electrons and electron impact ionization and three-body ionization. The subject of photon interactions in radiative transfer processes can become very complicated indeed. Fortunately, in many cases one can simplify this analysis considerably by assuming that the medium is in thermodynamic equilibrium (or at least in partial equilibrium).

4.3.3. EQUILIBRIUM MODELS

The simplest equilibrium model assumes that both the medium and the radiation field are in thermodynamic equilibrium at a temperature T. This is the case of black-body radiation. In this model, the radiative energy emitted per unit volume in frequency interval $d\nu$ about ν is exactly equal to the energy absorbed per unit volume in this frequency interval. The equilibrium radiation is isotropic and is independent of the specific properties of the medium. Detailed balance arguments[43, 46] can be used to show that the equilibrium radiation energy density is given by

$$u_\nu^P = \frac{8\pi h \nu^3}{c^3} \frac{1}{e^{h\nu/kT} - 1} \qquad (4.20)$$

This spectral energy density function is known as the Planck distribution (the radiation is said to be in "Planckian" equilibrium). It should be noted that the Planck distribution is a function only of the frequency ν and the temperature T. The most probable energy in the Planck distribution is 2.822 kT. The frequency-integrated radiant energy density is given by

$$U^P = \int_0^\infty u_\nu^P \, d\nu = \frac{4}{c}\sigma T^4 = 7.57 \times 10^{-15} T^4 \text{ ergs/cm}^3$$

where T is in K and σ is the Stephan-Boltzmann constant, $\sigma = 5.67 \times 10^{-5}$ ergs/cm^2 s K^4. Hence the total radiant energy density is proportional to the fourth power of the temperature for equilibrium radiation. Clearly, at high temperatures such as those characterizing ICF thermonuclear burn, the radiation energy density is quite large.

As an example, consider D-T fuel burning at a density of 200 g/cm^3 and a temperature of 50 keV.[53] The plasma thermal energy of the fuel is

$$E_{\text{plasma}} = \tfrac{3}{2}(1+Z)nkT = 1.224 \times 10^{19} \text{ ergs/cm}^3$$

The corresponding radiant energy density is

$$U^P = \frac{4}{c}\sigma T^4 = 8.57 \times 10^{20} \text{ ergs/cm}^3$$

Hence at typical thermonuclear burn conditions the radiation contributes more to the energy density of the fuel than does the thermal energy of the D-T plasma. This example is not totally valid because the D-T fuel is actually thin to the radiation, and the equilibrium energy density approximation is not correct. However, the example does illustrate how important radiation effects can become.

The radiant flux integrated over frequency is

$$S^P = \int_0^\infty S_\nu^P \, d\nu = \sigma T^4$$

We can also calculate the emission and absorption coefficients for Planckian equilibrium. To calculate the emission coefficient, we recall that both spontaneous and stimulated emission must be taken into account. When this is done, the emission term can be written as

$$j_\nu \left(1 + \frac{c^2}{2h\nu^3} I_\nu\right)$$

For equilibrium, the emission and absorption coefficients can be related by detailed balance[43] to find

$$\frac{j_\nu}{\kappa_\nu} = \frac{I_\nu}{1 + (c^2/2h\nu^2)I_\nu} \tag{4.21}$$

or in terms of the radiant energy density

$$J_\nu = 4\pi j_\nu = c u_\nu^P \kappa_\nu (1 - e^{-h\nu/kT})$$

Thus far we have assumed that both the radiation field and the material are in thermodynamic equilibrium. A somewhat less restrictive model assumes that the material, but *not* the radiation field is in local thermodynamic equilibrium (LTE) which is maintained by electron collision processes. That is, the radiation field is assumed to be sufficiently dilute that electron excitation and ionization exceed photon-induced processes to yield an equilibrium condition in which the state populations are related by the Boltzmann factor:

$$\frac{N_j}{N_i} \rightarrow \frac{g_i}{g_j} \exp(-h\nu_{ij}/kT)$$

Then the photon emission processes are essentially independent of the radiation field and are given by the Planck distribution

$$\frac{\varepsilon_\nu'}{\kappa_\nu'} = \frac{2h\nu^3/c^2}{\exp(h\nu/kT) - 1} \tag{4.22}$$

(This relationship is known as Kirchhoff's law.)

A somewhat different model takes the opposite extreme by assuming that both the free electron density and the radiation field are dilute enough to permit excited atoms to emit spontaneously and ionized atoms to recombine by photorecombination. In this corona equilibrium model there is no collisonal excitation or deexcitation; therefore to achieve equilibrium, the electron impact ionization and photorecombination processes are balanced.

In many instances these quasi-equilibrium models are inadequate and one is forced to a fully non-LTE calculation in which coupled rate equations for state population densities must be solved simultaneously with the equation of radiation transfer (and perhaps also with the hydrodynamics equations characterizing the motion of the host medium).[52]

4.3.4.   MODELS OF RADIATIVE TRANSFER

We can rewrite the equation of radiative transfer for a medium in local thermodynamic equilibrium as

$$\frac{1}{c}\frac{\partial I_\nu}{\partial t} + \hat{\Omega} \cdot \nabla I_\nu = \kappa_\nu'(I_\nu^P - I_\nu)$$

Since this equation is similar to those arising in other transport phenomena (e.g., neutron transport or gas dynamics), one can apply well-known methods to its analysis.[44-49]

Perhaps the most popular approach is to take angular moments of the radiative transfer equation. Integrating the equation over solid angle, one finds a continuity equation for the radiant energy density

$$\frac{\partial u_\nu}{\partial t} + \nabla \cdot \mathbf{S}_\nu + c\kappa'_\nu u_\nu = c\kappa'_\nu u_\nu^P$$

If the transfer equation is multiplied by $\hat{\Omega}^n$ and then integrated over solid angle, one arrives at moment equations of higher order.[45] However, it is customary to truncate this hierarchy by assuming that the radiant intensity $I_\nu(\mathbf{r}, \hat{\Omega}, t)$ is nearly isotropic so that the radiant energy density can be related to the energy flux by a diffusion approximation:

$$\mathbf{S}_\nu = -\frac{c}{3\kappa'_\nu} \nabla u_\nu$$

This leads to a diffusion equation for the spectral radiant energy density

$$\frac{\partial u_\nu}{\partial t} - \nabla \cdot D \nabla u_\nu + c\kappa'_\nu u_\nu = J_\nu \qquad (4.23)$$

This equation can be solved by any of the standard methods used for diffusion problems. One popular method is multigroup flux-limited diffusion theory.[9]

There is an important difference from conventional diffusion problems such as those encounted in heat conduction or neutron transport problems. The emission coefficient, the absorption coefficient and the diffusion coefficient, are strong functions of the plasma temperature.[50,51] Hence the radiation diffusion equation is highly nonlinear. Furthermore, unlike other particle diffusion processes in which the particles diffuse through a background, suffering scattering collisions which tend to randomize their distribution, the photons "diffuse" through a sequence of absorption and reemission processes. Radiation diffusing into a cold region raises its temperature, thereby changing the properties of the diffusion process. For this reason the radiation diffusion equation cannot be treated by itself but rather must be coupled to the hydrodynamic description of the medium to arrive at a self-consistent picture. This is most often accomplished by including coupling terms between the radiation diffusion and electron temperature equations. In the radiation diffusion model the relevant terms are

$$J_\nu - c\kappa'_\nu u_\nu$$

while in the electron temperature equation the terms of importance are

$$-\int_0^\infty J_\nu \, d\nu + c\int_0^\infty \kappa'_\nu u_\nu \, d\nu$$

The frequency dependence is generally treated in a multigroup approximation so that the integrals are replaced with sums.

A somewhat more accurate approximation to the radiative transfer equation involves obtaining the next moment equation

$$\frac{1}{c}\frac{\partial \mathbf{S}_\nu}{\partial t} + \nabla \cdot \mathbf{P}_\nu + \kappa'_\nu \mathbf{S}_\nu = 0$$

where $\mathbf{P}_\nu$ is the radiation pressure tensor, defined by

$$\mathbf{P}_\nu(r, t) = \frac{1}{c}\int_{4\pi} \hat{\Omega}\hat{\Omega} I_\nu(\mathbf{r}, \hat{\Omega}, t)\, d\hat{\Omega}$$

To close this set of equations, one can introduce an approximation known as the variable Eddington factor approximation by setting

$$P_\nu(\mathbf{r}, t) = f_\nu(\mathbf{r}, t) u_\nu(\mathbf{r}, t) \tag{4.24}$$

Here, $f_\nu(\mathbf{r}, t)$ is known as the Eddington factor. A variety of prescriptions have been given for calculating this quantity.[52,53]

A somewhat simpler approximation involves assuming local thermodynamic equilibrium so that we can replace u_ν with U^P. This approximation is known as the radiation conduction approximation[43] since it leads to a thermal conduction equation for the combined plasma-radiation fluid with a conduction law of

$$\mathbf{S} = -\frac{lc}{3}\nabla U^P = -\frac{16\sigma l T^3}{3}\nabla T \tag{4.25}$$

where l is the Rosseland mean free path, defined as

$$l = \frac{\displaystyle\int_o^\infty (\kappa_\nu)^{-1}(du_\nu^P/dT)\, d\nu}{\displaystyle\int_0^\infty (du_\nu^P/dT)\, d\nu} \tag{4.26}$$

This approximation is good for optically thick materials where gradients in the material temperature are small. The earlier diffusion approximation only requires the constraint that the material is optically thick so that gradients in u_ν are small. This is the most appropriate approximation for ICF targets where strong material temperature gradients can exist. If the simpler radiative conduction model is used, then we again have a nonlinear temperature equation, just as in the case of electron thermal conduction considered in Section 4.1. However, in this case the effective thermal conductivity has a different temperature dependence

$$\kappa \sim l(T)T^3$$

Many of the same analytical methods can be used to study this nonlinear thermal conduction equation—with many of the same results (e.g., wavelike as opposed to diffusive behavior).[3]

REFERENCES

1. L. D. Landau and E. M. Lifshitz, *Fluid Mechanics* (New York, Pergamon, 1959).

2. L. Spitzer, *Physics of Fully Ionized Gases*, 2nd ed. (New York, Wiley, 1962).

3. Ya. Zel'dovich and Y. P. Raizer, *Physics of Shock Waves and High Temperature Hydrodynamic Phenomena* (New York, Academic 1966).

4. R. E. Kidder and J. H. Zink, "Decoupling of Corona and Core of Laser-Heated Pellets," *Nucl. Fusion* **12**, 325 (1972).

5. G. A. Moses, "Laser Fusion Hydrodynamics Calculations", *Nucl. Sci. Eng.* **64**, 49 (1977).

6. R. J. Bickerton, "Thermal Conduction Limitations in Laser Fusion," *Nucl. Fusion* **13**, 457 (1973).

7. H. Saltzmann, "The Applicability of Fourier's Theory of Heat Conduction on Laser Produced Plasmas," *Phys. Lett.* **41A**, 363 (1972).

8. G. B. Zimmerman, "Calculational Results Concerning Some Laser Initiated Fusion Proposals," APS Plasma Science Division Meeting, Madison, Wi (Nov.,1971), postdeadline paper.

9. G. B. Zimmerman and H. L. Kruer, "Numerical Simulation of Laser Initiated Fusion," *Comments on Plasma Physics and Controlled Fusion* **2**, 51 (1975).

10. A. Shavit and Y. Zirin, "Macroscopic Phenomenological Relations for Nonlinear Processes in Kinetic Theory," *J. Stat. Phys.* **11**, 291 (1974).

11. B. Ripin, P. Burkhalter, F. Young, J. McMahon, D. Colombant, S. Bodner, R. Whitlock, D. Nagel, D. Johnson, N. Winsor, C. Dozier, R. Bleach, J. Stamper, and E. McLean, "X-Ray Emission Spectra from High-Power-Laser-Produced Plasmas," *Phys. Rev. Lett.* **34**, 1313 (1975).

12. J. Stamper and B. Ripin, "Faraday-Rotation Measurements of Megagauss Magnetic Fields in Laser-Produced Plasmas," *Phys. Rev. Lett.* **34**, 138 (1975).

13. J. Stamper, E. McLean, B. Ripin, "Studies of Spontaneous Magnetic Fields in Laser-Produced Plasmas by Faraday Rotation", *Phys. Rev. Lett.* **40**, 1177 (1978).

14. A. W. Ehler, "High-Energy Ions from CO_2 Laser-Produced Plasma," *J. Appl. Phys.* **48**, 2464 (1975).

15. R. C. Malone, R. L. McCrory, and R. L. Morse, "Indications of Strongly Flux-Limited Electron Thermal Conduction in Laser-Target Experiments", *Phys. Rev. Lett.* **34**, 721 (1975).

16. B. Yaakobi and T. C. Bristow, "Measurement of Reduced Thermal Conduction in (Layered) Laser-Target Experiments", *Phys. Rev. Lett.* **38**, 350 (1977).

17. F. Young, R. Whitlock, R. Decoste, B. Ripin, D. Nagel, J. Stamper, J. McMahon, and S. Bodner, "Laser-Produced-Plasma Energy Transport through Plastic Films," *Appl. Phys. Lett.* **30**, 45 (1977).

18. W. R. Mead, et al., "Observation and Simulation of Effects on Parylene Disks Irradiated at High Intensities with a 1.06 μm Laser," *Phys. Rev. Lett.* **37**, 489 (1976).

19. M. D. Rosen, et al., "The Interaction of 1.06 μm Laser Radiation with High-Z Disk Targets," *Phys. Fluids* **22**, 2020 (1979).

20. J. D. Hares, J. D. Kilkenny, M. H. Key, and J. G. Lunney, "Measurement of Fast Electron Energy Spectra and Preheating in Laser-Irradiated Targets," *Phys. Rev. Lett.* **42**, 1216 (1979).

21. K. B. Mitchell and R. P. Godwin, "Energy Transport Experiments in 10 μm Laser Produced Plasmas," *J. Appl. Phys.* **49**, 3851 (1978).

22. S. Braginskii, *Review of Plasma Physics*, Vol. 1 (New York, Consultants Bureau, 1965), p. 205.

23. D. Forslund, "Instabilities Associated with Heat Conduction in the Solar Wind and Their Consequences," *J. Geophys. Res.* **75**, 17 (1970).

24. R. J. Bickerton, "Thermal Conduction Limitations in Laser Fusion," *Nucl. Fusion* **13**, 457 (1973).

25. W. M. Manheimer, M. Lampe, R. W. Clarke, P. C. Liewer, and K. R. Chu, "Plasma Multistreaming and the Viability of Fluid Codes," *Phys. Fluids* **19**, 1788 (1976).

26. D. R. Gray and J. D. Kilkenney, "The Measurement of Ion Acoustic Turbulence and Reduced Thermal Conductivity Caused by a Large Temperature Gradient in a Laser Heated Plasma," *Plasma Phys.* **22**, 81 (1980).

27. C. Max, Lectures Presented at Les Houches Summer Workshop, 1980.

28. J. Albritton, et al.,"Transport of Long Mean Free Path Electrons in Laser Fusion Plasmas," *Phys. Rev. Lett.* **39**, 1536 (1977).

29. E. Valeo and I. Bernstein, "Fast Ion Generation in Laser Plasma Interactions," *Phys. Fluids* **19**, 1348 (1976).

30. A. Bell, R. Evans, and D. Nicholas, "Electron Energy Transport in Steep Temperature Gradients in Laser-Produced Plasmas," *Phys. Rev. Lett.* **46**, 243 (1981).

31. R. J. Mason, "Double-Diffusion Hot Electron Transport in Self-Consistent E and B Fields," *Phys. Rev. Lett.* **42**, 239 (1979).

32. R. J. Mason, "Monte Carlo (Hybrid) Suprathermal Electron Transport," *Phys. Rev. Lett.* **43**, 1975 (1979).

33. J. Kephart, R. Godwin, and G. McCall, "Bremsstrahlung Emission from Laser Produced Plasmas," *Appl. Phys. Lett.* **25**, 108 (1974).

34. C. Armstrong, B. Ripin, F. Young, R. Decoste, R. Whitlock, and S. Bodner, "Emission of Energetic Electrons from a Nd-Laser-Produced Plasma," *J. Appl. Phys.* **50**, 5233 (1979).

35. B. Yaakobi, I. Pelah, and J. Hoose, "Preheat by Fast Electrons in Laser-Fusion Experiments," *Phys. Rev. Lett.* **13**, 836 (1976).

36. R. Decoste and B. Ripin, "High Energy Ions From a Nd-Laser-Produced Plasma," *Appl. Phys. Lett.* **31**, 68 (1977).

37. R. Decoste and B. Ripin, "High-Energy Ion Expansion in Laser Plasma Interaction," *Phys. Rev. Lett.* **40**, 34 (1978).

38. E. McLean, R. Decoste, B. Ripin, J. Stamper, H. Griem, J. McMahon, and S. Bodner, "Spectroscopic Observation of Lost Ions from Laser Produced Plasma," *Appl. Phys. Lett.* **31**, 9 (1977).

39. R. L. Morse and C. W. Nielson, "Occurrence of High Energy Electrons and Surface Expansion in Radiantly Heated Target Plasmas," Los Alamos Scientific Laboratory Report LA-4986-HS (1972).

40. D. W. Forslund, J. M. Kindel, and K. Lee, "Theory of Hot Electron Spectra at High Laser Intensity," *Phys. Rev. Lett.* **39**, 284 (1977).

41. J. D. Lindl, "Effect of Superthermal Electron Tail on the Yield Ratio Obtained from DT-Targets Illuminated with a Shaped Laser Pulse," *Nucl. Fusion* **14**, 511 (1974).

42. T. Mehlhorn and J. J. Duderstadt, "Discrete Ordinates Solution of the Fokker-Planck Equation," *J. Comp. Phys.* **38**, 86 (1980)

43. Ya. Zel'dovich and Y. P. Raizer, *Physics of Shock Waves and High Temperature Hydrodynamic Phenomena* (New York, Academic, 1966), Chap. 6.

44. J. J. Duderstadt and W. R. Martin, *Transport Theory* (New York, Wiley, 1979).

45. G. Pomraning, *Radiation Hydrodynamics* (New York, Plenum, 1972).

46. S. Chandrasekhar, *Stellar Structure* (New York, Dover, 1939), Chap. 5.

47. S. Pai, *Radiation Gas Dynamics* (Berlin, Springer-Verlag, 1966), Chap. 3.

48. W. G. Vincenti and C. H. Kruger, *Introduction to Physical Gas Dynamics* (New York, Wiley, 1965).

49. H. Hottel and A. Sarofin, *Radiative Transfer* (New York, McGraw-Hill, 1967).

50. D. Bates, *Atomic and Molecular Processes* (New York, Academic, 1962).

51. H. Griem, *Plasma Spectroscopy* (New York, McGraw Hill, 1964).

52. G. Magelssen and G. Moses, "Pellet X-Ray Spectra for Laser Fusion Reactor Designs," *Nucl. Fusion* **19**, (1979).

53. P. Campbell, "A Variable Eddington Method for Radiation Transport in Dense Fusion Plasmas", KMS Fusion, Inc., Report KMSF-U458, Jan. 1976.

FIVE

Driver Energy Deposition

We now turn our attention to the absorption of the incident driver beam energy in the target. The detailed absorption mechanism depends on the driver type. Energy deposition by intense laser beams involves a host of complex processes characterizing the interaction of light (electromagnetic waves) with plasmas. Electron and ion beam drivers involve the slowing down of energetic charged particles in a plasma.

The classical mechanisms for both laser and charged particle beam energy deposition in plasmas involve charged particle collision processes. The electric field of an incident laser beam causes electrons in the target plasma to oscillate. This oscillation energy is converted into thermal energy as the electrons collide with ions. In a similar sense, energetic charged particles incident on the target slow down via charged particle collisions with the background plasma.

However, driver energy deposition in inertial confinement fusion targets is not restricted to such collisional processes. Indeed, since charged particle collision frequencies decrease rapidly with increasing temperature, classical absorption mechanisms can become rather ineffective at thermonuclear temperatures. Rather the driver energy deposition mechanisms of most interest involve various "anomalous" (or inadequately understood) processes that arise as a consequence of the high beam intensity or incident particle energy. For example, an incident laser beam or charged particle beam can couple to collective modes in the target corona plasma and drive the plasma into a turbulent state. This turbulence can lead to enhanced absorption. It can also act to reflect or scatter the incident beam from the target.

In this chapter we consider a variety of processes that are thought to characterize the interaction of the driver beams with the target and lead to driver energy deposition. Since a detailed study of such driver-target interaction mechanisms can become rather involved (e.g., venturing into imposing subjects such as plasma turbulence and relativistic beam-plasma interactions), we will confine ourselves for the most part to a qualitative discussion.

5.1. LASER LIGHT ABSORPTION IN PLASMAS

Laser light can interact in a variety of ways with a plasma. The plasma can refract, reflect, and/or absorb incident laser light. The simplest energy deposition mechanism is inverse bremsstrahlung or collisional absorption of the light. We recall that bremsstrahlung corresponds to the emission of radiation (photons) when a charged particle is decelerated, for example, when an electron emits a photon in a collision with an ion. Inverse bremsstrahlung occurs when an incident photon is absorbed by an electron in the Coulomb field of an ion.

A simpler way to think of inverse bremsstrahlung is to consider the motion of an electron in the oscillating electric field of an incident electromagnetic wave (the incident light beam). As the electrons oscillate in this field, they collide with ions, thereby converting the directed energy of the oscillation into the random energy of thermal motion. In other words, the oscillating electrons correspond to a current induced in the plasma by the incident light beam that then leads to resistive heating of the plasma due to charged particle collisions. In this way, the incident light energy is deposited in the form of increased electron thermal energy (temperature). Since this process depends on electron-ion collisions, we might expect that the absorption coefficient would scale as the electron-ion collision frequency ν_{ei}. We also know that ν_{ei} scales with temperature as $T_e^{-3/2}$. Hence as the plasma temperature increases, the collisional or inverse bremsstrahlung process becomes less effective. For example, the absorption length for 1.06-μm light propagating in a plasma characterized by a temperature $T_e \sim 1$ keV and an effective charge $Z \sim 3$ is about 100 μm. An increase in the plasma temperature to 10 keV would increase the absorption length to 3000 μm—the plasma would become essentially transparent to the incident light. Furthermore, for high beam intensities, the absorption coefficient can become intensity dependent (so-called nonlinear bremsstrahlung) and decrease as a power of the beam intensity. Thus for high temperature plasmas and high intensity laser light, the classical inverse bremsstrahlung absorption mechanism can become quite ineffective.

Fortunately there are other absorption mechanisms present that involve the coupling of the incident light into waves in the plasma. More specifically, the oscillation of the electrons in the electric field of the incident light across a variation in the plasma density drives a charge density fluctuation. If the incident light frequency is comparable to the electron plasma frequency, then this coupling can resonantly drive electron plasma waves. There are two primary sources of a plasma density gradient that lead to this coupling. The plasma blowoff resulting from the ablation process leads to an absorption process involving obliquely incident laser light known as resonance absorption. The density fluctuations from other plasma waves such as ion acoustic waves

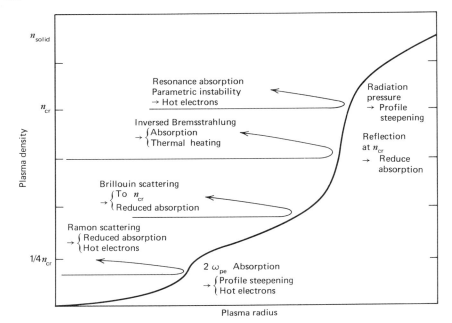

Figure 5.1. Laser plasma interaction mechanisms of importance in laser driven inertial confinement fusion.

can lead to the parametric excitation of plasma instabilities and thereby enhanced absorption.

The interaction of laser light with the plasma surrounding the target can become quite complex. Not only can such anomalous processes enhance absorption, but they can also lead to processes that focus the beam into narrow filaments (self-focusing and filamentation) or reflect the beam from the target (stimulated scattering). The coupling of light energy into the plasma waves (by either resonance absorption or parametric excitation processes) can produce high energy or suprathermal electrons. The incident light can also produce a ponderomotive force that modifies the density profile of the blowoff plasma, steepening this profile or rippling its surface. Some of the mechanisms that may arise in laser-plasma interactions in inertial confinement fusion targets are indicated schematically in Figure 5.1.

In this section we discuss both collisional and collective light absorption processes in plasmas. However, it is useful to begin with a brief discussion of the various ways in which intense light can interact with matter in general.

5.1.1. THE INTERACTION OF INTENSE LASER RADIATION WITH MATTER

The electromagnetic energy density at the focal spot of a high-powered pulsed laser reaches incredible magnitudes. By way of orientation, the parameters

Table 5.1   Focal Spot Parameters Characterizing an Incident Nd Laser Beam of 10 kJ Energy, 100 ps Pulse Length, and 10^{-3} cm² Focal Spot Area

Focused intensity	10^{18} W/cm²
Energy density	3×10^7 J/cm³
Photon density	3×10^{25} photons/cm³
E_{rms}	10^{10} V/cm
B_{rms}	50 MG
Average kinetic energy of electron oscillating in laser field	5 keV

characterizing the focal spot of a 10,000-J Nd glass laser beam, delivered in a 100-ps pulse on a focal spot area of 10^{-3} cm² are given in Table 5.1. The energy density in the focal spot, 3×10^7 J/cm³ is well above the energy density of electron binding to an atomic nucleus, 4×10^5 J/cm³ (although it is also well below the nuclear energy densities of 10^{11} J/cm³).

Therefore, if this very intense laser light is focused on a solid target, rather major transformations in the target surface occur. Not only is the solid surface vaporized, but it is ionized as well, producing a high density plasma that continues to absorb the incident laser light. The very rapid temperature increase at the surface of the solid causes the laser-produced plasma to blow off or ablate towards the laser beam. This, in turn, drives an intense hydrodynamic shock wave into the solid.[1]

Heating without Phase Change.   For sufficiently low light intensities, $I \gtrsim 10^7$ W/cm², the incident laser light merely heats the target surface without melting or vaporizing it. The incident light is absorbed (and partially reflected) within a skin depth of the surface. This absorbed energy is then transferred into the interior of the target by thermal conduction.

Laser Induced Melting and Evaporation.   For somewhat higher intensities, $I \sim 10^6 - 10^9$ W/cm², the incident laser energy will actually melt or evaporate the solid surface. One usually finds vaporization of the surface as opposed to melting. The mechanism for vaporization depends on the laser light intensity (the pulse width).

For longer pulses at low intensities, $I \sim 10^6 - 10^7$ W/cm², the laser light produces deep, narrow holes in the surface. There is very little blowoff of the vapor produced at the surface. Typically a 10J pulse delivered in 1 ms will produce a crater of about 1 mm in depth.

For higher intensity at shorter pulse lengths, $I \sim 10^9$ W/cm², only a small amount of the target material is vaporized. However, this vapor cloud or blowoff can interact with the incident laser light, absorbing the light and shielding the surface. Since higher intensities produce high pressures that drive the vapor away from the surface at high velocities, the back reaction from the blowoff drives strong shock waves into the solid target itself. For example, a 10 J pulse delivered in 30 ns will ablate only 1 to 3 μm of surface material.

Laser Induced Ionization and Gas Breakdown. It has been known for some time that sufficiently intense laser light can ionize materials. In particular, the focal spot of a large pulsed laser is capable of creating a "spark" in air, that is, producing gas breakdown.[2, 3] The threshold of this phenomenon is usually around $I \sim 10^{10}$ W/cm^2.

There appear to be two essential processes involved in such breakdown phenomena: the production of an initial ionization and then the subsequent growth of this ionization. The second process is usually taken as an avalanche mechanism. If there are free electrons in the gas, they will be rapidly accelerated to high energies by the electric field of the incident laser light. In a very short time they will have achieved sufficiently high energy to ionize other atoms and produce more free electrons—and so on in a cascading ionization process.

The initial or "priming" ionization is more difficult to understand, since the photons present in the incident laser beam have energies $h\nu$ many times less than the ionization potential of most atoms. For example, it would take the simultaneous absorption of 43 CO_2 laser photons to ionize a lithium atom. Such very high order multiphoton processes are very improbable. Nevertheless, in the very high photon densities of the laser focal spot, such a mechanism can become significant. Other explanations for the initial ionization include the presence of impurities, and also a distortion of the atomic electron energy levels in the very high intensity laser electric field that effectively lowers the ionization potential.

The calculation of the laser light threshold at which ionization occurs in a gas is rather complicated, and experimental verification is difficult. Fortunately at the very high intensities used in most applications (in excess of 10^{12} W/cm^2), the ionization of the target can be regarded as essentially instantaneous, regardless of the actual mechanism involved. (Here we need only recall that the kinetic energy of a free electron oscillating in the electric field of an incident laser beam is typically a keV or more.) Hence one usually considers the interaction of high intensity laser light with an ICF target to initially involve a dense, low temperature ($\sim$5eV) plasma.

5.1.2. COLLISIONAL (INVERSE BREMSSTRAHLUNG) ABSORPTION OF LASER LIGHT IN PLASMAS

The classical mechanism for laser light absorption in plasmas is *inverse bremsstrahlung* or *free-free absorption* in which a photon is absorbed by a free electron in the field of an ion. However, a more intuitive description of the absorption process would be to recognize that the incident electric field of the laser light oscillates free electrons. (The ions also oscillate, but their motion can be neglected because of their much larger mass.) This directed energy of electron motion is randomized and hence converted into heat energy by electron collisions with ions. In other words, the incident laser electric field drives electron currents, and the resistivity represented by electron-ion collisions leads to "Joule heating" of the plasma.

To analyze this process in more detail,[4, 5] consider the propagation of a linearly polarized plane electromagnetic wave in a medium as described by Maxwell's equations. We take the propagation vector along the z axis and the electric field vector along the x axis (see Figure 5.2) so that Maxwell's equations become

$$\frac{\partial E}{\partial z} = \frac{\partial B}{\partial t}$$

$$\frac{\partial B}{\partial z} = \mu J + \frac{1}{c^2} \frac{\partial E}{\partial t}$$

Here J represents the current induced by the electromagnetic wave in the x direction. This current can be calculated if we consider the motion of an electron in the electric field as described by its equation of motion for the x coordinate of velocity, u,

$$m \frac{du}{dt} + m \nu_{ei} u = -eE$$

Here, the electron-ion collision frequency has been introduced as a friction term in this equation. The induced current density can then be calculated in terms of u as

$$J = n_e e u$$

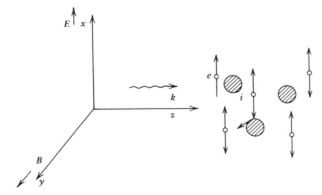

Figure 5.2. Coordinate system for analyzing inverse bremsstrahlung.

To solve these equations, we seek a plane wave solution in the form

$$\mathbf{E} = E\hat{\mathbf{e}}_x e^{i(kz+\omega t)}$$

We can then solve for the electron velocity as

$$u = \frac{eE}{m}\left(\frac{i\omega - \nu_{ei}}{\omega^2 + \nu_{ei}^2}\right) e^{i(kz+\omega t)}$$

and hence for the current density

$$J = \left(\frac{e^2 n_e}{\varepsilon_0 m}\right)\left(\frac{i\omega - \nu_{ei}}{\omega^2 + \nu_{ei}^2}\right) \varepsilon_0 E e^{i(kz+\omega t)}$$

where it is convenient to define the plasma frequency ω_p

$$\omega_p^2 \equiv \frac{e^2 n_e}{\varepsilon_0 m}$$

If we substitute this current into the field equations, we arrive at a relation between the propagation wave number k and the laser light frequency ω, that is, a dispersion relation for the laser light propagation in the plasma:

$$\left(\frac{kc}{\omega}\right)^2 = 1 - \left(\frac{\omega_p^2}{\omega^2 + \nu_{ei}^2}\right)\left(1 + i\frac{\nu_{ei}}{\omega}\right)$$

Since k is, in general, complex, the incident light is attenuated as it propagates. In particular, the index of refraction n_i is given by

$$n_i \equiv \mathrm{Re}\left\{\frac{kc}{\omega}\right\} = \left[\frac{\beta}{2} + \frac{1}{2}\left[\beta^2 + (1-\beta)^2\left(\frac{\nu_{ei}}{\omega}\right)^2\right]^{1/2}\right]^{1/2}$$

while the energy absorption coefficient is given by

$$\kappa \equiv 2\,\mathrm{Im}\{k\} = 2\frac{\omega}{c}\left[-\frac{\beta}{2} + \frac{1}{2}\left[\beta^2 + (1-\beta)^2\left(\frac{\nu_{ei}}{\omega}\right)^2\right]^{1/2}\right]^{1/2}$$

where

$$\beta \equiv 1 - \frac{\omega_p^2}{\omega^2 + \nu_{ei}^2}$$

Light Propagation in Plasmas.   Before continuing to examine the absorption process, it is important to note that for $\nu_{ei} \ll \omega$, the index of refraction simplifies to

$$n_i = \left[1 - \left(\frac{\omega_p}{\omega} \right)^2 \right]^{1/2}$$

When the plasma frequency becomes larger than the light frequency, $\omega_p > \omega$, the index of refraction n_i becomes imaginary. This means that light cannot propagate in the plasma when $\omega_p > \omega$. The blowoff plasmas characteristic of inertial confinement fusion targets have a density gradient similar to that shown in Figure 5.3. Since the plasma frequency scales as density, $\omega_p \sim n_e^{1/2}$, it is apparent that laser light incident from the lower density ("underdense") region can only propagate up to the density at which ω_p becomes equal to ω. This limit is usually referred to as the *critical density*, and it is defined more explicitly as

$$n_{ec} \equiv \frac{\varepsilon_0 m \omega_0^2}{e^2}$$

where ω_0 is the frequency of the incident laser light. The incident laser light will then be reflected at the critical density. From a more physical point of view, in the underdense region where the light frequency exceeds the plasma frequency, the electron inertia is sufficient to keep the material current in phase with the displacement current. In the overdense region where the light frequency is less than the plasma frequency, the material current opposes the displacement current in the light field, and the wave cannot propagate.

Several other comments are of use at this point. Since the index of refraction n_i is less than one, the light is refracted away from regions of higher density. In a later section we discuss the fact that most light absorption in a plasma occurs at or near the critical density n_{ec}.

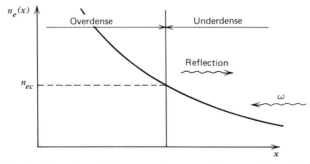

Figure 5.3.   Incident laser light cannot propagate beyond the critical density.

Table 5.2 The Critical Density for Various Laser Wavelengths

Laser Type	$\lambda_0(\mu m)$	$h\nu$ (eV)	ρ_c (g/cm^3)	n_{ec} (cm^{-3})
UV Excimer Nd ($\times 4$)	0.25	5.4	7.0×10^{-2}	2.2×10^{22}
Nd ($\times 2$)	0.50	2.5	1.5×10^{-2}	4.5×10^{21}
Nd	1.06	1.17	3.3×10^{-3}	10^{21}
CO$_2$	10.6	0.117	3.3×10^{-5}	10^{19}

Another important parameter that characterizes the plasma's response to the incident light is the "quiver velocity" with which the electrons oscillate in the light wave's electric field:

$$u_0 = \frac{e}{m_e} E_n \left| \frac{i\omega - \nu_{ei}}{\omega^2 + \nu_{ei}^2} \right| \sim \frac{eE}{m_e \omega}$$

The corresponding kinetic energy of the oscillating electron is given as

$$\langle KE \rangle_{\text{osc}} = \frac{1}{2} m_e u_0^2 = \frac{\bar{I}}{2 n_i n_e c}$$

where $\bar{I}$ is the average intensity of the incident light. We have tabulated the critical density n_{ec}, the quiver velocity u_0, the quiver energy $\langle KE \rangle_{\text{osc}}$, and several other parameters characterizing laser light of different wavelengths in Table 5.2.

Classical Absorption. In the underdense region of the blowoff plasma where the laser light absorption occurs, the collision frequency is much smaller than the frequency of the incident laser light, $\nu_{ei} \ll \omega_0$. Then we can simplify our dispersion relation to

$$\left(\frac{kc}{\omega_0} \right) \cong 1 - \frac{\omega_p^2}{\omega_0^2} + \frac{\omega_p^2}{\omega_0^2} \left(\frac{i\nu_{ei}}{\omega_0} \right)$$

Hence we can solve for $k(\omega_0)$

$$k \cong \frac{\omega_0}{c} \left(1 - \frac{\omega_p^2}{\omega_0^2} \right)^{1/2} \left\{ 1 + i \left(\frac{\nu_{ei}}{2\omega_0} \right) \left(\frac{\omega_p^2}{\omega_0^2} \right) \frac{1}{1 - \left(\omega_p^2/\omega_0^2 \right)} \right\}$$

We can then calculate the absorption coefficient as

$$\kappa = 2 \operatorname{Im}\{k\} \cong \left(\frac{\nu_{ei}}{c}\right)\left(\frac{\omega_p^2}{\omega_0^2}\right)\frac{1}{\left[1-\left(\omega_p^2/\omega_0^2\right)\right]^{1/2}}$$

If we now substitute in an explicit form for the electron-ion collision frequency ν_{ei}, we can write the absorption coefficient characterizing inverse bremsstrahlung [4, 6] as

$$\kappa = (2\pi)^{1/2}\left(\frac{16\pi}{3}\right)\frac{Zn_e^2 e^6 \ln \Lambda}{c(m_e kT_e)^{3/2}\omega_0^2\left(1-\left(\omega_p^2/\omega_0^2\right)\right)^{1/2}}$$

where

$$\Lambda = \frac{v_{\text{th}}}{\omega_p p_{\min}}, \quad p_{\min} \equiv \max\left\{\frac{Ze^2}{kT_e}, \frac{\hbar}{(m_e kT_e)^{1/2}}\right\}$$

Let us examine this expression in more detail.[7] For fixed plasma density n_e and temperature T_e, it appears that $\kappa \sim 1/\omega_0^2 \sim \lambda_0^2$, and hence we might expect that longer wavelength radiation is absorbed more effectively. However, this is a misleading comparison, since most absorption occurs near the critical density n_{ec} corresponding to a plasma frequency $\omega_p = \omega_0$. It is more illuminating to rewrite the absorption coefficient in an alternative form

$$\kappa \cong \frac{\phi^2}{(1-\phi)^{1/2}}\kappa_{0c}$$

where $\phi \equiv n_e/n_{ec}$ and κ_{0c} is the absorption coefficient characterizing the critical density,

$$\kappa_{0c} \equiv \frac{1}{c}\nu_{ei}(n_e = n_{ec})$$

Actually, κ_{0c} is the most appropriate measure of the effective absorption coefficient for laser light of a given wavelength. Since the collision frequency scales as $\nu_{ei} \sim n_e$, it is apparent that $\kappa_{0c} \sim \omega_0^2$. Hence the effective absorption length decreases rapidly as the wavelength of the incident laser light decreases (see Figure 5.4).

Effect of Plasma Density Gradients. The plasma blowoff cloud surrounding an ICF target is characterized by a density variation $n_e(x)$ (assuming a plane geometry for the moment). Hence the effective absorption of the incident beam

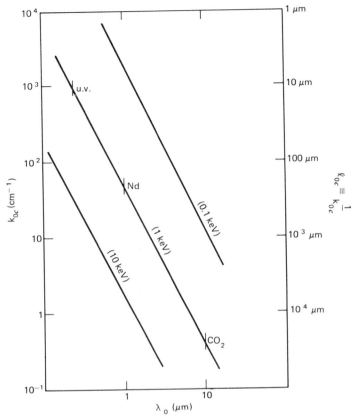

Figure 5.4. The dependence of the effective absorption length characterizing inverse bremsstrahlung upon wavelength and temperature.

must take into account the variation of the absorption coefficient, $\kappa(x)$, caused by this spatial density gradient.[5, 8] More specifically, the absorption would be given by

$$\exp\left[-\int \kappa(x)\, dx\right]$$

where we can write

$$\int \kappa(x)\, dx \sim \kappa_{0c} \int \frac{\left[n_e(x)/n_{ec}\right]^2}{\left[1 - n_e(x)/n_{ec}\right]^{1/2}}\, dx$$

For a linear density gradient, $n_e(x) = n_{ec} x/L$, this expression yields the beam absorption (including absorption of both the incident and reflected beam)

$$\text{Absorption} = 1 - \exp\left(-\tfrac{32}{15}\kappa_{0c} L\right)$$

For an exponential density profile, $n(x) = \beta n_{ec} \exp(-x/L)$, the coefficient in the above expression is modified to

$$\tfrac{32}{15} \to \tfrac{4}{3}\left[1 - \left(1 + \frac{\beta}{2}\right)(1-\beta)^{1/2}\right], \quad (\beta < 1)$$

It should be apparent that the absorption depends essentially on the dimensionless factor, $\kappa_{0c}L$, that is, the ratio of the "scale height" L of the density profile to the absorption length at the critical density, κ_{0c}^{-1}. For appreciable inverse bremsstrahlung absorption to occur, the plasma cloud must be characterized by a scale height on the order of an absorption length in the neighborhood of the critical density n_{ec}.

Nonlinear Bremsstrahlung. There are many modifications that can arise in this simple picture of light absorption in plasmas. For sufficiently low temperatures, bound-bound and bound-free absorption can occur. At higher intensities, the strong electric field of the light will distort the distribution of electron thermal velocities, hence modifying the collision frequency ν_{ei} and leading to a dependence of the absorption coefficient κ on the light intensity I (hence the name *nonlinear bremsstrahlung*). This latter effect occurs at light intensities at which the energy of oscillation of the electrons in the light beam electric field is comparable to their thermal energy,

$$\langle KE \rangle_{\text{osc}} \sim kT_e$$

Although the detailed calculation of the effect of intense light on the electron distribution function $f_e(\mathbf{v})$ is cumbersome, several qualitative features of such an effect will be discussed. In the simplest model, the electron distribution function $f_e(\mathbf{v})$ is assumed to remain roughly Maxwellian, but the electron-ion collision frequency ν_{ei} is modified because the electron velocity v must now include the quivering component u_0.[9-15] In this case one finds that the absorption coefficient is modified to

$$\kappa \to \frac{\kappa}{1 + \tfrac{3}{2}(u_0/v_{\text{th}})^2}$$

Since $(u_0/v_{\text{th}})^2 = \langle KE \rangle_{\text{osc}}/kT$, it is apparent that for large incident light intensities, the absorption coefficient will decrease as I^{-1}.

A more detailed calculation[16, 17] which takes into account the fact that $f_e(v)$ will be perturbed from a Maxwellian finds that

$$\kappa \to \kappa \exp\left(-\frac{v_W^2}{2v_{\text{th}}^2}\right)$$

where

$$\tfrac{1}{2}m_e v_W^2 \equiv 8\left[\left(\frac{n_e}{n_{ec}}\right)\left(\frac{Z}{\lambda_0}\right)\left(\frac{\ln\Lambda}{10}\right)\right]^{2/3}\text{eV}$$

The absorption coefficient predicted by this estimate can be reduced by as much as a factor of two over that for the usual ("linear") inverse bremsstrahlung process.

Although there is little doubt that such effects can be present, direct experimental evidence is scant because of the presence of a variety of other processes that set in at lower beam intensities (e.g., resonance absorption and stimulated scattering).

5.1.3.   RESONANCE ABSORPTION

When light is incident on a spatially inhomogeneous plasma, electrostatic waves are generated whenever the light has a component of the electric field **E** along the density gradient ∇n. In particular, when p-polarized light (with E parallel to the plane of incidence) is incident obliquely on the density gradient, then the component of the light E field parallel to the plasma density gradient can drive electron plasma waves (see Figure 5.5). Near the critical density the electric field becomes very large and will resonantly excite these waves. Hence one finds an energy transfer mechanism from the light into the waves, and eventually through the damping of the waves into the electron temperature:

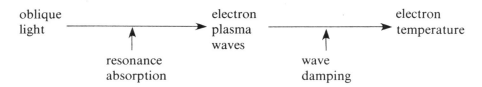

(Notice that if the light is s-polarized, with **E** out of the plane of incidence, there will be no coupling to the plasma waves.) This process is known as

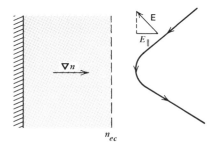

Figure 5.5. Resonance absorption occurs when the light is obliquely incident upon a density gradient.

resonance absorption.[8-25] It is now felt that this is a dominant mechanism involved in the absorption of laser light in ICF targets.

A Simple Model of Resonance Absorption.[22, 26] Consider a nonuniform plasma driven by a uniform electric field of strength E_d and frequency ω_0. We can combine Maxwell's equations,

$$\nabla \cdot \mathbf{E} = 4\pi\rho$$

$$\frac{\partial \rho}{\partial t} + \nabla \cdot \mathbf{J} = 0$$

to find

$$\nabla \cdot \left(4\pi\mathbf{J} + \frac{\partial \mathbf{E}}{\partial t}\right) = 0$$

or

$$\frac{\partial \mathbf{E}}{\partial t} + 4\pi\mathbf{J} = \left\langle \frac{\partial \mathbf{E}}{\partial t} + 4\pi\mathbf{J}\right\rangle$$

where the average represents the spatially independent component. We neglect ion motion and linearize to write

$$J = -en_0(z)u$$

where u is the oscillation velocity. If we differentiate with respect to time and use the linearized equation of motion, we find

$$\frac{\partial^2 E}{\partial t^2} + \omega_p^2(z)E + \nu_{ei}\frac{\partial E}{\partial t} = -\left[\omega_p^2(z) - \langle\omega_p^2(z)\rangle\right]E_d\cos\omega_0 t$$

If we assume a field $E \sim \exp(i\omega_0 t)$, we find a response

$$E = \frac{\omega_p^2(z)E_d}{\omega_0^2 - \omega_p^2(z) + i\nu_{ei}\omega_0}$$

If we assume a linear density gradient, $n(z) = n_{ec}(z/L)$, we can compute the power absorbed from the laser driver field:

$$P_{abs} = \int \frac{\nu_{ei}|E|^2}{8\pi}dz = \frac{\omega_0 L E_d^2}{8}$$

Notice in particular that the collision frequency ν_{ei} cancels out in this expression. That is, the amount of resonance absorption is independent of the detailed wave damping mechanism.

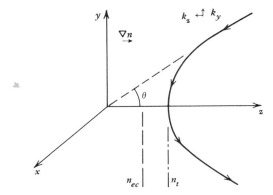

Figure 5.6. Coordinate system for analyzing resonance absorption.

Oblique Incidence.[26] In the more general case, the light is incident obliquely upon a plasma density gradient as shown in Figure 5.6. The dispersion relation characterizing the light wave is

$$\omega_0^2 = \omega_p^2 + \omega_0^2 \sin^2\theta + k_z^2 c^2$$

The maximum distance of penetration occurs where $k_z = 0$, or at

$$n_t = n_{ec} \cos^2\theta$$

For p-polarized light (in the $y - z$ plane), there is a component of the electric field vector, $E_\parallel$, along the density gradient. To analyze this, we need to compute this component and then use it for E_d in our earlier modeled problem result. Kruer[22] shows that the component of the electric field which drives the resonant process is

$$E_d = E_\parallel \sim \frac{E_0}{\left(\dfrac{\omega_0 L}{c} \right)^{1/16}} \sin\theta \exp\left[-\frac{2}{3}\left(\frac{\omega_0 L}{c} \right) \sin^3\theta \right]$$

where L is the scale height (assuming a linear density gradient). The fractional absorption can then be calculated as

$$f = \tfrac{1}{2}\phi^2(\tau)$$

where

$$\tau = (k_0 L)^{1/3} \sin\theta$$

$$\phi(\tau) = 2.31\tau \exp\left(-\tfrac{2}{3}\tau^3\right)$$

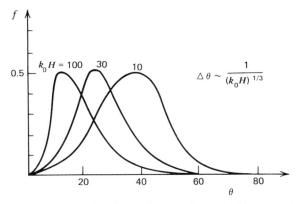

Figure 5.7. Fractional absorption for various angles of incidence and scale lengths.

The fractional absorption is plotted in Figure 5.7 as a function of incident angle.

For resonance absorption to occur, the electric field of the incident wave must tunnel from the turning point at $n_{ec}\cos^2\theta$ to the critical density n_{ec}. If the angle of incidence, θ, is too large, the light will be turned away too far from the critical density. If the angle θ is too small, the component of the electric field parallel to the density gradient, $E_{\parallel}$, is too small at the turning point. In both of these instances, there will be little resonance absorption. The angle of incidence for maximum absorption is given by

$$\sin\theta \cong 0.8\left(\frac{c}{\omega_0 L}\right)^{1/3}$$

The range of angles, $\Delta\theta$, for which there is appreciable absorption depends on the scale height L. For scale heights large compared to the laser wavelength, $L \gg \lambda_0$, there is absorption for only a narrow range of incident angles, $\Delta\theta$. For shorter scale heights, $L \gtrsim 10\lambda_0$, resonance absorption occurs for a broad range of angles $\Delta\theta$. For example, for current experiments on small targets, $L = 1\ \mu m$ with $\lambda_0 = 1.06\ \mu m$, the optimum angle of incidence is 20. Reactor grade targets for which $L \sim 1000\ \mu m$ would require an angle of incidence of as small as 3°. It is difficult to see how such small angles of incidence could be produced under conditions of uniform illumination. Fortunately, the ponderomotive forces produced by the incident light modify the blowoff plasma density profile, steepening it and shortening the scale height in such a manner as to increase resonance absorption. In any event, however, the fraction of incident light that can be absorbed by the resonance process is limited to roughly 50%.

Nonlinear Effects and Hot Electron Production. Although the resonance absorption process is basically linear, there are some nonlinear effects that become important at high light intensity. Two dimensional plasma simulations

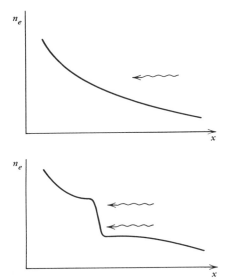

Figure 5.8. Modification of the density gradient profile by the ponderomotive force exerted by the incident light.

have shown that the resonantly driven wave field grows to sufficient intensity that electrons can be accelerated through in one oscillation period. This phenomenon is known as "wavebreaking." It leads to electron trapping by the localized oscillating field and the production of very high energy (suprathermal) electrons.[27-30] A more detailed discussion of these suprathermal electrons and their implications for target dynamics has been provided in Section 4.2.

Profile Modification. A second nonlinear effect of some importance is the influence of the incident light on the plasma density gradient. The ponderomotive force exerted by the beam (the light pressure) can dam up the plasma flow, thereby steepening the density profile and reducing the scale height[31-37] (see Figure 5.8). This can have a significant effect on the range of angles over which significant resonance absorption will occur. It furthermore tends to reduce the energy deposited in the form of suprathermal electrons.

Other types of profile modification can become important. For example, computer simulations have suggested that the incident beam can ripple the critical surface. This also broadens the incident angle range over which appreciable resonance absorption occurs. There is some experimental evidence for this type of profile modification from the measurement of backscattered light.[38]

5.1.4. ACOUSTIC TURBULENCE

A second anomalous (nonclassical) process that might lead to enhanced absorption involves the presence of *ion-acoustic turbulence* driven by the heat flux from the energy deposition region into the ablation front.[37-42] The general idea is that strong absorption in the underdense region of the plasma, due, for example, to resonance absorption, will produce a large heat flux toward higher density regions of the target. The heat flux can drive ion-acoustic turbulence (via a two-stream mechanism due to the counterflowing cold and hot electron components or the hot electrons and the ion stream due to plasma blowoff). Since the ion-acoustic turbulence increases the effective electron-ion collision frequency, ν_{ei}, it would lead to enhanced absorption (analogous to collisional absorption) of the incident light. This process can only occur for densities between the critical density and roughly one-third the critical density, $0.3n_{ec} \gtrsim n_e < n_{ec}$. However, in this region, light absorption could be enhanced considerably, corresponding to an additional absorption of the incident beam of roughly 20%. Although this absorption mechanism does produce suprathermal electrons, they are far less energetic than those produced by resonance absorption.

5.1.5. PARAMETRIC PROCESSES IN ICF PLASMAS

Incident laser light can couple together the natural collective modes or waves in an ICF target plasma in such a way as to drive these modes unstable.[43-45] These instabilities will then grow until they saturate in a turbulent state. This turbulent state of the plasma will then be characterized by enhanced values of transport coefficients such as those characterizing absorption, thermal and electrical conduction, and electron-ion energy transfer.

For example, the incident light can couple together electrostatic modes such as electron plasma waves and ion acoustic waves in such a way as to lead to enhanced absorption of the incident light in the vicinity of the critical surface. The light can also couple into electromagnetic modes and excite instabilities that lead to an enhanced reflectivity of the plasma.

These phenomena are examples of a *parametric excitation* process.[46-58] More precisely, parametric excitation involves the amplification of the oscillation of a natural mode of a system due to a periodic modulation of a parameter that characterizes the system. Perhaps the most common example of this is a child on a swing. The natural frequency of the oscillating motion of the swing is determined by the mass and the length of the rope (see Figure 5.9). But the child can influence this motion by kicking its feet in such a way as to change its center of mass—that is, the child can change the effective length of the swing and thereby the frequency of the swing in a periodic fashion. If the child

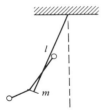

Figure 5.9. The child on a swing is an example of a single-mode parametric process. $d^2\theta/dt^2 + \Omega^2\theta = 0$; $\Omega^2 = (g/l)(1 + \varepsilon\cos\omega t)$.

kicks at a frequency just double that of the natural frequency of oscillation, then it can amplify the motion of the swing. In this case, the parametric excitation appears through a modulation of the frequency parameter.

In a plasma, there are a variety of different natural modes, such as electron plasma waves and ion acoustic waves, that depend on parameters such as density and temperature. Since these latter parameters can be modified—indeed, modulated—by the electromagnetic field of the incident light wave, it is not surprising to find that parametric excitation can play an important role in the interaction of laser light with plasmas. A variety of different coupling processes can occur involving both electrostatic and electromagnetic modes. These are indicated schematically in Figure 5.10. All of these processes involve three waves. There are also four-wave and higher order processes that can occur, but these are usually of secondary importance in laser-plasma interactions.

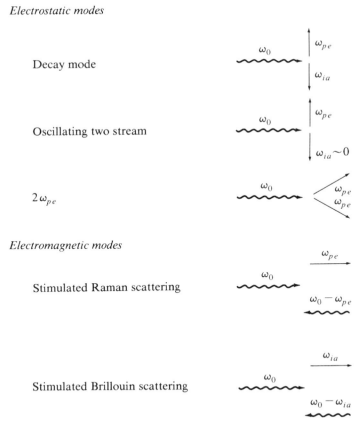

Figure 5.10. There are a variety of three-wave parametric processes that can occur in laser-plasma interactions.

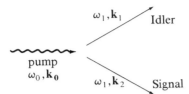

ω_1, k_1 Idler

pump
ω_0, k_0

ω_1, k_2 Signal   **Figure 5.11.** A three-mode parametric process.

The most general type of three-wave parametric interaction process can be represented schematically as shown in Figure 5.11. Here we have employed the usual terminology of parametric amplifiers in electrical engineering by referring to the driving force (e.g., the incident light wave) as the "pump," the lower frequency natural mode (e.g., an ion acoustic wave) as the "idler," and the higher frequency natural mode (e.g., an electron plasma wave) as the "signal." Such parametric processes are characterized by several general properties:

1. A matching or "resonance" condition among frequency and wave numbers must be obeyed for strong coupling:

$$\omega_0 \sim \omega_1 + \omega_2$$

$$k_0 \sim k_1 + k_2$$

2. Since all natural modes of oscillation are damped, the driver or pump amplitude must exceed a certain threshold intensity to drive the modes unstable. These modes then grow in amplitude with a certain growth rate as they absorb energy from the pump.

3. The final frequency of the amplified oscillation is determined by the pump frequency rather than the natural frequency of the modes. This is referred to as "frequency locking."

The general approach to analyzing parametric coupling processes in laser-plasma interactions is to first recast the equations characterizing the modes of the plasma into a form in which the parametric excitation is explicit. These equations are then analyzed in the linear limit to determine the thresholds on the pump amplitude for the onset of instabilities and the growth rates of these instabilities. The more complex analysis of the nonlinear evolution of the parametrically driven modes and their eventual saturation in a turbulent state requires the use of plasma simulation codes.

Single Mode Analysis. To be more explicit, consider the case of a single mode parametric process, that is, in which both the idler and the signal are the same mode. The oscillator amplitude $X(t)$ satisfies

$$\frac{d^2 X}{dt^2} + 2\Gamma \frac{dX}{dt} + (\Omega^2 + \Gamma^2) X(t) = 0$$

where

$$\Omega = \text{frequency of oscillation}$$

$$\Gamma = \text{damping coefficient}$$

The parametric excitation is introduced as a modulation in the frequency:

$$\Omega^2 = \Omega_0^2(1 - 2\varepsilon \cos \omega_0 t)$$

where we will refer to Ω_0 as the natural frequency of the system in the absence of the modulation, and ω_0 is the pump frequency.

If we use a variable substitution

$$X(t) = e^{-\Gamma t}Y(t)$$

we find

$$\frac{d^2Y}{dt^2} + \Omega_0^2(1 - 2\varepsilon \cos \omega_0 t)Y(t) = 0$$

This is a differential equation with periodic coefficients and corresponds to a special case of Hill's equation known as the Mathieu equation. A theorem due to Floquet indicates that the solution to this equation will have a periodic form

$$Y\left(t + \frac{2\pi}{\omega_0}\right) = e^{i\mu}Y(t)$$

The usual approach to the analysis of such parametric processes involves perturbation theory. If we assume (1) a weak pump such that $\varepsilon \ll 1$ (so that the equation can be linearized) and (2) small damping such that $\Gamma/\Omega_0 \ll 1$, then we can simplify the analysis considerably. We begin by introducing a Fourier transform:

$$X(t) = \frac{1}{2\pi}\int e^{-i\omega t}X(\omega)\,d\omega$$

If we now transform the differential equation, we find

$$D(\omega)X(\omega) = \varepsilon\Omega_0^2\left[X(\omega - \omega_0) + X(\omega + \omega_0)\right]$$

where $D(\omega)$ is a dispersion relation for the natural oscillation of the system

$$D(\omega) = -\omega^2 - 2i\Gamma\omega + \left(\Omega_0^2 + \Gamma^2\right)$$

We can now study two cases suggested by Floquet's theorem:

Case 1: $\omega_0 \cong 2\Omega_0$, where ω_0 is the pump frequency.

If we look at frequencies $\omega \sim \Omega_0$, then $\omega - \omega_0 \sim -\Omega_0$ and is resonant with the natural oscillation frequency Ω_0. But $\omega + \omega_0 \sim 3\Omega_0$ and is off-resonance. Hence we can neglect $X(\omega + \omega_0)$ as an off-resonance term and consider only the coupling of $X(\omega)$ and $X(\omega - \omega_0)$

$$D(\omega)X(\omega) = \varepsilon\Omega_0^2 X(\omega - \omega_0)$$

In a similar fashion, we find that $X(\omega - \omega_0)$ satisfies

$$D(\omega - \omega_0)X(\omega - \omega_0) = \varepsilon\Omega_0^2 X(\omega)$$

where we have neglected $X(\omega - \omega_0) \sim X(-3\Omega_0)$ as off-resonance. Combining these, we find the dispersion relation

$$D(\omega)D(\omega - \omega_0) = \varepsilon^2\Omega_0^4$$

If we now factor the dispersion relation for the natural mode:

$$D(\omega) = -(\omega + \Omega_0 + i\Gamma)(\omega - \Omega_0 + i\Gamma)$$

and use the fact that $\omega \sim \Omega_0$ and $\Gamma/\Omega_0 \ll 1$, we can simplify this to

$$D(\omega) \sim -2\Omega_0(\omega - \Omega_0 + i\Gamma)$$

$$D(\omega - \omega_0) \sim 2\Omega_0(\omega - \omega_0 + \Omega_0 + i\Gamma)$$

so that our dispersion relation becomes

$$(\omega - \Omega_0 + i\Gamma)(\omega - \Omega_0 - \Delta + i\Gamma) + \varepsilon^2\frac{\Omega_0^2}{4} = 0$$

where we have introduced the frequency mismatch

$$\Delta \equiv \omega_0 - 2\Omega_0$$

Let us now write

$$\omega = (\Omega_0 + x) + iy$$

Here, a positive value of y would imply an instability in the mode. If we separate real and imaginary parts such that

$$x(x - \Delta) - (y + \Gamma)^2 + \varepsilon^2\Omega_0^2/4 = 0$$

$$(2x - \Delta)(y + \Gamma) = 0$$

we can identify two different types of solutions:

1. $y = -\Gamma$ which yields a damped oscillation with frequency $x = \frac{1}{2}[\Delta \pm [\Delta^2 - \varepsilon^2 \Omega_0^2]^{1/2}]$. This solution makes sense only when $\Delta^2 > \varepsilon^2 \Omega_0^2$. But we recall that $\Delta = \omega_0 - 2\Omega_0$. Hence we require $(\omega_0 - 2\Omega_0)^2 > \varepsilon^2 \Omega_0^2$, that is, ε sufficiently small or frequency mismatch Δ sufficiently large.

2. $x = \Delta/2$: Now we find $\text{Re}\{\omega\} = x + \Omega_0 = \omega_0/2$. Notice that this is a "frequency-locked" situation, since it does not depend on the natural frequency. We also find that the growth rate is $y = -\Gamma \pm \frac{1}{2}[\varepsilon^2 \Omega_0^2 - \Delta^2]^{1/2}$, which is valid in the region $\varepsilon^2 \Omega_0^2 > \Delta^2$ or $\varepsilon^2 \Omega_0^2 > (\omega_0 - 2\Omega_0)^2$. From our expression for y we see that one of the modes is less damped than the natural oscillation, the other mode is more heavily damped. The less damped mode becomes unstable when

$$\varepsilon^2 > \frac{\Delta^2 + 4\Gamma^2}{\Omega_0^2}$$

Thus we have arrived at a threshold condition on the pump intensity for instability. Notice that for zero frequency mismatch, $\Delta = 0$, we have the minimum threshold condition, $\varepsilon_{\min} = 2\Gamma/\Omega_0$. The maximum growth rate also occurs for $\Delta = 0$, $y_{\max} = -\Gamma + \varepsilon/2\Omega_0$.

Case 2: $\omega_0 \cong \Omega_0$. This analysis can be repeated to find two types of solution once again, one damped and one "frequency locked" with an oscillation frequency no longer dependent on the natural frequency but rather on the frequency mismatch.

Coupled Mode Parametric Excitation. The situation of more direct interest to three-wave processes is the coupling of two natural modes by a pump mode, as described by

$$\frac{d^2 X}{dt^2} + 2\Gamma_1 \frac{dX}{dt} + \left(\omega_1^2 + \Gamma_1^2\right) X(t) = \lambda Z(t) Y(t)$$

$$\frac{d^2 Y}{dt^2} + 2\Gamma_2 \frac{dY}{dt} + \left(\omega_2^2 + \Gamma_2^2\right) Y(t) = \mu Z(t) X(t)$$

where the pump is given by

$$Z(t) = 2Z_0 \cos \omega_0 t$$

It is customary to assume, without loss of generality, that $\omega_1 \ll \omega_2$.

The analysis of these coupled oscillators was first given by Nishikawa[37] and applied to analyze the interaction of electromagnetic waves with plasmas. The perturbation analysis of this problem, while quite similar to that of our previous single mode example, is cumbersome, and so we only discuss results here. For the frequency matching condition

$$\omega_0 \sim \omega_1 + \omega_2$$

we can again Fourier transform the coupled mode equations to find the dispersion relation

$$D_1(\omega) = \lambda \mu Z_0^2 \left[\frac{1}{D_2(\omega + \omega_0)} + \frac{1}{D_2(\omega - \omega_0)} \right]$$

where

$$D_s(\omega) = -\omega^2 - 2i\Gamma_s \omega + (\omega^2 + \Gamma_s^2), \qquad s = 1, 2$$

A perturbation analysis of this dispersion relation indicates two classes of solution. Both cases can be driven unstable, but in one of these cases there is a nonoscillatory solution. We study these solutions in more detail in the next section when we consider the particular application of the coupled-mode equations to electrostatic waves in a plasma.

Application to Electrostatic Waves in Plasmas.[36, 58]   To apply these results to the parametric excitation of electrostatic waves in a plasma by an incident laser beam, we must first transform the relevant equations describing the laser-plasma interaction into the form of the coupled parametric oscillators. It is typically assumed that the laser beam is represented by a uniform electric field oscillating at a frequency comparable (but greater than) the plasma frequency. The dynamics of the plasma are represented by the coupled hydrodynamics equations:

$$\frac{\partial n_\sigma}{\partial t} + \mathbf{u}_\sigma \cdot \frac{\partial n_\sigma}{\partial \mathbf{r}} + n_\sigma \frac{\partial}{\partial \mathbf{r}} \cdot \mathbf{u}_\sigma = 0$$

$$n_\sigma \left[\frac{\partial \mathbf{u}_\sigma}{\partial t} + \mathbf{u}_\sigma \cdot \frac{\partial \mathbf{u}_\sigma}{\partial \mathbf{r}} \right] + \frac{1}{m_\sigma} \frac{\partial p_\sigma}{\partial \mathbf{r}} = \frac{e_\sigma}{m_\sigma} n_\sigma \mathbf{E} - \nu_\sigma n_\sigma \mathbf{u}_\sigma$$

$$\frac{\partial}{\partial \mathbf{r}} \cdot \mathbf{E} = 4\pi \sum_\sigma e_\sigma n_\sigma, \qquad \sigma = e, i$$

These equations are then linearized about a spatially homogeneous part (oscillating with the applied field) and averaged over the high frequency motion of the electrons. The resulting set of equations then takes the form

$$\frac{\partial^2 n_e}{\partial t^2} + \nu_e \frac{\partial n_e}{\partial t} + \omega_{pe}^2(k) n_e(t) = \frac{ie}{m_e} \mathbf{k} \cdot \mathbf{E}_0 n_i(t)$$

$$\frac{\partial^2 n_i}{\partial t^2} + \nu_i \frac{\partial n_i}{\partial t} + \omega_{ia}^2(k) n_i(t) = -\frac{ie}{m_i} \mathbf{k} \cdot \mathbf{E}_0 n_e(t)$$

which is identical to that of the coupled oscillator problem. The results of the analysis of that problem can be applied directly.

More specifically, the dispersion relation for three-wave parametric coupling of electrostatic waves becomes

$$\left(\omega^2 + i\nu_i\omega - \omega_{ia}^2\right) = \frac{1}{4}\omega_{pi}^2\omega_{pe}^2 k^2 d_0^2\left[\frac{1}{(\omega-\omega_0)^2 - \omega_R^2 + i\nu_e(\omega-\omega_0)}\right.$$

$$\left. + \frac{1}{(\omega+\omega_0)^2 - \omega_R^2 + i\nu_e(\omega+\omega_0)}\right]$$

where

$$d_0 \equiv \frac{eE_0}{m\omega_0^2} = \text{electron "quiver" displacement}$$

$$\omega_R^2 \equiv \omega_{pe}^2 + \frac{\gamma_e k_B T_e}{m_e} k^2 = \text{electron plasma wave frequency}$$

$$\omega_{ia} \equiv \left(\frac{k_B T_e}{m_i}\right)^{1/2} k = \text{ion-acoustic wave frequency}$$

In deriving this dispersion relation, the assumptions of a weak pump ($kd_0 \ll 1$) and near critical density ($\omega_0 \sim \omega_R$) have been used. If we confine our attention to low frequency modes, $\omega \ll \omega_0$ and solve the dispersion relation for a frequency $\omega \equiv \omega_r + i\gamma$, we can determine the threshold for instability by setting $\gamma = 0$. There are two cases of interest:

1. *Decay mode instability.* $\omega_r \gg \nu_e$, $\omega_0 \sim \omega_r + \omega_R$.
   Then we find

$$\nu_i\omega_{ia}\nu_e\omega_R \sim \frac{1}{4}\omega_{pi}^2\omega_{pe}^2 k^2 d_0^2$$

If we note that the light intensity is given by

$$I = \frac{1}{2}cn_i\varepsilon_0 E_0^2$$

we can calculate the threshold for the parametric decay mode as

$$I_{\text{DM}} = 4cn_in_e(k_B T_e)\left(\frac{\nu_e}{\omega_{pe}}\right)\left(\frac{\nu_i}{\omega_{ia}}\right) = 2\left(\frac{\nu_e}{\omega_{pe}}\right)\left(\frac{\nu_i}{\omega_{ia}}\right)I_{\text{th}}$$

where we have defined the laser light intensity at which the quiver energy is equal to the thermal energy as

$$I_{\text{th}} = 2n_i cn_e(k_B T_e)$$

2. *Oscillating two-stream instability.* $\omega_r=0, \gamma=0$. Then

$$-\omega_{ia}^2 \sim \frac{1}{4}\,\omega_{pi}^2\omega_{pe}^2 k^2 d_0^2 \left[\frac{2(\omega_0^2-\omega_R^2)}{(\omega_0^2-\omega_R^2)^2+\nu_e^2\omega_0^2}\right]$$

(Notice that this implies that we must have $\omega_0<\omega_R$.) The minimum threshold occurs at $\omega_0^2-\omega_R^2=\nu_e\omega_0$ which yields

$$I_{2S}=2\left(\frac{\nu_e}{\omega_{pe}}\right)I_{th}$$

The thresholds for both the decay mode and two-stream instability depend on the damping of the waves. For the ion acoustic wave, one can write

$$\frac{\nu_i}{\omega_{ia}} \sim \left(\frac{m_e}{m_i}\right)^{1/2}+\left(\frac{T_e}{T_i}\right)^{3/2}\exp\left(-\frac{1}{2}\frac{T_e}{T_i}\right)$$

while for the electron plasma wave

$$\frac{\nu_e}{\omega_{pe}} \sim \frac{\nu_{ei}}{\omega_{pe}}+\left(\frac{\pi}{8}\right)^{1/2}(k_{max}\lambda_D)^{-3}\exp\left[-\frac{1}{2}(k_{max}\lambda_D)^{-2}\right]$$

Here, k_{max} is the wave number of the fastest growing ion acoustic mode, given approximately by

$$k_{max}\lambda_D \sim \left[\frac{2}{3}\left(\frac{\omega_0}{\omega_{pe}}-1\right)+\left(\frac{1.7}{3}\right)^2\frac{m_e}{m_i}\right]^{1/2}-\left(\frac{1.7}{3}\right)\left(\frac{m_e}{m_i}\right)^{1/2}$$

For the case of equal electron and ion temperatures, $T_e=T_i$, we find $\nu_i/\omega_{ia}\sim 0.6$ so that the thresholds for the two stream and decay mode instabilities are comparable, $I_{2S}\sim I_{DM}$. In the situation more typical of the blowoff plasma, $T_e\gg T_i$ and $\nu_i/\omega_{ia}\sim(m_i/m_e)^{1/2}\gg 1$ so that the threshold for the two stream instability is significantly greater than that for the decay mode, $I_{2S}\gg I_{DM}$.

The dispersion relations can be used to calculate the corresponding growth rates for these instabilities:

Decay mode

$$\gamma_{max}=\frac{3^{1/2}}{2}\,\omega_{pe}\left[\frac{1}{14}\frac{m_e}{m_i}(kd_0)^2\right]^{1/3}$$

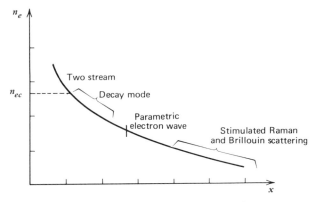

Figure 5.12. Density windows for parametric processes.

Two-stream

$$\gamma_{\max} = \tfrac{1}{2}\,\omega_{pe}\left[\frac{m_e}{m_i}\,(kd_0)^2\right]^{1/3}$$

Notice that in both cases, the growth rates scale as $\gamma_{\max} \sim I^{1/3}$.

The implications of these results are important. Consider laser light incident upon a density gradient as shown in Figure 5.12. The above analysis indicates that parametrically driven electrostatic wave instabilities can only occur near the critical surface. The minimum threshold for the excitation of these instabilities occurs for $T_e \gg T_i$ and $n_e \sim n_{ec}$. Then the damping is collisional in nature, and we find that the decay mode instability has the lowest threshold which scales as $I_{DM} \sim n_{ec}^{3/2}/T_e^{1/2}$.

Stimulated Scattering Processes.[58, 59] The incident light wave can also parametrically excite electromagnetic waves in the plasma leading to a stimulated scattering of light, that is, an enhanced reflection. For example, *stimulated Brillouin scattering* involves the parametric coupling of the incident light with an ion-acoustic wave and a backscattered electromagnetic wave. The *stimulated Raman scattering* process involves coupling of the incident light wave with an electron plasma wave and a backscattered electromagnetic wave. Frequency-matching conditions imply that these processes can occur in the underdense region if there is a sufficiently large scale height in the blowoff plasma.

Such stimulated scattering processes are potentially quite serious since they could scatter the incident laser light back off of the target plasma before it can penetrate into the critical surface where the most effective absorption (resonance or parametric) can occur. Brillouin scattering is the most serious, since the frequency matching condition $\omega_0 \sim \omega_{\text{scat}} + \omega_{ia}$ allows it to occur over a large region of the underdense plasma. Plasma simulations and experiments have indicated that nearly all of the energy of the incident light can be transferred to the scattered wave if conditions are right.

Several remedies have been proposed to circumvent this process. For example, one can avoid building up a large scale height plasma corona surrounding the target. An alternative approach is to modulate the frequency of the incident light (e.g., generating a random frequency by passing the light through a filter before focusing it on the target) in such a way as to increase the effective threshold intensity for the stimulated Brillouin process.[60-62]

5.2.   RELATIVISTIC ELECTRON BEAM ENERGY DEPOSITION

When electron beam drivers are used in inertial confinement fusion, the incident electron beam deposits its energy in the electron component of the ablation material of the target. These hot electrons then equilibrate with the target material ions. Several energy absorption mechanisms are of interest: (i) collisional absorption, (ii) modifications due to beam magnetic fields, and (iii) anomalous energy deposition due to collective (beam-plasma) effects.

5.2.1.   CLASSICAL (COLLISIONAL) ABSORPTION

At low target temperatures and low beam current, electron energy loss is primarily through binary collisions with target material electrons. As the target material becomes a plasma, it is important to account for changes due to electron shielding at close range and the excitation of plasmas waves (Cerenkov radiation).

More precisely, the possible interactions of an electron beam of energy $E \sim 1$ MeV incident upon a target is tabulated for both low and high Z target materials in Table 5.3. Here we have ranked the various energy deposition mechanisms in order of importance and given crude estimates of their relative contribution to the overall beam energy deposition.

We can write the formula characterizing collisional energy deposition as[63,64]

$$\frac{dE}{dx} = \frac{dE}{dx}\bigg|_{\text{ionization}} + \frac{dE}{dx}\bigg|_{\substack{\text{scattering} \\ \text{(electrons)}}} + \frac{dE}{dx}\bigg|_{\text{bremsstrahlung}}$$

where

$$\frac{dE}{dx}\bigg|_{\text{ionization}} = 4\pi r_0^2 \frac{m_0 c^2}{\beta} NZ \left[\ln \beta \left(\frac{E + m_0 c^2}{I} \right) \left(\frac{E}{m_0 c^2} \right)^{1/2} - \frac{\beta^2}{2} \right]$$

$$\frac{dE}{dx}\bigg|_{\text{scattering}} = 4\pi r_0^2 \frac{m_0 c^2}{\beta} NZ \left[\ln \left(\frac{m_0 c \lambda_D (\gamma - 1)}{2\hbar} \right)^{1/2} + \frac{1}{4} \ln \left(\frac{\beta^2 m_0 c^2}{kT} \right) \right]$$

$$\frac{dE}{dx}\bigg|_{\text{bremsstrahlung}} = 4\pi r_0^2 (E + m_0 c^2) 4Z^2 \left[\ln 2 \left(\frac{E + m_0 c^2}{m_0 c^2} \right) - \frac{1}{3} \right]$$

Table 5.3.   Energy Deposition Mechanisms for Relativistic Electron Beams

Mechanism	Relative contribution (%)	
	Low-Z Targets	High-Z Targets
Ionization	~ 90	<20
Elastic backscatter by nucleus	<10	>50
Inelastic backscatter by atomic electrons	<10	<10
Bremsstrahlung	~ 0	8%

Here, $m_0 c^2 = 0.511$ MeV, $\beta = v/c$, and $4\pi r_0^2 = 10^{-24}$ cm^2. In particular, for relativistic electrons slowing down in a 1-keV plasma, this yields roughly

$$\frac{dE}{dx} \sim 2.5\rho \frac{Z}{A} \qquad \text{MeV/cm}$$

For solid density, $dE/dx \sim 1$ MeV cm^2/g. Hence the range of a 1-MeV electron in solid density D-T is several centimeters. Since typical target designs are on the order of millimeters in diameter, it is apparent that classical absorption mechanisms are insufficient to absorb the incident driver beams.

5.2.2.   MAGNETIC FIELD EFFECTS AND ANOMALOUS ABSORPTION

An intense relativistic electron beam produces a magnetic field. If this field can penetrate the target, it can influence energy deposition.[65,66] For example, the electron Larmor radius in a megagauss field ranges from 0.01 to 0.1 mm. Electrons can be turned around and trapped in a layer of the order of the Larmor radius. But for this effect to occur, the magnetic field of the incident beam must penetrate the target. We can estimate the distance the magnetic field can diffuse into the target during a time τ as

$$\delta \sim 10^6 (\tau \ln \Lambda)^{1/2} T^{-3/4}$$

For example, if $\tau \sim 10^{-10}$ s, then the penetration distance is only 10^{-4} cm, far too short to lead to appreciable absorption.

If anomalous resistivity is present (due to plasma turbulence), the penetration distance becomes larger. There is some thought that a two-stream instability induced by the return current in the target could have this effect. This process has been observed experimentally, but it does not appear to be a strong effect. In fact, the coupling of the incident electron beam appears to be only several times that of the classical coupling mechanism.

As with laser-plasma interactions, there is some possibility that the incident electron beam will excite plasma waves that will interact with the beam electrons and lead to more efficient energy absorption in the outer layers of the target. Unfortunately such beam plasma interactions have not yet been demonstrated to lead to a beam absorption efficiency sufficient for ICF target design.

5.2.3. HIGH Z ABSORPTION TARGETS

One remedy to the absorption problem is to use a layer of high Z material such as gold to shorten the energy deposition range. However, this leads to a new problem, since the bremsstrahlung generated by a relativistic electron beam incident upon a high Z target is appreciable. In Table 5.4 we have compared the fraction of the incident beam energy converted into bremsstrahlung for low and high Z targets. The importance of this radiation production is apparent when the mean free paths of these bremsstrahlung photons are recognized to be of the order of centimeters or longer (see Table 5.5). The bremsstrahlung produced during beam absorption in the high Z shell can penetrate into the target, preheating the inner shell and the fuel.

More detailed calculations indicate that roughly 8% of the incident energy in a beam of 1-MeV electrons would be converted into bremsstrahlung radiation. For 10-MeV electrons, this percentage increases to 33%. Several modifications in target design have been proposed to mitigate bremsstrahlung preheat. Layers of differing Z (e.g., carbon, gold, etc.) might be used. Target designs have also been proposed that facilitate the diffusion of the incident beam magnetic field in an attempt to shorten the electron range. There have also been target designs that produce internally generated magnetic fields to shorten electron range.

In summary, however, there appear to be serious problems in achieving the necessary beam energy deposition characteristics to facilitate efficient implosion of ICF targets using relativistic electron beams. The long electron energy deposition range suggests that massive target designs may be necessary to absorb the incident beam energy. Bremsstrahlung preheating appears to be a very significant problem. These features, coupled with the difficulties in delivering and focusing intense relativistic electron beams over some distance onto an ICF target have raised serious doubts as to the suitability of this type of ICF driver.

Table 5.4. Fraction of the Incident Relativistic Electron Beam Energy Converted into Bremsstrahlung Radiation

Beam Energy	Target Z			
	1	3	29	79
1 MeV	0.001	0.002	0.027	0.082
10 MeV	0.01	0.02	0.16	0.33

Table 5.5. Bremsstrahlung Mean Free Paths in Various Materials

Material	Bremsstrahlung Energy (cm)	
	0.1 MeV	1 MeV
Al	2.2	6
Pb	0.17	1.25
D-T	35	96

5.3.   ION BEAM ENERGY DEPOSITION

Focused ion beams present an attractive alternative to laser and relativistic electron beam drivers. Ion beams have the distinct advantage that they appear to produce no preheating radiation such as suprathermal electrons or hard X rays. Ion energy deposition is "classical," that is, based on well-known collision processes.[67] The plasma effects (turbulence, suprathermal electron generation, thermal conduction inhibition) that have plagued the laser driven approach to inertial confinement fusion do not arise in ion beam driven targets. This feature allows target designers to return to those thrilling days of yesteryear[68] when their only concerns were classical coupling and thermal transport mechanisms.

Ion beams have the added advantage that there is no critical density associated with the beam propagation or energy deposition. Ions can penetrate deeper into the target, thereby coupling their energy into high density target material and driving a more efficient implosion process. Since ion beams cannot be reflected from the target, the absorption efficiency is 100%. These two effects combine to allow ion beam drivers to achieve overall implosion efficiences as high as 15 to 20% compared to the 5 to 10% efficiencies characterizing laser driven implosions.

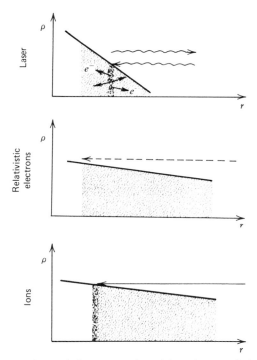

Figure 5.13.   A comparison of the energy deposition characteristics of laser, electron, and ion beams.

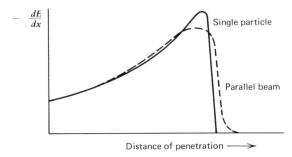

Figure 5.14. The energy loss profile, dE/dx versus x, for ions.

Yet another advantage for ion beam drivers arises from the spatial dependence of the driver beam energy deposition in the target. The energy deposition characteristics of ion, laser, and electron beams are compared in Figure 5.13. Ion beam energy deposition is characterized by a Bragg peaking phenomena that concentrates the energy deposition near the end of the ion range (see Figure 5.14). This effect allows the outer pellet material to serve as a tamper for the region where the ion energy is deposited. (See Chapter 9 for an example of such a target design.) By way of contrast, in laser driven implosions the outer material is below critical densities (e.g., 10^{21} cm^{-3}) and is blown off into a vacuum, hence serving no tamping function at all. In the case of relativistic electron beams, the stopping power is nearly constant over the electron range.

The more favorable energy deposition characteristics of ion beams give them a decided advantage over the more established approaches to inertial confinement fusion using laser or relativistic electron beam drivers. In this section we consider the range requirements for ion beams imposed by target implosion demands. We then examine theoretical models of ion energy deposition in targets and various ion range/energy characteristics appropriate for ICF applications.

5.3.1. REQUIREMENTS ON THE ION RANGE

To effectively drive an ICF target implosion, the specific energy (J/g) achieved by ion beam deposition in the target material must be sufficient to generate an ablation velocity of 10^7 cm/s. We recall then from Section 3.6 that this ablation will produce (through the rocket effect) an implosion velocity of 2×10^7 cm/s, the minimum required for thermonuclear ignition. The specific kinetic energy corresponding to target material velocity of 10^7 cm/s is 5×10^6 J/g. When the thermal energy of the ablator is added to this, the total specific energy is roughly 2×10^7 J/g. To infer an ion range R, let us assume a spherical target geometry so that

$$\frac{E}{m} = \frac{E}{4\pi r^2 \Delta r \rho} = \frac{E}{4\pi r^2 R} = 2 \times 10^7 \text{ J/g}$$

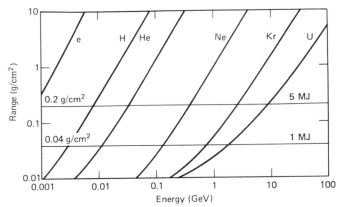

Figure 5.15. A comparison of the range of ions in cold material as a function of energy and mass.

where we have identified $\rho\Delta r = R$ as the range of the ions. For a typical target, $r \sim 0.3$ cm and $E \sim 1$ to 5×10^6 J, we require $R \sim 40$ to 200 mg/cm^2.

Figure 5.15 presents a plot of the range of different ions in cold material as a function of the ion energy. Using our range limits as a guide, we can quickly identify the acceptable energies of each ion species. For example, the ions listed in Table 5.6 correspond to a range of 100 mg/cm^2. From the perspective of a target designer, each of these ion species has the same range and is therefore equivalent. Of course this is not precisely true since the detailed deposition profile of the high Z ions will be different from that of low Z ions, and the implosion could be sensitive to this. However, to a first approximation, ion range is the most important factor in target design.

It should also be noted that while various ion species of differing energies but identical ranges may be regarded as essentially equivalent from the viewpoint of target design, they may present a considerable difference from that of driver design. High energy ions can be used to achieve a given driver beam intenstiy with a much lower current than low energy ions. Beam current is a significant factor in achieving the required beam focus on the target, as we see in Chapter 8.

Table 5.6.   Ions with a Range of 100 mg/cm^2

Ion	Energy (MeV)
H	4.8
He	19
Ne	240
Kr	1500
U	4400

5.3.2. CLASSICAL ION STOPPING IN COLD MATTER

Interest in ion stopping in solid or gaseous materials dates back to the early 1900s.[69] There has been considerable theoretical and experimental work on this subject.[70] The bulk of this work has dealt with the absorption of low intensity ion beams in cold matter. In this case, "low intensity" means that the ions do not interact with one another, and that they do not dynamically alter the properties of the matter, for example, through heating. Much of this information can be applied to the stopping of ions in the dense ablator material of ICF targets. Some modifications must be made to account for the high temperature of this material (~ 100 eV) and the corresponding free electron population.

The basic slowing down mechanisms involve excitation and ionization of bound atomic electrons through Coulomb interactions with the ion. This process is usually described by the Bethe equation[71] for ion stopping

$$\left(\frac{dE}{dx}\right)_{\text{Bethe}} = \frac{4\pi N_0 (Z_{\text{eff}})^2 \rho e^4 Z_2}{m_e c^2 \beta^2 A_2} \left[\ln \frac{2m_e c^2 \beta^2 \gamma^2}{\bar{I}} - \beta^2 - \sum_i \frac{c_i}{Z_2} - \frac{\delta}{2} \right]$$

where

$\quad\quad Z_1 =$ atomic number of projectile ion

$\quad\quad Z_{\text{eff}} =$ effective charge of the projectile ion

$\quad\quad N_0 =$ Avogadro's number

$\quad\quad \rho =$ density of stopping material

$\quad\quad A_2 =$ atomic weight of stopping material

$\quad\quad Z_2 =$ atomic number of stopping material

$\quad\quad \delta =$ polarization effect correction term

$\quad\quad \bar{I} =$ average ionization potential

$\quad\quad \sum_i \frac{c_i}{Z_2} =$ sum of the effects of shell correction terms

$$\gamma = \left(1 - \beta^2\right)^{-1/2}, \quad\quad \beta = v/c$$

The range of ions scales essentially as

$$R \sim \frac{A}{Z^2} \left(\frac{E}{A}\right)^{1.8}$$

Hence even for high energy ions (e.g., 10^3 MeV), the range is still well under 0.1 mm.

The average ionization potential is a very important parameter in the Bethe formula. It is formally defined by

$$Z \ln \bar{I} = \sum_n f_n \ln E_n$$

where E_n are the possible electronic states and f_n are the corresponding dipole oscillator strengths for the stopping material. In practice, this formal definition is not very useful and the average ionization is measured experimentally.[72]

The Bethe formula is valid as long as the average ionization $\bar{I}$ is less than $2m_e c^2 \beta^2 \gamma^2$. However, it diverges for higher ionizations unless atomic shell corrections and polarization effects are included. The shell correction terms can be included in the form of a five parameter least square fit to the available data for proton stopping[73]

$$\sum_i \frac{c_i}{Z_2} = a_0 + a_1 \ln E + a_2 (\ln E)^2 + a_3 (\ln E)^3 + a_4 (\ln E)^4$$

The inclusion of the shell corrections improves the Bethe model at low ion energy, but it remains invalid for very low ion energies. Here it is customary to use the LSS model of Linhard.[74] In this model a Thomas-Fermi description of the bound electrons is used, and the stopping power due to excitation and ionization is added to a contribution from nuclear elastic scattering of the projectile ion. The electronic contribution is given by

$$\frac{dE}{dx}\bigg|_{LSS} = C_{LSS} E^{1/2}$$

where

$$C_{LSS} = K (E_L / 1.602 \times 10^{-9})^{1/2} / (R_L \times 10^4)(keV^{1/2}/\mu m)$$

$$E_L = (1 + A) Z_1 Z_2 e^2 / A a$$

$$a = 0.468 c (Z_1^{2/3} + Z_2^{2/3})^{-1/2} \times 10^{-8} \qquad (cm)$$

$$R_L = (1 + A)^2 / 4\pi A N a^2$$

$$A = A_2 / A_1$$

N = target atom number density

$$K = \frac{0.0793 \, Z_1^{2/3} Z_2^{2/3} (1 + A)^{3/2}}{(Z_1^{2/3} + Z_2^{2/3})^{3/4} A_2^{1/2}}$$

A_1 = atomic weight of projectile ion

The range of validity of this formula is restricted by $Z_1^{1/3} \geq 137 \beta$.

The elastic nuclear scattering contribution can be expressed as [75, 76]

$$\frac{dE}{dR}\bigg|_{nuc} = C_n \varepsilon^{1/2} \exp\left[-45.2(C_n'\varepsilon)^{0.277}\right] \qquad (\text{MeV}/\text{g cm}^2)$$

where $R = \rho x$,

$$C_n = 4.14 \times 10^6 \left(\frac{A_1}{A_1 + A_2}\right)^{3/2} \left(\frac{Z_1 Z_2}{A_2}\right) \left(Z_1^{2/3} + Z_2^{2/3}\right)^{-3/4}$$

$$C_n' = \frac{A_1 A_2}{(A_1 + A_2)} \frac{1}{Z_1 Z_2} \left(Z_1^{2/3} + Z_2^{2/3}\right)^{-1/2}$$

$$\varepsilon = E/A_1$$

The stopping power of energetic ions in cold material can be estimated by

$$\frac{dE}{dx}\bigg|_{bound} = \min\left\{\frac{dE}{dx}\bigg|_{Bethe}, \frac{dE}{dx}\bigg|_{LSS}\right\} + \frac{dE}{dx}\bigg|_{nuc}$$

A final problem with the Bethe model is the value of Z_{eff}, the effective charge of the fast ion. The stopping of ions other than protons might be expected to scale as Z^2. However their measured values deviate from this dependence. This discrepancy is included in the theory by defining a value for Z_{eff} that matches the experimental results.[76-79] A suitable expression for this is given by Brown and Moak as[78]

$$\frac{Z_{eff}}{Z_1} = 1 - 1.034 \exp\left(-137.04 \, \beta/Z_1^{0.69}\right)$$

This completes our discussion of the stopping of ions via classical mechanisms in cold matter.[80] We now turn our attention to those modifications that occur in the high temperature plasmas characterizing ICF targets.

5.3.3.   HIGH TEMPERATURE (PLASMA) EFFECTS ON ION STOPPING

The Bethe and LSS models (and their modifications) adequately describe the stopping of ions in cold matter.[80] But in the case of ICF targets enough energy is deposited to heat the ablator material to several hundreds of electron volts. At these temperatures there is appreciable ionization. For example, gold at $T = 50$ eV and 10% of solid density will have $Z_{eff} \sim 10.3$. The free electrons produced by this ionization of target material will contribute significantly to the stopping power. Ionization also changes the properties of the remaining bound electrons and screens these from the projectile ion.

The effect of ionization on the average ionization potential of the atom can be roughly treated by replacing the ionization potential of the ionized atom with an expression such as the following for doubly charged oxygen:[81]

$$\bar{I}=(\text{oxygen}^{+2})=\left(\frac{Z_{\text{oxygen}}}{Z_{\text{carbon}}}\right)^2 \bar{I}\,(\text{neutral carbon})$$

Simple binary collision theory for impact parameters within a Debye length coupled to plasma wave excitation (collective phenomena) outside of a Debye length can be used as the basis for the stopping power of free plasma electrons and ions. The energy loss relation for free electrons is given by[82]

$$\left.\frac{dE}{dx}\right|_{\text{free}}=\frac{\omega_p^2(Z_{\text{eff}})^2 e^2}{c^2\beta^2}G(y_e)\ln\Lambda_{\text{free}}$$

where

$$G(\xi)=\text{erf}(\xi^{1/2})-2\frac{\xi^{1/2}}{\pi}e^{-\xi}$$

$$y_e=\beta/\beta_e=\left(\frac{m_e c^2\beta^2}{2kT_e}\right)^{1/2}$$

$$\omega_p^2=\frac{4\pi\rho\bar{Z}_2 e^2 N_0}{m_e A_2}$$

$$\Lambda_{\text{free}}=\frac{0.764\beta c}{p_{\text{min}}\omega_p}$$

$$p_{\text{min}}=\max\left\{\frac{e^2 Z_1}{\mu\bar{u}^2},\frac{h}{2\mu\bar{u}}\right\},\qquad \mu=\frac{m_1 m_2}{m_1+m_2}$$

$\bar{u}=$ average relative speed between the ion and target electron

A similar expression can be written for the plasma ion component

$$\left.\frac{dE}{dx}\right|_{\text{ion}}=\frac{Z_{\text{eff}}^2 Z_2 e^2}{\beta^2 c^2 A_2}\left(\frac{m_e}{m_p}\right)\omega_p^2 G(y_i)\ln\Lambda_i$$

where

$$y_i=\frac{A_2 E}{A_i kT_i}$$

$$\Lambda_i=p_{\text{max}}/p_{\text{min}}$$

$$p_{\text{max}}=\lambda_D=\text{Debye length}$$

$$p_{\text{min}}=\frac{\mu\beta_i^2}{Z_1 Z_2}\frac{m_p c^2}{e^2},\qquad \mu=\frac{A_1 A_2}{A_1+A_2}$$

Thus the stopping of ions in the hot, dense ablator of ICF targets is significantly more complex than the stopping in cold matter. The latter has the great advantage that much experimental data is available to normalize the theoretical models. There is essentially no data to verify the high temperature effects.

5.3.4. ION STOPPING: SOME TYPICAL RESULTS

A monoenergetic beam of carbon ions stopping in a solid density cold gold target is a good test of the classical theory. Figure 5.16 shows the stopping as predicted by the Bethe theory with shell corrections and the LSS model. We see that the Bethe theory diverges at low ion energy, but this is precisely where the LSS theory picks up. A combination of the two closely approximates the tabulated stopping powers of Northcliffe and Schilling.[83]

The important effects of the free electrons in a partially ionized plasma can be seen in Figure 5.17 where the individual components of the stopping power are plotted for gold at $T=200$ eV and 10% solid density. The characteristics of different ions are shown in Figures 5.18 and 5.19 where the deposition profiles for protons, carbon, and xenon are given for cold and high temperature gold. The range shortening that is evident in these figures is more explicitly displayed in the three-dimensional plot of Figure 5.20. The range of protons in gold at a temperature of 100 eV is only about one-half that of the cold range. At higher temperatures, as the free electron component becomes more dominant, the Bragg peak disappears.

These details are of considerable importance to the target designer since they determine the thickness of the ablator and the intensity of the beam necessary to achieve an energy deposition of 2×10^7 J/g. The most concentrated energy deposition occurs at 50 to 100 eV in the sharp Bragg peak. However, these temperatures only correspond to about 5 to 10×10^6 J/g of specific internal energy. The 2×10^7 J/g needed for the target implosion comes at temperatures of about 200 eV where the Bragg peak has smoothed out.

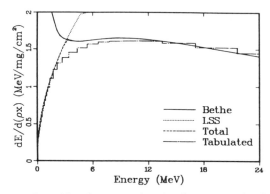

Figure 5.16. Energy deposition from a carbon ion beam stopping in solid density cold gold targets. (Courtesy of Sandia Laboratory.)

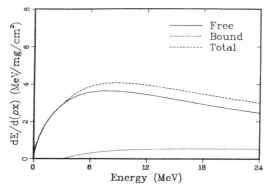

Figure 5.17. Energy deposition for carbon ions incident on a gold target plasma at a temperature of 200 eV. (Courtesy of Sandia Laboratory.)

5.3.5. THE VALIDITY OF THE LOW INTENSITY THEORY FOR ICF APPLICATIONS

The bound electron theory, the Bethe and LSS models, and the plasma stopping theory are all applicable to single test particle situations (i.e., a dilute beam of ions). But can such theories be used for very high intensity beams in the 100-TW range?

Consider, for example, the case of a 2-MeV proton beam of intensity 100 TW/cm^2 incident upon a target with an electron density of 10^{22} to 10^{24} cm^{-3}, typical of solid materials. The interparticle spacing in the ion beam is 10^{-6} cm, while the Coulomb shielding distance in the target is 10^{-8} cm. Hence, the 100 shielding lengths between the ions in the beam should imply that they interact as independent particles and that collective phenomena within the beam itself

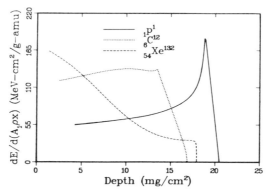

Figure 5.18. Proton, carbon, and xenon ion deposition profiles in cold, normal density gold. (Courtesy of Sandia Laboratory.)

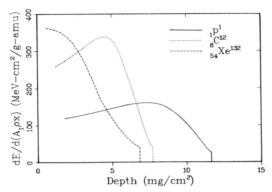

Figure 5.19. Proton, carbon, and xenon ion deposition profiles in 200-eV, 0.01 solid density gold. (Courtesy of Sandia Laboratory.)

is unlikely once it has penetrated the material. At higher ion energies, such as 10-GeV uranium, the number of beam ions is proportionately lower for the same intensity. In this case the interparticle spacing is 17 times as large as for 2-MeV protons, and there are 1700 shielding lengths between the ions.

Therefore these beams of ions are not really very "intense" once they have penetrated the ablator material, due to shielding effects.

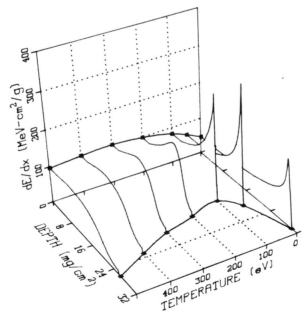

Figure 5.20. Deposition profiles for 2 MeV protons in gold at a density of 0.193 g/cm^3 as a function of temperature. (Courtesy of Sandia National Laboratory.)

REFERENCES

1. C. DeMichelis, "Laser interaction with Solids–A Bibliographical Review," *IEEE J. Quantum Electron.* **QE-6**, 630 (1970).

2. P. Nelson, P. Veyrie, H. Berry, and Y. Durand, "Experimental And Theoretical Studies of Air Breakdown by Intense Pulse of Light," *Phys. Lett.* **13**, 226 (1964).

3. A. J. Alcock, C. DeMichelis, V. V. Korobkin, and M. C. Richardson, "Preliminary Evidence for Self-Focusing in Gas Breakdown Produced by Picosecond Laser Pulses," *Appl. Phys. Lett.* **14**, 141 (1969).

4. J. W. Shearer, "A Survey of the Physics of Plasma Heating by Laser Light," Lawrence Livermore Laboratory Report UCID-15745 (1970).

5. J. W. Shearer, "Effect of Oblique Incidence on Optical Absorption of Laser Light by a Plasma," *Phys. Fluids* **14**, 501 (1971).

6. T. Johnston and J. Dawson, "Correct Values for High Frequency Power Absorption by Inverse Bremsstrahlung in Plasmas," *Phys. Fluids* **16**, 722 (1973).

7. J. W. Shearer and J. J. Duderstadt, "Wavelength Dependence of Laser-Light Absorption by a Solid Deuterium Target," *Nucl. Fusion* **13**, 401 (1973).

8. C. Max, 1980 Les Houches Lectures.

9. S. Rand, "Inverse Bremsstrahlung with High-Intensity Radiation Fields," *Phys. Rev.* **136**, B231 (1964).

10. V. P. Silin, "Non-linear High Frequency Plasma Conductivity," *Sov. Phys. JETP* **20**, 1510 (1965).

11. M. B. Nicholson-Flourence, "Intensity Dependence of Free-Free Absorption," *J. Phys.* **A4**, 574 (1971).

12. R. K. Osborn, "Nonlinear Bremsstrahlung," *Phys. Rev.* **A5**, 1660 (1972).

13. G. J. Pert, "Inverse Bremsstrahlung Absorption in Large Radiation Fields During Binary Collisions-Born Approximation I. Elastic Collisions," *J. Phys.* **A5**, 1221 (1972).

14. H. Brysk, "Multiphoton Inverse Bremsstrahlung," *J. Phys.* **A8**, 1260 (1975).

15. J. F. Seely and E. G. Harris, "Heating of a Plasma by Multiphoton Inverse Bremsstrahlung," *Phys. Rev.* **A7**, 1064 (1973).

16. L. Schlessinger and J. Wright, "Inverse Bremsstrahlung Absorption Rate in an Intense Laser Field," *Phys. Rev.* **A20**, 1934 (1979).

17. B. Langdon, "Non-linear Inverse Bremsstrahlung and Heated Electron Distributions," *Phys. Rev. Lett.* **44**, 575 (1980).

18. V. L. Ginzburg, *The Propagation of Electromagnetic Waves in Plasmas* (New York, Pergamon, 1964).

19. N. Denisov, "On a Singularity of the Field of an Electromagnetic Wave Propagated in an Inhomogeneous Plasma," *Sov. Phys JETP* **4**, 544 (1957).

20. A. Piliya, "Wave Conversion in an Inhomogeneous Plasma," *Sov. Phys.–Tech. Phys.* **11**, 609 (1966).

21. J. P. Freidberg, R. W. Mitchell, R. L. Morse and L. I. Rudsinski, *Phys. Rev. Lett.* **28**, 795 (1972).

22. W. Kruer, Lecture Notes for Scottish Summer School on Laser-Plasma Interactions, 1979.

23. J. Balmar and T. Donaldson, "Resonance Absorption of 1.06 μm Laser Radiation in Laser Generated Plasma," *Phys. Rev. Lett.* **39**, 1084 (1977).

24. K. Manes, V. Rupert, J. Auerbach, P. Lee, J. Swain, "Polarization and Angular Dependence of 1.06 μm Laser-Light Absorption by Planar Plasmas, *Phys. Rev. Lett.* **39**, 281 (1977).

25. J. Perlman and J. J. Thompson, "Polarization Dependent Energy Transport in Laser Produced Plasmas," *Appl. Phys. Lett.* **32**, 703 (1978).

26. W. Kruer and K. Estabrook, "Laser Light Absorption Due to Self-Generated Magnetic Fields," *Phys. Fluids* **20**, 1688 (1977).

27. K. Estabrook and W. Kruer, "Properties of Resonantly Heated Electron Distributions," *Phys. Rev. Lett.* **40**, 42 (1978).

28. D. Forslund, J. Kindl, and K. Lee, "Theory of Hot Electron Spectra at High Laser Intensity," *Phys. Rev. Lett*, **39**, 284 (1977).

29. B. Bezzerides, S. Gitomer, and D. Forslund, "Randomness, Maxwellian Distributions and Resonance Absorption," *Phys. Rev. Lett.* **44**, 651 (1980).

30. C. Armstrong, B. Ripin, F. Young, R. Decoste, R. Whitlock, and S. Bodner, "Emission of Energetic Electrons from a Nd-Laser-Produced Plasma," *J. Appl. Phys.* **50**, 5233 (1979).

31. H. Hora, "Nonlinear Confining and Deconfining Forces Associated with the Interaction of Laser Radiation with Plasma," *Phys. Fluids* **12**, 182 (1969).

32. J. D. Lindl and P. K. Kaw, "Ponderomotive Force on Laser-Produced Plasmas," *Phys. Fluids* **14**, 371 (1971).

33. E. Valeo and K. Estabrook, "Stability of the Critical Surface in Irradiated Plasmas," *Phys. Rev. Lett.* **34**, 1008 (1975).

34. E. Lindman, "Absorption and Transport in Laser Plasmas," *J. Phys. Colloq.* **C-6**, 9 (1977).

35. D. W. Phillion, et al., "Evidence for Profile Steepening in Laser Irradiated Plasmas," *Phys. Fluids* **20**, 1892 (1977).

36. J. J. Thomson, et al., "Theoretical Interpretation of Angle and Polarizaton Dependent Laser Light Absorption Measurements," *Phys. Fluids* **21**, 707 (1978).

37. H. Nishimura, et al., "Resonance Absorption and Surface Instability at a Critical Density Surface of a Plasma Irradiated by a CO_2 Laser," *Plasma Phys.* **21**, 69 (1980).

38. B. Ripin, "Laser Fusion Studies at NRL," Ed. S. Bodner, NRL Memo Report 3591, Oct. 1977, p. 128.

39. J. Dawson and C. Oberman, "High Frequency Conductivity and the Emission and Absorption Coefficients of a Fully Ionized Plasma," *Phys. Fluids* **5**, 517 (1962).

40. R. Faehl and W. Kruer, "Laser Light Absorption by Short Wavelength Ion Turbulence," *Phys. Fluids* **20**, 55 (1977).

41. W. Manheimer, D. Colombant, and B. Ripin, "Efficient Light Absorption by Ion-Acoustic Fluctuations in Laser Produced Plasmas," *Phys. Rev. Lett.* **38**, 1135 (1977).

42. W. Manheimer and D. Colombant, "Light Absorption by Ion-Acoustic Turbulence in Laser Produced Plasmas," *Phys. Fluids* **21**, 1818 (1978).

43. J. W. Shearer, et al., "Experimental Indications of Plasma Instabilities Induced by Laser Heating," Lawrence Livermore Laboratory Report UCRL-73489 (1971); *Phys. Rev.* **A6**, 764 (1972).

44. R. A. Haas, et al., "Irradiation of Parylene Disks with a 1.06 Micron Laser," *Phys. Fluids* **20**, 322 (1977).

45. W. L. Kruer, R. A. Haas, W. C. Mead, D. W. Phillion, and V. C. Rupert, "Collective Behavior in Recent Laser-Plasma Experiments," *Plasma Phys.*, **64** (1977).

46. K. Nishikawa, "Parametric Excitation of Coupled Waves," *J. Phys. Soc. (Japan)* **24**, 916, 1154 (1968).

47. J. Dawson and C. Oberman, "High Frequency Conductivity and the Emission and Absorption Coefficients of a Fully Ionized Plasma," *Phys. Fluids* **5**, 517 (1962).

48. V. P. Silin, "Parametric Resonance in a Plasma," *Sov. Phys.–JETP* **21**, 1127 (1965).

49. G. G. Comisar, "Theory of the Stimulated Raman Effect in Plasma," *Phys. Rev.* **141**, 200 (1966).

50. E. A. Jackson, "Parametric Effects of Radiation on a Plasma," *Phys. Rev.* **153**, 235 (1967).

51. D. F. Dubois, "Parametrically Excited Plasma Fluctuations," *Phys. Rev.* **164**, 207 (1967).

52. J. Dawson, P. Kaw, and B. Green, "Optical Absorption and Expansion of Laser-Produced Plasmas," *Phys. Fluids* **12**, 875 (1969).

53. J. R. Sanmartin, "Electrostatic Plasma Instabilities Excited By a High-Frequency Electric Field," *Phys. Fluids* **13**, 1533 (1970).

54. V. V. Pustovalov and V. P. Silin, "Anomalous Absorption of an Electromagnetic Wave," *Sov. Phys. JETP* **32**, 1198 (1971).

55. W. L. Kruer, J. Katz, J. Byers, and J. DeGroot, "Plasma Heating by Large-Amplitude, Low-frequency Electric Fields," *Phys. Fluids* **15**, 1613 (1972).

56. A. A. Galeev and R. Z. Sagdeev, "Parametric Phenomena in a Plasma," *Nucl. Fusion* **13**, 603 (1973).

57. S. Jorna, "Laser Induced Instabilities in Homogeneous Plasmas," *Phys. Fluids* **17**, 765 (1974).

58. C. S. Liu, "Parametric Instabilities in Homogeneous Unmagnetized Plasmas," *Adv. Plasma Phys.* **6**, 83 (1975) and "Parametric Instabilities in an Inhomogeneous Unmagnetized Plasma," *Adv. Plasma Phys.* **6**, 121 (1975).

59. D. W. Forslund, J. M. Kindel, and E. L. Lindman, "Theory of Stimulated Scattering Processes in Laser-Irradiated Plasmas," *Phys. Fluids* **18**, 1002 (1975).

60. J. J. Thomson, W. L. Kruer, and S. E. Bodner, "Parametric Instability Thresholds and Their Control," *Phys. Fluids* **17**, 849 (1974).

61. E. J. Valeo and C. R. Oberman, "Model of Parametric Excitation by an Imperfect Pump," *Phys. Rev. Lett.* **30**, (1973).

62. J. J. Thomson and J. I. Karush, "Effects of Finite-Bandwidth Driver on the Parametric Instability," *Phys. Fluids* **17**, 1608 (1974).

63. J. D. Jackson, *Classical Electrodynamics*, 3rd ed. (New York, Wiley, 1975), Chap. 13.

64. R. D. Evans, *The Atomic Nucleus*, (New York, McGraw Hill, 1969), Chap. 21.

65. D. Mosher and I. Bernstein, "Magnetic-Field-Induced Enhancement of Relativistic-Electron Beam Energy Deposition," *Phys. Rev. Lett.* **38**, 1483 (1977).

66. S. L. Bogolyubsky, B. P. Gerasimov, V. I. Liksonov, Yu. P. Topov, L. I. Pudakov, A. A. Samarskii, V. P. Smirnov, and L. I. Urutskoev, *Pis'ma Zh. Eksp. Teor. Fiz.* **24**, 202 (1976).

67. T. A. Mehlhorn, "A Finite Material Temperature Model for Ion Energy Deposition in Ion-Driven ICF Targets," Sandia National Laboratory Report SAND 80-0038 (May, 1980)

68. J. Nuckolls, L. Wood, A. Thiessen, and G. Zimmerman, "Laser Compression of Matter to Super-High Densities: Thermonuclear (CTR) Applications," *Nature* **239**, 139 (1972).

69. N. Bohr, "On the Theory of the Decrease of Velocity of Moving Electrified Particles on Passing Through Matter," *Philos. Mag.* **25**, 10 (1913).

70. S. P. Ahlen, "Theoretical and Experimental Aspects of the Energy Loss of Relativistic Heavily Ionized Particles," *Rev. Mod. Phys.* **52**, 121 (1980).

71. H. Bethe, "On the Theory of the Passage of Fast Particle Beams Through Matter," *Ann. Phys.* **5**, 325 (1930).

72. E. Williams, "Application of Ordinary Space-Time Concepts in Collision Problems and the Relation of Classical Theory to Born's Approximation," *Rev. Mod. Phys.* **17**, 217 (1945).

73. H. H. Andersen and J. F. Aiegler, *Hydrogen—Stopping Powers and Ranges in All Elements* (New York, Pergamon, 1977).

74. J. Linhard, M. Scharff, and H. E. Shiott, Kgl. Danske Videnskab. Selskab, *Mat. Fys. Medd.* **33**, (14) (1963).

75. P. C. Steward and R. W. Wallace, Lawrence Livermore Laboratory Report UCRL-19128 (1970).

76. J. Linhard and M. Scharff, "Energy Dissipation by Ions in the Kev Range," *Phys. Rev.* **124**, 128 (1964).

77. H. D. Betz, "Charge States and Charge-Changing Cross Sections of Fast Heavy Ions Penetrating Through Gaseous and Solid Media," *Rev. Mod. Phys.* **44**, 465 (1972).

78. M. D. Brown and C. D. Moak, *Phys. Rev.* **B6**, 90 (1972).

79. P. Steward, "Stopping Power and Range for Any Nucleus in the Specific Energy Interval 0.01 – 500 MeV/amu in Any Nongaseous Material," Lawrence Livermore Laboratory Report UCRL-18127 (1968).

80. J. F. Ziegler, *Stopping Cross Sections for Energetic Ions in All Elements*, Vol. 15 (New York, Pergamon, 1972).

81. D. Mosher, in *ERDA Summer Study of Heavy Ions for Inertial Fusion*, Lawrence Berkeley Laboratory Report LBL-5543 (1976), p. 39.

82. J. D. Jackson, *Classical Eelctrodynamics*, 3rd ed. (New York, Wiley, 1975), p. 643.

83. L. Northcliffe and R. Schilling, *Nucl. Data Tables* **A7**, 233 (1970).

SIX

Computer Simulation

The dynamics of an inertial confinement fusion target involve a variety of complex hydrodynamic and transport processes as well as driver beam-target interaction and fusion reaction kinetics. Although any single process can usually be described by simple models when decoupled from other processes, the complete description of pellet implosion and burn must be simulated using large computer codes that contain all of the relevant physics.[1-11] The successful implosion of ICF targets requires an accurate description of driver energy deposition in the outer layers of the target, the transport of this energy into the ablation surface via thermal conduction or particle transport, and finally the conversion of this thermal energy into hydrodynamic motion that leads to isentropic compression of the fuel to high density. As the imploding shock waves converge to the pellet center with a velocity of the order of 3×10^7 cm/s, they shock heat the central region of the compressed pellet to 4 to 10 keV and thermonuclear burning occurs. This burning self-heats the central fuel region to over 20 keV, and a supersonic burn wave propagates outward from the pellet center heating the surrounding cold fuel by fusion reaction product energy deposition and bringing it to ignition conditions.

The achievement of high target fusion energy gain depends sensitively on many details of the implosion and energy transport process. We can conveniently picture the imploding pellet as consisting of three regions as shown in Figure 6.1. Driver beam energy deposition occurs in the outermost region, and the physics of the energy absorption process and energy transport dominate the dynamics of this region.

To make this discussion more explicit, consider the particular case of laser fusion. For laser drivers, the energy deposition region is bordered by the outer edge of the pellet plasma or corona and the critical density surface where the electron plasma frequency is equal to the laser light frequency (10^{21} cm^{-3} for 1.06-μm and 10^{19} cm^{-3} for 10.6-μm light, respectively). In this region laser light is absorbed or reflected from the plasma, with most of the interaction

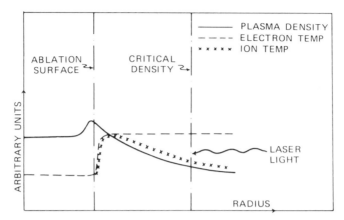

Figure 6.1. Plasma density and electron and ion temperature profiles during the implosion of a spherical target.

occurring near the critical density surface. Laser light absorption is usually modeled as a combination of classical (inverse bremsstrahlung) and resonance absorption. Here we recall that resonant absorption occurs because of a resonant coupling between the laser light electric field and electron plasma waves.[12, 13] This process requires the oblique incidence of laser light on the electron density gradient in the corona region near the critical density surface. Since this electron density gradient also refracts laser light, there will be an optimum angle at which resonant absorption is maximized.[14, 15] In actual practice this resonant absorption process cannot be self-consistently determined within a plasma hydrodynamics calculation since the time and spatial scales are far too small. Instead the results of numerical plasma particle simulations together with experimental results are used to provide a recipe for determining the amount of energy absorbed through resonant absorption.[16]

The resonant absorption process deposits much of the incident light energy in fast electrons.[17-19] These so-called suprathermal electrons may be 10 to 100 times hotter than the thermal plasma background (10 to 100 keV) and consequently possess mean free paths that are very long in comparison with the gradients in the thermal plasma. The transport of these suprathermal electrons in the corona region is presumed to give rise to fast ion emission observed in laser-plasma experiments.[20, 21] Those hot electrons streaming inward from the critical density surface can stream ahead of the ablation front and preheat the compressed pellet core.[22] This results in a degradation of the implosion process. Therefore it is very important to treat the origin and transport of these suprathermal electrons properly in any simulation of the pellet implosion process.

In addition to this nonhydrodynamic coupling of laser light energy into the plasma electrons, there is also a contribution to the hydrodynamic pressure due to the presence of laser radiation near the critical density. This so-called ponderomotive force,[23, 24] due to the intense electromagnetic field of the incident light, can significantly alter the electron density gradient profile near the critical density surface and, thus, can affect both classical and resonant absorption processes. This makes the task of properly modeling the coupling of laser light into the plasma even more difficult. To date no laser fusion hydrodynamics code can consistently predict the amount of laser energy deposited in the plasma. Typically the amount of energy deposited is determined by a prescription that comes from the particular experiment being modeled, and the remainder of the simulation is performed under this assumption.

The second region of interest is bordered by the energy deposition region (e.g., the critical density surface for laser fusion) on the outside and the ablation front on the inside. This region has a fluid velocity in the outward direction. However, the dominant energy flow is inward via the thermal electron conduction process, suprathermal electron streaming, and radiative transfer. Thus the transport of energy dominates the dynamics of this region, underscoring the fact that inertial confinement fusion calculations should not be considered as hydrodynamic calculations in the classical sense that envision mass flow as the dominant process. In this middle region energy flow is an equally important process, and the transport of nonthermal energy in the form of suprathermal electrons and radiation is of considerable importance. Later we will find that the use of standard procedures for computing hydrodynamic transport coefficients leads to erroneous results unless ad hoc "fix-ups" are provided in the equations to ensure plausible results.

The sensitivity of the implosion process to uncertainties in the electron thermal conduction process is displayed in Figure 6.2, where thermal conductivity is scaled from its nominal value, while all other parameters, including the

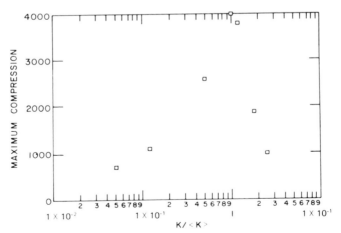

Figure 6.2. Results of hydrodynamic simulations showing the dependence of the maximum compression of the pellet core on the ratio of electron thermal conductivity to its nominal value.

laser input pulse shape, are held fixed.[25] These results are for a solid D-T sphere and a 60-kJ laser pulse with an optimum power law shape. The maximum compression of the pellet core varies by over an order of magnitude for changes in κ_e within the uncertainty of its correct value.

Important reductions of κ_e (so-called thermal flux inhibition) can result from plasma instabilities[26] and locally intense self-generated magnetic fields.[27] Ion-acoustic instabilities can result from the interaction between the inwardly directed thermal conduction electrons, the counter-streaming cold electron return current that is induced to maintain charge neutrality, and the background ion motion in the outward direction. These instabilities can lead to an increase in the effective electron-ion collision frequency and thereby reduce thermal conduction. Again, these effective transport coefficients cannot be calculated in a self-consistent fashion within a hydrodynamics description, but rather they are usually approximated by using a prescription such as ratios of electron and ion temperatures and the magnitude of the electron heat flux.[28] Self-generated magnetic fields can result from thermoelectric currents that are produced in an inhomogeneous plasma.[29] One typical situation is nonparallel density and temperature gradients, $\nabla n \times \nabla T$, that result from nonuniform laser illumination and / or the development of two-dimensional flow during the implosion process. Two-dimensional flow can result from a non-symmetric target configuration such as the ball-on-disk target or from the development of fluid instabilities during a symmetric implosion. In either case the thermal transport coefficient characterizing conduction transverse to the magnetic field lines is reduced by the factor $(1+\Omega^2\tau^2)^{-1}$, where Ω is the electron-cyclotron frequency, and τ is the electron-ion collision frequency. A two-dimensional hydrodynamic description is necessary to predict this important effect, even for a presumably symmetric implosion. This is of great significance because two-dimensional calculations are extremely expensive, and accurate treatment of the transport processes, which are important in addition to the hydrodynamics, can become prohibitively expensive in two dimensions.

The third region of interest is that portion of the target within the ablation surface. In this dense, cold plasma, hydrodynamics is presumably an accurate model. However, care must be taken to properly evaluate the associated plasma equations of state. At high compressions, cold electrons behave as a Fermi degenerate gas, and their properties must be treated accordingly. The other hydrodynamic effect of interest is fluid instability. Conditions at the ablation front can correspond to Rayleigh-Taylor instability conditions, and growth rates can be fast enough to destroy the symmetric compression. Again, an accurate description of this phenomenon necessitates the use of multidimensional models. Perturbation techniques[30] as well as full two-dimensional calculations[31] are used to study this problem.

For all three of these regions, radiation emission, transport, and absorption also must be considered.[32] High energy photons created in the laser absorption region can stream inward to preheat the pellet core just as the suprathermal electrons. More importantly, in the current experiments, radiation is one of the principal diagnostic tools. The radiation spectrum is used to determine

Table 6.1. Target Physics

Energy Deposition Region	Energy Transport Region	Pellet Core Region
Hydrodynamics	Hydrodynamics	Hydrodynamics
Driver energy deposition	Electron thermal conduction	Hydrodynamic instabilities
Laser-plasma interaction	Suprathermal electron	Electron and ion preheat
Resonant absorption	transport	of cold fuel
Stimulated scattering	Photon transport	Isentropic compression
Density profile modification	Plasma instabilities	Equation of state near
Suprathermal electron generation	Magnetic fields	Fermi degeneracy
Suprathermal electron transport	Equation of state	Shock convergence and timing
Fast ion generation		Shock heating of central
Plasma sheath formation		hot spot
		Thermonuclear reaction rates
		Fusion reaction product
		transport
		Thermonuclear burn wave
		propagation
		Target disassembly

electron temperature, suprathermal electron temperature, and the number of suprathermal electrons.[33] The spatial distribution of X rays provides information regarding the compression of glass microballoon targets. X-ray line emission can be used to diagnose the presence of suprathermal electron preheat of compressed D-T fuel.

In addition to these three spatial regions, there is a final stage of the inertial confinement fusion process that involves the burning of the compressed fuel core. Ignition of the central hot spot causes the burn process to proceed more rapidly that the implosion process. This process again involves the transport of nonthermal particles, in this case, the charged particle fusion reaction products as well as neutral particles such as fast neutrons.[1, 34] These particles slow down in the dense pellet core, giving up their energy to bootstrap heat the core to optimum burn temperatures. In current laser fusion experiments, a negligible amount of charged particle energy is redeposited in the D-T fuel. Many of the particles escape, and these can be used as a diagnostic to determine the temperature of the burning fuel. In this case, charged particle transport must be adequately modeled to predict the amount of energy loss and spectrum broadening that results from interactions with the surrounding tamping zones, so that experimental results can be translated into D-T ion temperatures. Suitable models of neutron transport in the dense pellet core must also be provided.

The essence of the preceding discussion is summarized in Table 6.1 where the fusion target is divided into spatial regions, and the most significant physics in each region is listed. From this table we can clearly see that classical plasma hydrodynamics by itself is inadequate to model these plasmas. Indeed, even with the addition of particle transport models and magnetic field effects, the present computer code models of inertial confinement fusion targets still omit a number of important physical processes.

6.1. HYDRODYNAMICS CODES

Essentially all inertial confinement plasma hydrodynamics codes use a two-temperature, one-fluid model of the plasma. In this model, electrons and ions are assumed to flow as one fluid, implying no charge separation (at least on the length scale of interest). However, each species maintains its own characteristic temperature due to weak energy coupling between the two populations. Radiation can be included as either a third temperature equation, assuming a local Planckian distribution,[4, 35] or as an energy-dependent treatment of the photon distribution function.[2] We consider only the treatment of the electrons and ions here, since the radiation is most typically treated in an energy-dependent fashion and is coupled to the electron equation through emission and absorption terms. The basic equations of the two-temperature, one-fluid model of the plasma (minus the radiation coupling) take the form:

$$\frac{\partial \rho}{\partial t} + \nabla \cdot \rho \mathbf{u} = 0$$

$$\rho \left(\frac{\partial}{\partial t} + \mathbf{u} \cdot \nabla \right) \mathbf{u} = - \nabla p \tag{6.1}$$

$$\rho c_{ve} \left(\frac{\partial}{\partial t} + \mathbf{u} \cdot \nabla \right) T_e = \nabla \cdot \kappa_e \nabla T_e - p_e (\nabla \cdot \mathbf{u}) - \omega_{ei}(T_e - T_i) + S_e$$

$$\rho c_{vi} \left(\frac{\partial}{\partial t} + \mathbf{u} \cdot \nabla \right) T_i = \nabla \cdot \kappa_i \nabla T_i - p_i (\nabla \cdot \mathbf{u}) + \omega_{ei}(T_e - T_i) + S_i$$

6.1.1. LAGRANGIAN COORDINATES[36]

When we write the hydrodynamics equations in terms of $\mathbf{r}$ and t, we are using an Eulerian description in terms of a fixed coordinate frame. For example, the Eulerian form of the equations of continuity and motion are

$$\frac{\partial \rho}{\partial t} + \nabla \cdot \rho \mathbf{u} = 0$$

$$\frac{\partial \mathbf{u}}{\partial t} + \mathbf{u} \cdot \nabla \mathbf{u} = - \frac{1}{\rho} \nabla p \tag{6.2}$$

However, when the fluid is rapidly expanding or contracting or reactions are occurring, it is useful to move to a reference frame that moves with the local flow velocity of the fluid. Such a frame is known as a Lagrangian description.

Example: *Plane Flow*   Consider a fluid particle at position x and the coordinate of a reference particle at x_1 (see Figure 6.3). The mass (per unit area) between x and x_1 is given by

$$m = \int_{x_1}^{x} \rho(x')\, dx' \tag{6.3}$$

(Note that m just labels which particle we are considering—namely that particle that has a certain mass between it and x_1.) Hence we can treat m as a new variable, noting that

$$dm = \rho\, dx$$

The fluid mass separating one particle from another, m, is therefore taken as the appropriate Lagrangian variable.

To convert the hydrodynamics equations describing plane fluid flow to Lagrangian form, we first write their Eulerian form as

$$\frac{\partial \rho}{\partial t} + \frac{\partial}{\partial x}(\rho u) = \frac{\partial \rho}{\partial t} + u\frac{\partial \rho}{\partial x} + \rho\frac{\partial u}{\partial x} = 0$$

or

$$\frac{D\rho}{Dt} + \rho\frac{\partial u}{\partial x} = 0 \tag{6.4}$$

where we have introduced the substantial derivative (in the reference frame of the flow)

$$\frac{D}{Dt} = \frac{\partial}{\partial t} + u\frac{\partial}{\partial x}$$

If we now make a variable substitution, $dm = \rho\, dx$, we find that the equation of continuity can be written as

$$\frac{DV}{Dt} = \frac{\partial u}{\partial m}$$

where we have introduced the specific volume, $V = 1/\rho$. But in Lagrangian

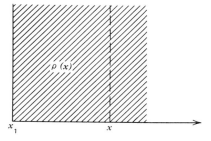

$\rho(x)$

x_1 x

Figure 6.3.   Relationship between space and mass coordinates.

Table 6.2 A Comparison of Eulerian and Lagrangian Coordinates in One-Dimensional Plane Flow

	Eulerian	Lagrangian
Continuity equation	$\dfrac{D\rho}{Dt} = -\rho\dfrac{\partial u}{\partial x}$	$\dfrac{\partial V}{\partial t} = \dfrac{\partial u}{\partial m}$
Equation of motion	$\dfrac{Du}{Dt} = -\dfrac{1}{\rho}\dfrac{\partial p}{\partial x}$	$\dfrac{\partial u}{\partial t} = -\dfrac{\partial p}{\partial m}$

coordinates moving with the flow,

$$\frac{D}{Dt} \to \frac{\partial}{\partial t}$$

Hence the continuity equation in Lagrangian coordinates becomes

$$\frac{\partial V}{\partial t} = \frac{\partial u}{\partial m} \qquad (6.5)$$

where we now regard m and t as the independent variables, $V(m,t)$ and $u(m,t)$ as dependent variables. A comparison of the Eulerian and Lagrangian form of the continuity and motion equations in one dimensional plane flow is given in Table 6.2.

Suppose we have solved for $V(m,t)$ and $u(m,t)$. Then to get back to Eulerian variables, $\rho(x,t)$ and $u(x,t)$, we first integrate

$$dx = V(m,t)\,dm$$

to find

$$x(m,t) = \int_0^m V(m',t)\,dm' + x_1(t) \qquad (6.6)$$

We can then use this to find $\rho(x,t) = [V(x,t)]^{-1}$ and $u(x,t)$.

Example: *Spherical Geometry* We now define the Lagrangian mass coordinate in spherical geometry by

$$dm = \rho 4\pi r^2\,dr$$

Then the Lagrangian form of the hydrodynamics equations becomes

$$\frac{\partial V}{\partial t} = \frac{\partial}{\partial m}(4\pi r^2 u)$$

$$\frac{\partial u}{\partial t} = -4\pi r^2 \frac{\partial p}{\partial m}$$

Frequently one omits the 4π factor and defines the dependent variables per steradian. Then the Lagrangian form of the hydrodynamics equations in spherical geometry become

$$\frac{\partial V}{\partial t} = \frac{\partial}{\partial m}(r^2 u)$$

$$\frac{\partial u}{\partial t} = -r^2 \frac{\partial p}{\partial m}$$

The most general Lagrangian form is

$$\frac{\partial V}{\partial t} = V \nabla \cdot \mathbf{u}$$

$$\frac{\partial \mathbf{u}}{\partial t} = -V \nabla p$$

However, it should be noted at this point that Lagrangian coordinates are most easily implemented in coordinate systems with one-dimensional symmetry. In two and three dimensions they lead to some complications in defining an appropriate numerical mesh structure that avoids singularities due to distortion.

The general Lagrangian form of the hydrodynamics equations for one-dimensional symmetry is

$$\frac{\partial V}{\partial t} = \frac{\partial}{\partial m}(r^{\delta-1}u)$$

$$\frac{\partial u}{\partial t} = -r^{\delta-1}\frac{\partial p}{\partial m} \qquad (6.7)$$

where

$$\delta = \begin{cases} 1 \text{ plane (mass/unit area)} \\ 2 \text{ cylindrical (mass/unit length-radian)} \\ 3 \text{ spherical (mass/steradian),} \end{cases}$$

We will write the general Lagrangian form of the two-temperature, one-fluid plasma model as

$$\frac{\partial \mathbf{u}}{\partial t} = -\frac{1}{\rho} \nabla p$$

$$\rho c_{ve} \frac{\partial T_e}{\partial t} = \nabla \kappa_e \nabla T_e - p_e(\nabla \cdot \mathbf{u}) - \omega_{ei}(T_e - T_i) + S_e$$

$$\rho c_{vi} \frac{\partial T_i}{\partial t} = \nabla \kappa_i \nabla T_i - p_i(\nabla \cdot \mathbf{u}) + \omega_{ei}(T_e - T_i) + S_i \qquad (6.8)$$

The key to Lagrangian coordinates involves the fact that they move with the fluid. That is, they are like chalk lines on a rubber sheet. If you stretch or distort the sheet, the chalk lines distort with it. When the hydrodynamics equations are discretized for numerical solution, the mass (rather than the volume) of the spatial zones in a Lagrangian mesh cell is conserved.

6.1.2. NUMERICAL SOLUTION OF THE HYDRODYNAMICS EQUATIONS

To be explicit, let us consider the Lagrangian form of the hydrodynamics equations in one-dimensional spherical geometry:

$$\frac{\partial V}{\partial t} = \frac{\partial}{\partial m}(r^2 u)$$

$$\frac{\partial u}{\partial t} = -r^2 \frac{\partial p}{\partial m} \tag{6.9}$$

Since these equations are hyperbolic, they have solutions with a characteristic propagation speed. Furthermore these equations (in the absence of dissipative terms) admit discontinuous solutions (shock waves) that can create difficulties for finite difference schemes. Hence we need some tricks to handle their numerical solution.

Artificial Viscosity (von Neumann q). It is difficult to handle shock waves described by the inviscid hydrodynamics equations numerically, since a step discontinuity arises. Although suitable numerical schemes can be devised for very specialized problems, the general simulation of the complex flows encountered in ICF targets requires a more universal "fixup." Von Neumann[37] proposed that one introduce a phony or artificial viscosity, which he labeled by q, to spread out the shock wave over a few zones of the numerical spatial mesh. Since the artificial viscosity is set equal to zero on either side of the shock, it does not affect the flow transition across the shock which is determined by the conservation laws as manifested in the Rankine-Hugoniot relations.

The von Neumann q must be chosen very carefully to yield a constant thickness for all shocks. This requires quadratic terms in velocity gradients (equivalent to using a small viscosity for weak shocks and a large viscosity for strong shocks). For example, one can define

$$q = \frac{1}{V}\left[b\Delta m \frac{\partial u}{\partial m}\right]^2, \qquad \frac{\partial V}{\partial t} < 0 \qquad \text{(compression)}$$

$$= 0, \qquad \frac{\partial V}{\partial t} \geqslant 0 \qquad \text{(expansion)} \tag{6.10}$$

where it is customary to choose $b = 2^{1/2}$.

In summary, then, the artificial viscosity is chosen so that it dissipates energy in a shock to a few surrounding finite difference zones while preserving the Rankine-Hugoniot relations across the shock. This preserves the essential features of the shock while reducing the gradients across the shock to values that allow the treatment by general finite difference methods. (A useful hint in analyzing data generated by hydrodynamics codes is to print out the value of q along with other hydrodynamic variables. Then one can easily determine those regions where shock waves are forming by noting where q is nonzero. Furthermore, the strength of the shock is measured by the size of q.)

Recent developments in flux-conserving transport based numerical methods have largely removed the need for introducing an artificial viscosity.[38] However, most of the hydrodynamics codes used in ICF applications continue to use this older method.

Differencing Schemes. The hydrodynamics equations are coupled, nonlinear partial differential equations since the transport coefficients depend strongly upon temperature, for example, $\kappa_e \sim T_e^{5/2}$ and $\omega_{ei} \sim T_e^{-3/2}$. These equations are usually solved using standard finite difference techniques. The first equation (the equation of motion) is hyperbolic and has a characteristic propagation speed. It also admits discontinuous solutions (shocks) that create difficulties for general finite difference schemes. Because of these properties, this equation is usually solved using an explicit differencing technique:[39]

$$\frac{u^{n+1/2} - u^{n-1/2}}{\Delta t^n} = -\frac{1}{\rho^n} \nabla (p^n + q^n) \qquad (6.11)$$

where

$$q^n = q(t^n) \quad \text{for} \quad \dot{V} < 0 \quad \text{(compression)}$$

$$= 0 \qquad \text{for} \quad \dot{V} > 0 \quad \text{(expansion)}$$

The spatial difference was not specified here since it varies depending on whether the equation is solved in one or two dimensions. This equation requires a stability condition given by[39]

$$\frac{c_s \Delta t}{\Delta x} < 1 \qquad (6.12)$$

where c_s is the sound speed in the plasma. This is known as the Courant condition. It prevents a disturbance from propagating across a finite difference zone in less than one time step.

To be more explicit in the differencing scheme, let us consider the case of a simple one-dimensional geometry. We begin by breaking up the Lagrangian grid into J zones. That is, we discretize the variable m as shown in Figure 6.4. It is common to choose equal increments in m, that is, equal mass zoning. Notice that the radius of each zone, $r_j(t)$, is actually a function of time. In simple one-dimensional codes one might choose from 30 to 100 zones. The

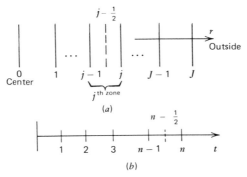

Figure 6.4. Time and spatial zoning mesh structures. (a) Spatial zoning. (b) Time zoning.

time variable is also discretized (again, see Figure 6.4). Here we will use the standard notation

$$u(m_j, t_n) = u_j^n$$

If we use an explicit differencing scheme, we can write

$$\frac{u_j^{n+1/2} - u_j^{n-1/2}}{\Delta t^n} = -\left(r_j^n\right)^2 \left[\frac{(p_{j+1/2}^n - p_{j-1/2}^n)}{\Delta m_j} + \frac{(q_{j+1/2}^n - q_{j-1/2}^n)}{\Delta m_j} \right]$$

where

$$\Delta t^n = \tfrac{1}{2}\left(\Delta t^{n+1/2} + \Delta t^{n-1/2}\right)$$

$$\Delta m_j = \tfrac{1}{2}\left(\Delta m_{j+1/2} + \Delta m_{j-1/2}\right)$$

$$q_{j-1/2} = \frac{1}{V_{j-1/2}} \left[b\Delta m \left(\frac{u_j - u_{j-1}}{\Delta m} \right) \right]^2$$

Next, we use

$$r_j^{n+1} - r_j^n = u_j^{n+1/2} \Delta t^{n+1/2}$$

to give the change in the location of the jth zone as a function of time. The continuity equation can be bypassed by calculating

$$V_{j-1/2}^{n+1} = \frac{\Delta r_{j-1/2}^{n+1}}{\Delta m_{j-1/2}} \left[r_j^{n+1} r_{j-1}^{n+1} + \tfrac{1}{3}\left(\Delta r_{j-1/2}^{n+1}\right)^2 \right]$$

We also need to use

$$V_{j-1/2}^{n+1/2} = \tfrac{1}{2}\left(V_{j-1/2}^{n+1} + V_{j-1/2}^{n} \right)$$

$$\dot{V}_{j-1/2}^{n+1/2} = \frac{1}{\Delta t^{n+1/2}}\left(V_{j-1/2}^{n+1} - V_{j-1/2}^{n} \right)$$

Temperature Equations (Heat Conduction).   The coupled temperature equations are parabolic, which should imply that they have an infinite propagation speed. However, due to nonlinear thermal conductivity, they exhibit a behavior similar to that of a wave equation and, in fact, do have a finite propagation (recall Section 4.1). Nevertheless, these equations are typically solved using an implicit differencing technique, since their characteristic time scale is usually much more rapid than that characterizing the equation of motion and hence an explicit scheme would impose a much stricter time step constraint. The temperature equations need not be solved simultaneously, since their coupling is weak.

To be more precise, consider the simple linear heat conduction equation

$$\frac{\partial T}{\partial t} = \kappa \frac{\partial^2 T}{\partial x^2}$$

An explicit numerical solution scheme would discretize this equation as

$$\frac{T_j^{n+1} - T_j^n}{\Delta t} = \kappa \left[\frac{T_{j+1}^n - 2T_j^n + T_{j-1}^n}{(\Delta x)^2} \right]$$

(The node structure is shown in Figure 6.5.) Here we note that the spatial

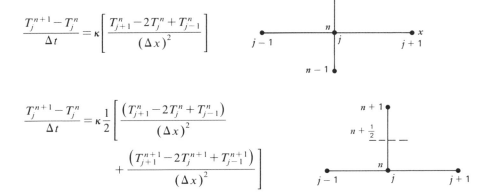

$$\frac{T_j^{n+1} - T_j^n}{\Delta t} = \kappa \left[\frac{T_{j+1}^n - 2T_j^n + T_{j-1}^n}{(\Delta x)^2} \right]$$

$$\frac{T_j^{n+1} - T_j^n}{\Delta t} = \kappa \frac{1}{2}\left[\frac{\left(T_{j+1}^n - 2T_j^n + T_{j-1}^n\right)}{(\Delta x)^2} + \frac{\left(T_{j+1}^{n+1} - 2T_j^{n+1} + T_{j-1}^{n+1}\right)}{(\Delta x)^2} \right]$$

Figure 6.5.   Differencing schemes for thermal conduction equations.

derivative is evaluated at the earlier time step t_n. Since all T^n are known, we can explicitly solve for the temperature T_j^{n+1}. However, stability problems arise if $\kappa \Delta t / (\Delta x)^2 > \frac{1}{2}$.

An alternative and more satisfactory approach is to use the Crank-Nicholson differencing scheme:[39]

$$\frac{T_j^{n+1} - T_j^n}{\Delta t} = \frac{\kappa}{2} \left[\left(\frac{T_{j+1}^n - 2T_j^n + T_{j-1}^n}{(\Delta x)^2} \right) + \left(\frac{T_{j+1}^{n+1} - 2T_j^{n+1} + T_{j-1}^{n+1}}{(\Delta x)^2} \right) \right]$$

(6.13)

This is now an unconditionally stable difference scheme. But T_{j+1}^{n+1}, T_j^{n+1}, and T_{j-1}^{n+1} on the right-hand side are unknown. Hence we must solve this equation implicitly (in this case, this corresponds to a linear tridiagonal system of equations for the T_j^{n+1}).

Although this is a straightforward scheme for linear problems, it does present problems when we have a nonlinear conduction problem of the form

$$\frac{\partial T}{\partial t} = \frac{\partial}{\partial x} \kappa(T) \frac{\partial T}{\partial x}$$

Then we must use either extrapolation or iteration techniques. For example, we might write

$$\frac{T^{n+1} - T^n}{\Delta t^{n+1/2}} = \nabla \cdot \kappa^n \nabla \left[\theta T^n + (1 - \theta) T^{n+1} \right] + \cdots$$

where, once again, a specific spatial differencing scheme has not been included to keep the notation simple. The parameter θ is frequently taken as $\frac{1}{2}$, for if the equations were linear (which they are not), this value would give a second-order accurate differencing scheme. In light of nonlinearities and the variable time and space mesh steps, this can at best be considered an imperfect solution. The nonlinear thermal conductivity has been shown with an explicit evaluation

$$\kappa^n = \kappa \left[T^n(r) \right]$$

in the equation of thermal conduction, which is an approximation difficult to improve on in a two-dimensional calculation. In a one-dimensional calculation, several other possibilities exist. Past temperatures at each mesh point can be saved and used to extrapolate ahead in time to evaluate κ at $n + \frac{1}{2}$,

$$\tilde{\kappa}^{n+1/2} = \kappa(\tilde{T}^{n+1/2}), \qquad \tilde{T}^{n+1/2} = F(T^{n-2}, T^{n-1}, T^n)$$

or the temperature equations can be iterated by reevaluating the nonlinear coefficients until the temperatures converge pointwise. However, even in one dimension, each of these methods has drawbacks. Experience has shown that neither is as stable as the single explicit evaluation method, and they often require more stringent conditions on the time step than the explicit evaluation. Furthermore, an "uninteresting" part of the plasma is usually the source of the smaller time step (e.g., the blowoff region far beyond the critical density). In laser fusion calculations, the plasma electrons usually increase in temperature due to thermal conduction from a source of heat. This is a stable process for these two schemes; however, when electrons decrease in temperature, it is often because of an expansion of the plasma. The strong thermal conduction continually "feeds" these loss zones and an instability results.

For linear diffusion equations, the Crank-Nicholson $(\theta = \frac{1}{2})$ differencing scheme ensures unconditional stability. However this is not the case for nonlinear equations where no rigorous estimate of the stability condition can be made. Experience has shown that a time step constrained to ensure $\max(\Delta T/T) < \frac{1}{4}$ usually provides stable solutions, but values as small as $\frac{1}{10}$ are necessary for some calculations. Quite often it is this ad hoc stability condition that sets the time step in the hydrodynamics calculation rather than the Courant condition.

Some Comments on Two-Dimensional Lagrangian Codes. The most sophisticated Lagrangian hydrodynamics codes are two dimensional in nature (typically $r - z$ cylindrical geometry). Such codes allow the simulation of non-symmetric hydrodynamic motion and heat conduction due to nonuniform illumination of the target.[40-48]

Figure 6.6 shows a symmetric target nonuniformly illuminated by two laser beams. Figure 6.7 presents a prediction of the glass fuel interface at the time of peak thermonuclear burn (along with temperature and density contours) using the two-dimensional LASNEX code.[2] Figure 6.8 shows a LASNEX calculation of an unstable implosion using very fine zoning.[41]

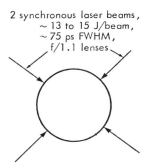

2 synchronous laser beams,
    ~ 13 to 15 J/beam,
    ~ 75 ps FWHM,
    f/1.1 lenses

DT-filled glass microsphere,
80 μm diameter,
0.6 μm wall thickness,
2 mg/cm³ fill

Figure 6.6. Schematic of a glass microsphere implosion with strong irradiation asymmetry. (Courtesy of Lawrence Livermore Laboratory.)

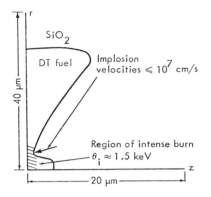

Figure 6.7. Conditions predicted by LASNEX at the time of peak thermonuclear burn. (Courtesy of Lawrence Livermore Laboratory.)

The point to be made by these examples is that the processes involved in practical inertial confinement fusion applications rarely are amenable to a one-dimensional simulation. The validity of one-dimensional results depend upon the symmetry of the problem and the careful interpretation of the output. In fact, even two-dimensional calculations cannot fully describe many situations that are actually three dimensional in nature. Some important effects such as fluid instabilities and magnetic field generation are inherently two-dimensional in nature and cannot be modeled with one-dimensional codes.

Two-dimensional calculations are very expensive and are feasible only on the fastest of computers. Even with this great expense, the numerical algo-

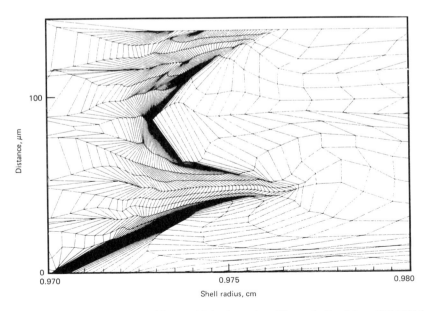

Figure 6.8. Implosion of a 100-μm thick, 1-cm beryllium shell with a 200-TW laser indicating strong asymmetries in compression.

rithms are usually very simple extensions of one-dimensional methods. Alternating direction methods are often used to solve the implicit heat transport equations, while explicit methods are again used for the equation of motion. However these methods can lead to gross inaccuracies when the finite difference mesh distorts, as we saw in Figure 6.8. To solve these problems more sophisticated full matrix inversion schemes such as the incomplete Cholesky-Conjugate Gradient method[49] have been implemented in codes such as LASNEX. This approach involves a full matrix inversion scheme rather than the standard method of operator splitting that reduces the problem to tridiagonal matrix inversion.

Zone distortion is an important problem in multidimensional Lagrangian codes because the mesh is imbedded in the fluid. This may not be a problem in two dimensions if the motion is well-behaved. However shear flows can often develop, and this can prove troublesome.[50] Vortices also cannot be handled by Lagrangian codes. Figure 6.9 demonstrates the problem with a very simple example of a strong force pushing diagonally on the corner of a quadrilateral zone. If the other mesh points of the zone remain fixed, the zone develops a concave boundary and becomes a "banana zone." The temperature equations are usually solved in two dimensions by making two sweeps through the mesh, (operator splitting) once along the k-index lines and then along the l-index lines, while holding the diffusion in the respective transverse direction fixed during each sweep (Figure 6.10). The problems of a banana zone become clear when the sweep path through the zone is as indicated in Figure 6.9. Such badly distorted zones also demand very small time steps to maintain stability.

Many different "fixups" were devised to minimize or correct this zone distortion problem. A more sophisticated generalization of the artificial viscos-

LAGRANGIAN COORDINATES

   – MESH IS IMBEDDED IN FLUID

CLASSIC PROBLEM

   – ZONE DISTORTION

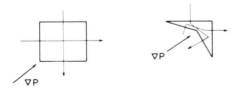

SOLUTION

   – REZONING

   – MIXED EULERIAN- LAGRANGIAN

ADVANTAGES

   – BETTER RESOLUTION

Figure 6.9. A schematic diagram of the problems created by the distortions of two-dimensional Lagrangian finite difference zones.

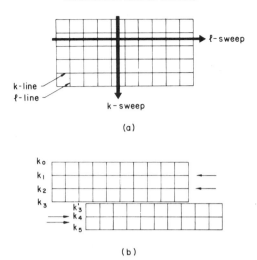

Figure 6.10. A schematic of two-dimensional zoning showing the use of slide lines.

ity can be added to the code that will mitigate zone distortion while maintaining the conservation properties of the fluid. When shear flow can be identified a priori, use of "slide lines" may be helpful. In this scheme, as shown in Figure 6.10, specified k or l lines are allowed to decouple and slide against one another. This idea is conceptually simple; however, in practice it is an extremely complex bookkeeping task to implement. A somewhat different approach involves stopping the calculation once zones have distorted sufficiently and rezoning the problem before continuing. This is commonly done with most large two-dimensional codes, and sophisticated computer graphics programs have been developed to allow the user to automatically reposition mesh points with a light pen. Here again, one attempts to maintain conservation laws. Another method of hydrodynamic computation, known as mixed Eulerian-Lagrangian, can also alleviate the zone distortion problem somewhat.

Despite all of these drawbacks, there are nevertheless good reasons for using Lagrangian coordinates. As noted earlier, when the mass of the fluid originally falls within the borders of a zone, it remains within that zone throughout the calculation. This is very important when materials with very different properties are adjacent to one another, such as would be the case in a D-T gas-filled glass microballoon. In Eulerian calculations, the mesh remains fixed and the fluid flows through it, so that as time progresses, zones near the original D-T/glass interface contain both glass and D-T gas. Very fine Eulerian zoning would be necessary near the interface to maintain spatial resolution; the average properties of the D-T/glass mixture would be difficult to determine. There is no mixing in Lagrangian coordinates and hence no averaging must be done. Since mesh points follow the fluid in a Lagrangian calculation, there tend to be more mesh points in steep density gradients and fewer points where the gradient is small, just as there should be to maintain good resolution. These considerations most often outweigh the disadvantages of zone distortion.

Time Step Size.   Yet another important factor in the numerical solution of the plasma hydrodynamics equations is the choice of the time step size, Δt. We have already noted that this choice is constrained by stability considerations (i.e., the Courant condition). The time step size will vary throughout the hydrodynamic calculation by orders to magnitude. While the implosion process occurs on the time scale of nanoseconds, the thermonuclear burn phase is through in a few picoseconds. During the final stages of compression and thermonuclear burn, the time step size can become as small as 10^{-14} s. The computer code must contain logic to determine the optimum time step size as the calculation proceeds.

Table Interpolation.   Another important feature of ICF target design codes involves interpolation of equation of state and opacity data in multidimensional tables. This data is generated at great expense using complex computer codes that calculate the atomic properties of materials at extreme pressures and temperatures. The results of these calculations are tabulated on a density-temperature grid for use in hydrodynamics calculations. As an example of such data, we have shown the specific internal energy for Xe in Figure 6.11. The derivative of the specific energy with respect to temperature yields the specific heat, $c_v = (\partial e / \partial T)_v$, which is a key parameter in a hydrodynamics calculation.

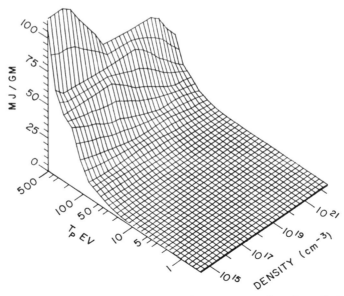

Figure 6.11.   The specific internal energy of xenon as a function of density and temperature. (Courtesy of University of Wisconsin.)

Both the temperature and density grid as well as the data itself range over several orders of magnitude. Therefore it is often convenient to tabulate the logarithm of the data on a logarithmic temperature-density grid. Bilinear interpolation of the logarithmic data followed by exponentiation has been found to yield quite accurate results. Another method of data storage involves using analytic functions to fit the data, and then storing the fitting coefficients. While this latter method requires more storage space and is not cost effective for many target design calculations, it does allow for a more accuration determination of derivatives of the data (such as the specific heat).

6.1.3. WAZER: AN EXAMPLE OF AN ICF PLASMA HYDRODYNAMICS CODE

One of the early plasma hydrodynamics codes developed for inertial confinement fusion studies was the WAZER code developed at Lawrence Livermore Laboratories.[3] We summarize the essential features of this code, since it is typical of many of the codes in use today. The code describes the hydrodynamic behavior of a single fluid, two-temperature plasma in one-dimensional geometry (in plane, cylindrical, or spherical coordinates). Specific processes included in the code include:

1. Heat flow due to thermal diffusion.

2. Heat exchange between electrons and ions.

3. Heat loss from electrons as radiation.

4. Hydrodynamic effects of all energy transformations.

5. Addition of energy through absorption of laser light.

6. Production of energy by thermonuclear reactions.

7. Gain or loss of energy by heat conduction across boundaries or work done on or by the system at boundaries.

Hydrodynamic Equations. The one-dimensional hydrodynamic equations in Lagrangian coordinates used in the code (written in spherical coordinates) are:

$$\frac{\partial V}{\partial t} = \frac{\partial}{\partial m}(r^2 u)$$

$$\frac{\partial u}{\partial t} = -r^2 \frac{\partial p}{\partial m}$$

Heat Transfer. The temperature equations for the electrons and ions are taken as

$$c_{vi}\frac{\partial T_i}{\partial t} + T_i p_{Ti}\frac{\partial V}{\partial t} = \frac{\partial}{\partial m}\left(\kappa_i r^2 \frac{\partial T_i}{\partial r}\right) - \omega_{ei}(T_i - T_e) + S_i$$

$$c_{ve}\frac{\partial T_e}{\partial t} + T_e p_{Te}\frac{\partial V}{\partial t} = \frac{\partial}{\partial m}\left(\kappa_e r^2 \frac{\partial T_e}{\partial r}\right) - \omega_{ei}(T_e - T_i) + S_e - S_r$$

where the thermal conductivities are given by

$$\kappa_i = \frac{40(2/\pi)^{3/2} k (kT_i)^{5/2}}{m_i^{1/2} (Ze)^4 \ln \Lambda_i}$$

$$\kappa_e = \frac{40(2/\pi)^{3/2} k (kT_e)^{5/2}}{m_e^{1/2} e^4 \ln \Lambda (Z + 3.44 + 0.26 \ln Z)}$$

and the electron-ion energy exchange coefficient is

$$\omega_{ei} = \frac{(2\pi/k)^{1/2} (N_e Ze^2)^2 V \ln \Lambda}{m_i m_e [(T_e/m_e) + (T_i/m_i)]^{3/2}}$$

Radiation.   Heat loss through bremsstrahlung is modeled by a sink term in the electron temperature equation:

$$S_r = 2\omega_r T_e$$

where

$$\omega_r = \left[\frac{2^5 N_i Z^3 e^6 V}{3hmc^3} \right] \left(\frac{2\pi k}{m_e T_e} \right)^{1/2}$$

Equation of State.   A variety of equation of state models can be used, ranging from ideal gas ($p = NkT$) to more complex Thomas-Fermi-Dirac or tabulated data.

Absorption of Laser Light.   If ϕ is the incident photon energy flux, then the energy source term due to laser light energy absorption that appears in the electron temperature equation is

$$S_e = \kappa V \phi$$

where κ is the light absorption coefficient. We can break the light flux into two components, an inward-directed component ψ^- and an outward-directed component ψ^+

$$\phi = \frac{1}{r^2} (\psi^- + \psi^+)$$

The light flux enters the system from the outer boundary such that

$$\psi^- (r_{max}) = \Psi(t)$$

where $\Psi(t)$ is a prescribed function characterizing the time dependence of the incident laser pulse. The light then passes through the zones of the plasma as governed by

$$\frac{d\psi^-}{dr} = \kappa\psi^-$$

When it reaches the critical surface (or the origin of an underdense spherical plasma), it is reflected and propagates back outward as governed by

$$\frac{d\psi^+}{dr} = -\kappa\psi^+$$

The light absorption coefficient can be modeled in several ways. If only classical (inverse bremsstrahlung) absorption is present, we would take

$$\kappa_{IB} = AN_eN_iZ^2\left(\frac{I_0}{\hbar\omega}\right)^{7/2}\frac{F(\hbar\omega/kT_e)}{\left[1-(\omega_p/\omega)^2\right]^{1/2}}$$

However it is more common to model the absorption assuming an anomalous process such as resonance absorption or parametric excitation occurs in the neighborhood of the critical surface. Typically one simply assumes that a certain fraction of the incident laser light is dumped into the first overdense mass zone.

Thermonuclear Reactions. In a D-T fueled pellet, the dynamics of the thermonuclear reactions are described by the rate equations

$$\frac{\partial N_T}{\partial t} = -\langle v\sigma_{DT}\rangle N_T N_D + \tfrac{1}{4}\langle v\sigma_{DD}\rangle N_D^2$$

$$\frac{\partial N_D}{\partial t} = -\langle v\sigma_{DD}\rangle N_D^2 - \langle v\sigma_{DT}\rangle N_T N_D$$

A variety of options can be used to describe the fusion energy release. In very dilute fuels, one can simply assume that the alphas and neutrons carry away the reaction energy. However in dense fuels, one usually assumes that the alphas deposit their energy immediately as

$$S_{DT} = (3.6\ \text{MeV})\langle v\sigma_{DT}\rangle N_D N_T V$$

In partitioning of the alpha energy deposition among electrons and ions, one can use a crude model

$$S_i = \frac{T_e S_{DT}}{32+T_e} + \frac{T_e S_{DD}}{120+T_e}$$

$$S_e = \frac{32 S_{DT}}{32+T_e} + \frac{120 S_{DD}}{120+T_e}$$

where temperatures are in keV.

Other Features.   Most such codes contain the capability to calculate total energy balances as a check on the accuracy of the code. The time steps used in the code are usually variable. They are restricted by the Courant condition that demands $\Delta x / \Delta t$ be larger than the rate at which a disturbance would be propagated through the medium hydrodynamically. They are also frequently restricted by demanding that the fractional density or temperature change of any zone in a given time step be less than some specified limit (say, 10%).

6.2.   PARTICLE TRANSPORT

In addition to plasma hydrodynamics, particle transport processes play an important role in inertial confinement fusion targets.[51, 52] These processes include radiative transfer, suprathermal electron transport, and charged particle fusion reaction product transport. Fast neutron transport can also become important for very high ρ-R targets.[53] Each of these particles possesses a mean free path that is considerably longer than the characteristic scale length of the hydrodynamic background plasma in some regions of the target. This behavior rules out the possibility of treating these species within a hydrodynamics framework.

These particles must be treated as separate species that interact with the plasma fluid through gain and loss terms in the temperature equations and through a momentum source term in the equation of motion. Furthermore, to characterize those particles "created" and "destroyed" by thermonuclear reactions, as well as the heating of thermal electrons up into a suprathermal distribution by driver interaction, the continuity equation must also include a gain and loss term. In practice, mass is automatically conserved by each zone in a Lagrangian code, so that this gain and loss computation is actually just a matter of bookkeeping.

The transport of radiation is not discussed here (see Chapter 5 for details) except to point out some of the standard methods of solution of the equations of radiative transfer. If the radiation mean free path is short enough, the photon distribution can be assumed to be the local thermal Planckian distribution (local thermodynamic equilibrium), and the radiation can be characterized by its own temperature and temperature diffusion equation.[4] This treatment is consistent with the hydrodynamic treatment of the plasma fluid and leads to a "three-temperature" fluid model (electrons, ions, and photons). The photon transport equation itself can be solved more directly using discrete ordinate (S_N)[54] or Monte Carlo techniques.[55] However these are generally very expensive to implement in time-dependent problems and are used only in special cases. More typically the photon transport equation is expanded in angular moments to arrive at an approximate diffusionlike equation (using a prescription known as the Eddington or variable Eddington method).[32] The frequency dependence is discretized into a number of groups to lead to a multigroup diffusion equation description. This set of equations is frequently modified using the "flux-limiting" prescription described later in this section to account for long photon mfp's.

More peculiar to inertial confinement fusion plasmas is the transport of suprathermal electrons and charged particle reaction products. Suprathermal electrons play a significant role in the pellet implosion process, since they are created by driver (laser light) interaction in the outer regions of the pellet and stream inward carrying their energy into the dense fuel core. With energies in excess of 100 keV, these hot electrons may not be stopped until they stream through the ablation front into the cold compressed fuel ahead of the front. This heats up the cold plasma, resulting in a nonisentropic compression and a degradation of the implosion process. To remedy this situation, pellets are designed with high-Z material surrounding the D-T fuel that serves both as a tamper/pusher to enhance the implosion process and as a shield against the suprathermal electrons and high-frequency X-rays.

Fusion reaction product charged particles are principally the alphas ($^4_2$He), tritons ($^3_1$T), and helium-3 nuclei ($^3_2$He) occurring as products of D-T and D-D fusion reactions. The temperature of the burning D-T fuel is typically 20 to 100 keV, so that each of the reaction products is highly nonthermal (ranging in energy from 1 to 3 MeV).

6.2.1. MULTIGROUP FLUX-LIMITED DIFFUSION THEORY

The method that has received the greatest use to date for the transport of these nonthermal particles in inertial confinement fusion hydrodynamics codes is multigroup flux-limited diffusion theory.[56-58] This method is used in LASNEX, a two-dimensional Lagrangian hydrodynamics code developed at the Lawrence Livermore Laboratory. Flux-limited diffusion theory uses a diffusion equation to model particle transport with a flux-corrected diffusion coefficient to account for long mfp situations. The particle slowing down is described by a discrete particle energy (multigroup) structure:

$$\frac{\partial N_g}{\partial t} - \nabla \cdot D_g \nabla N_g + L_g N_g = S_g \qquad g=1,\ldots, G \qquad (6.14)$$

where $N_g(\mathbf{r}, t)$ is the density of particles with energies between E_g and E_{g-1}, while L_g is a slowing down operator that removes and adds particles from energy group g. To define the coefficients in the equation, one can refer to the Fokker-Planck equation for the distribution function $f_\alpha(\mathbf{r}, \mathbf{v}, t)$ characterizing particles of species α

$$\frac{\partial f_\alpha}{\partial t} + \mathbf{v} \cdot \frac{\partial f_\alpha}{\partial \mathbf{r}} + \frac{\mathbf{F}_\alpha}{m} \cdot \frac{\partial f_\alpha}{\partial \mathbf{v}} = \left(\frac{\partial f_\alpha}{\partial t} \right)_c \qquad (6.15)$$

where the force term is given by

$$\mathbf{F}_\alpha = q_\alpha(\mathbf{E} + \mathbf{v} \times \mathbf{B})$$

while the collision term is

$$\left(\frac{\partial f_\alpha}{\partial t} \right)_c = \frac{\partial}{\partial \mathbf{v}} \cdot \left[\boldsymbol{\kappa} \frac{\partial f_\alpha}{\partial \mathbf{v}} - \mathbf{L} f_\alpha \right]$$

where **K** and **L** are the velocity space diffusion and drag terms. Due to the long range nature of the Coulomb potential, charged particles slow down through many small angle scattering events. The collision term in the Fokker-Planck equation models this process.

Solution of the Fokker-Planck equation in an infinite medium yields several characteristic transport times of interest. The time required for a fast particle to deflect through an angle of 90°, τ_D, is given by[59]

$$\tau_D = \frac{(m/2)^{1/2} E^{3/2}}{2\pi Z^2 e^4 \ln \Lambda_e \sum_i Z_i^2 N_i}$$

where m, E, and Z are the mass, energy, and charge of the fast particle, and Z_i and N_i are the charge and density of the ith ionic species. Using this deflection time, a macroscopic transport cross section can be defined as

$$\Sigma_{tr} = (v\tau_D)^{-1} = \lambda_{tr}^{-1}$$

while the multigroup diffusion coefficient is then given as

$$D_g = \frac{v_g}{3\Sigma_{trg}}$$

The deflection time τ_D is obtained from the Fokker-Planck equation by considering the K term in the collision model. This term modifies the distribution by relaxing it toward isotropy.

The slowing down of particles depends on both the K and L terms in the Fokker-Planck equation. For fast ion transport at high energy, the L term dominates and is due to interactions with electrons. As the fast ion slows down, the K term dominates and slowing is due to larger angle collisions with thermal ions. In Figure 6.12 we have plotted dv/ds for the thermal electron and ion contributions as a function of fast particle velocity. The domination of the electron component at high energy and the ion component at low energy is apparent.

If we let

$$\frac{dv}{dt} = v\frac{dv}{ds}$$

and define

$$\tau_g = \frac{v_g}{\dfrac{dv_g}{dt}}$$

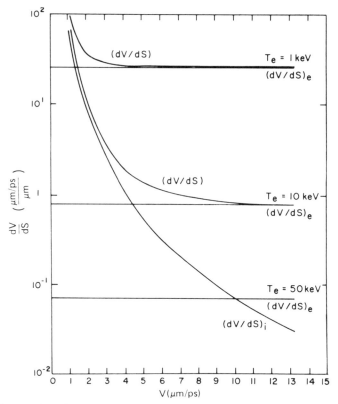

Figure 6.12. A plot of dv/ds versus the velocity of a fast alpha particle slowing down in a D-T plasma, for electron temperatures of 1, 10, and 50 keV.

then the slowing down operator can be expressed as

$$L_g N_g = \frac{N_g}{\tau_g} - \frac{N_{g-1}}{\tau_{g-1}}$$

The multigroup diffusion equations can then be written as

$$\frac{\partial N_g}{\partial t} - \nabla \cdot D_g \nabla N_g + \frac{1}{\tau_g} N_g - \frac{1}{\tau_{g-1}} N_{g-1} = S_g \qquad (6.16)$$

$$g = 1, \ldots, G$$

where we assume particles can only scatter into group g from the next higher energy group. When particles reach thermal energies, they are added back into the hydrodynamic equations characterizing the thermal plasma.

Unfortunately, diffusion theory is not a very good approximation to the transport of fast charged particles. This approximation can be "fixed up" by the procedure of flux limiting.[60] Recall that the diffusion equation is obtained from the particle continuity equation

$$\frac{\partial N}{\partial t} + \nabla \cdot \mathbf{J} + LN = S$$

by using a diffusion or Fick's law type of approximation to express the particle current in terms of the gradient of the density

$$\mathbf{J} \cong -D \nabla N$$

For those situations in which $|N/\nabla N| < \lambda_{mfp}$ (that is, in the transport regime), the magnitude of the particle current is overestimated by the diffusion approximation. We can correct for this by redefining the particle current density as an interpolation between its diffusion form and the free streaming limit (refer to Chapter 5 for more details):

$$\mathbf{J} = \frac{-D \nabla N}{1 + |\widetilde{D \nabla N}/J_{max}|} = -\tilde{D} \nabla N \qquad (6.17)$$

where J_{max} is chosen in some appropriate way to correspond to the current in the free-streaming limit.

The expression $\widetilde{D \nabla N}$ is designated in this way because the gradient operator does not enter the difference equations in an explicit manner. This entire expression is simply estimated from values of D and N from the previous time step, and the flux-limited diffusion coefficient $\tilde{D}$ is computed using this estimate. This ad hoc diffusion coefficient ensures physically plausible transport of the particles when diffusion theory holds:

$$\mathbf{J} \to -D \nabla N \qquad \text{for} \quad D \nabla N \ll J_{max}$$

When the mean free path is very long,

$$\mathbf{J} \to \mathbf{J}_{max} \qquad \text{for} \quad D \nabla N \gg J_{max}$$

An accurate treatment of the slowing down of charged particles by multigroup methods can be a problem. This is a continuous slowing down process with a strong coupling between the energy, space, and time variables. This can result in a great deal of numerical diffusion in energy space, with as many as 100 groups required to give reasonable accuracy.

The multigroup equations also are of only first-order accuracy in the energy variable and usually are of only first-order accuracy in time as well, since a fully implicit differencing scheme must be used to allow the hydrodynamic time step to be used:

$$\frac{N_g^{n+1} - N_g^n}{\Delta t^{n+1/2}} = \nabla \cdot D_g \nabla N_g^{n+1} - \frac{1}{\tau_g} N_g^{n+1} - \frac{1}{\tau_{g-1}} N_{g-1}^{n+1} + S_g^{n+1/2}$$

Of course, this is a general scheme that can be used to model almost any kind of particle transport. It produces reasonably good results for nondiffusive problems, and it is very straightforward to implement numerically, even in two dimensions. It embodies the essentials of most transport problems and should give reasonable answers for integrated quantities, such as total reaction rates. However, care should be taken when interpreting detailed results of the flux-limited diffusion treatment of transport-dominated problems. Over less than one mean free path, flux-limited diffusion can entirely miss details for certain types of problems.

6.2.2.   PARTICLE TRACKING METHODS

An alternative approach to charged particle transport is to use techniques that are specifically adapted to the behavior of the particular charged particles. In Figure 6.13 we see that an ion beam streaming through a background plasma has far less dispersion in velocity than an electron beam.[61] The electron beam becomes much more isotropic as it slows down because of large-angle scattering; however, the ion beam simply slows down without much scattering out of the beam trajectory. The electron behavior tends to reinforce the applicability of flux-limited diffusion theory as a description of electron transport, for that treatment demands that the distribution function be nearly isotropic. In the case of thermonuclear reaction products, however, ion beam results indicate

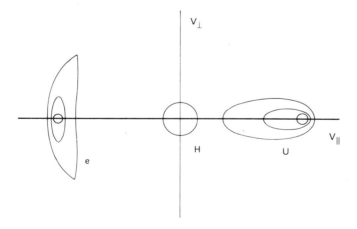

Figure 6.13.   A plot of velocity space at different times for beams of electrons and ions in a hydrogen background plasma.

that isotropy is not reached until near the end of the particle path to thermalization. This behavior can be used to great advantage in modeling the reaction product transport for it implies that a simple straight line trajectory will be a good approximation. This is the basic assumption used in particle tracking algorithms.

In a one-dimensional Lagrangian hydrodynamic treatment, a spherical plasma is divided into concentric shells or zones, and the plasma behavior is represented by the finite difference solution to the hydrodynamics equations for these zones. To model the charged particle fusion reaction products, we assume that those created in each zone stream along a finite number of rays that originate in the center of the zone.[61] As they stream, they experience the Coulomb drag force represented by the L terms in the Fokker-Planck equation. A range-energy relation can be expressed in the following form:

$$-\frac{dv}{ds} = A + \frac{B}{v^3} = K(v) \tag{6.18}$$

where

$$A = A_0(Z^2/m)(\ln \Lambda_e)N_e/T_e^{3/2}$$

$$B = B_0(Z^2/m)(\ln \Lambda_i)\sum_i (Z_i^2 N_i/\rho)$$

where A is due to scattering on thermal electrons and B corresponds to scattering from thermal ions. (These expressions have been used in Figure 6.14.) The straight line trajectory is valid during most of the slowing down process; however, near the end, the fast particle loses its energy in large-angle collisions with thermal ions (straggling). In the transport of fusion reaction products, it is the energy redeposition back into the thermal electrons that is of greatest importance. Near the end of a reaction product trajectory, it has very little energy remaining; hence the error in estimating redeposition ignoring straggling is small.

We solve the range-energy equation along a ray as the particle passes from one zone boundary to the next by integrating along its exact path length:

$$\Delta s = \int_{v_0 - \Delta v}^{v_0} g(v)\, dv$$

where Δs is the distance across the zone along the particle trajectory, v_0 is the particle velocity on entering the zone, Δv is the velocity loss in crossing the zone, and $g(v) = [K(v)]^{-1}$. This equation is an integral equation for Δv that can be solved by using a Taylor expansion to find

$$\Delta v = \frac{K(v_0)\Delta s}{\left[1 + \frac{1}{2}K'(v_0)\Delta s\right]}$$

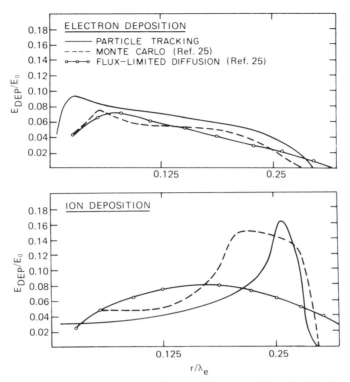

Figure 6.14. Plots of the fraction of alpha particle energy deposited in electrons and ions as a function of the distance traveled by the alpha particle measured in units of its mfp to a collision with an electron.

This procedure is accurate for $\Delta v/v_0 \ll 1$; however, should $\Delta v/v_0 \gtrsim 1$, then the particles have thermalized within the zone so again the error to energy redeposition is not serious. Only the partitioning of the energy to the electrons and ions is important. The total energy lost in a zone is simply

$$\Delta E = \tfrac{1}{2} m \left[v_0^2 - (v_0 - \Delta v)^2 \right] N$$

where N is the number of ions streaming along a given ray. The fraction of this energy going to the electrons is

$$\Delta E^{(e)} = A \left(\frac{\Delta s}{\Delta v} \right) \Delta E$$

Should the particles slow to thermal energy in the zone, then the fraction of energy going to the electrons can be obtained from the results of an infinite medium calculation, tabulated as a function of electron temperature. Then the loss to the ions in either case is

$$\Delta E^{(i)} = \Delta E - \Delta E^{(e)}$$

In addition to energy redeposition, the nonthermal ions also impart momentum to the zone

$$m \Delta v N \cos \alpha = M \Delta u$$

where M is the zone mass, m is the nonthermal mass, and α is the angle between the trajectory of the ions and the outward radial direction. The algorithm to compute the energy redeposition from the reaction product ions involves a computation for each reaction product, in each finite difference zone, and along each straight line ray originating in that zone. This computation tracks the ion trajectories from creation to thermalization or escape and would tally the amount of energy and momentum deposited in each zone along their path. This algorithm must be executed for each reaction product that is transported from each finite difference zone along each straight line ray.

The particle tracking algorithm discussed above has taken an "adiabatic" approach to the time dependence of the slowing down process. That is, a slowing down calculation is performed for each time step in the hydrodynamics calculation. Fast ions are forced to thermalize or escape the plasma during the time step in which they are created. This assumption can break down, however, and in these cases one can implement a time-dependent particle tracking method that develops an energy-time relationship analogous to the range-energy relationship used in the adiabatic method.[1,8] However this generalization requires a rather complex programming logic.

Particle tracking methods differ significantly from flux-limited diffusion. The former presumes streaming as the dominant physical process and then corrects for large-angle scattering, while the latter presumes diffusion and corrects for streaming. From Fokker-Planck studies of the detailed behavior of charged particles, it is apparent that flux-limited diffusion methods are most applicable to suprathermal electron transport, while particle tracking methods are most useful for nonthermal ion transport (e.g., reaction products).

6.2.3. DIRECT SOLUTION OF THE FOKKER-PLANCK EQUATION

The most common description of the cumulative effect of many random, small angle collisions on the particle distribution function is provided by the Fokker-Planck equation.[62] We have already noted that the Fokker-Planck equation is used as the basis for approximate theories of charged particle transport including flux-limited diffusion theory and particle tracking methods. More direct numerical methods for solving the Fokker-Planck equation have been developed for magnetic fusion systems[63] (particularly magnetic mirror devices), but generally the spatial dependence is suppressed, and only the velocity space diffusion is retained.

It is possible to solve the Fokker-Planck equation in full generality including time, space, and velocity dependence by using discrete ordinates methods.[51,52] Discrete ordinates methods are particularly attractive for this purpose since they have been highly developed for the solution of neutral particle (neutron and photon) transport problems. Indeed, there are a variety of sophisticated and efficient one and two dimensional computer codes available that have planar, cylindrical, and spherical geometry options (as well as toroidal and triangular meshes for nonorthogonal geometries) and allow for a variety of boundary and source conditions. The incorporation of the Fokker-Planck collision term into a multigroup discrete ordinates formalism provides for an easy extension to a variety of geometries and source configurations.

For example, the TIMEX discrete ordinates code, originally developed for time-dependent neutron transport, has been modified to describe charged particle transport by solving the Fokker-Planck equation directly.[52] This approach utilizes a discrete ordinates treatment of the spatial and angular variables, a multigroup treatment of the energy variable, and an explicit time differencing scheme. The resulting computer code allows the simulation of charged particle transport as described by the Fokker-Planck equation allowing for full spatial, velocity, and time dependence.

6.3. PLASMA SIMULATION (PARTICLE) CODES

Plasma physics is an inherently nonlinear phenomenon which can exhibit a complex structure on a microscopic scale. Hence conventional hydrodynamic descriptions of the plasma state have only a limited validity. A more thorough analysis and prediction of plasma behavior generally requires the use of plasma simulation or particle codes in which the equations of motion for the particles (ions and electrons) comprising the plasma are integrated directly.[64,65] In this sense, particle simulation codes represent a microscopic simulation of the plasma dynamics.

The basic approach in plasma simulation is shown in Figure 6.15. The coordinates of the particles are used to determine a macroscopic charge and current density. These densities are then used in Maxwell's equations to

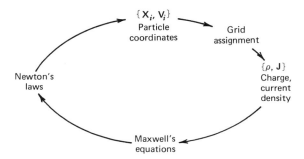

Figure 6.15. The basic strategy used in plasma particle dynamics simulation calculations.

determine the corresponding self-consistent electric and magnetic fields. These fields determine the forces acting on the particles, and the corresponding equations of motion are then integrated with these forces to determine the coordinates at the next time step.

There are several difficulties with this straightforward approach, however. The dynamics of the plasma occur on a time scale characterized by the plasma frequency, hence implying that very small time steps must be taken. Furthermore, the long range of the Coulomb interaction implies that many particles interact simultaneously. For example, the effective range of interaction is determined by the Debye length. In a typical fusion plasma, there are perhaps 10^5 to 10^6 particles in a Debye sphere. Therefore one cannot simply solve the equations of motion for each particle,

$$m_i \frac{d^2 \mathbf{r}_i}{dt^2} = \mathbf{F}_i$$

Rather, one must average or smear out the detailed structure over a relatively coarse grid (on the scale of the Debye length rather than the interparticle spacing). Plasma simulation codes attempt to describe the collective behavior of the plasma rather than the microscopic fields associated with particle discreteness. A spatial grid is chosen that is sufficiently fine to resolve this collective behavior while still being sufficiently coarse to ignore the microscopic field structure on length scales comparable to the interparticle spacing.

In effect plasma simulation codes determine the self-consistent fields in the plasma only on a macroscopic scale characterizing collective behavior. Since a great many of the plasma particles then experience the same forces arising from these self-consistent fields (since the variations in the field are not resolved on the scale of interparticle spacing), the computer simulation model actually treats a group of particles as an effective or "finite-size" particle for the purposes of the calculation.

To be more precise, consider an electrostatic plasma simulation code in which Maxwell's equations reduce to

$$\nabla \cdot \mathbf{E} = 4\pi\rho$$

Our first task is to assign the particles to a spatial grid. Suppose we consider the one-dimensional spatial grid shown in Figure 6.16. For convenience we

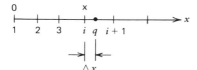

Figure 6.16. The grid structure used in a one-dimensional simulation calculation.

have taken the cell size $\delta = 1$. There are two schemes used to assign the charge to the grid:

1. *Nearest grid point (NGP)*

$$\rho(i) = q$$

2. *Particle in cell (PIC)*. One interpolates as

$$\rho(i) = q(1 - \Delta x)$$
$$\rho(i+1) = q \Delta x$$

Next we determine the electric field from this charge density using Poisson's equation. We can either finite-difference Poisson's equation directly

$$\frac{\partial E}{\partial x} = -4\pi(\rho_e - \rho_i) = -4\pi\rho$$
$$E(i+1) = E(i) - 4\pi \frac{\rho(i+1) + \rho(i)}{2} \delta$$

or Fourier transform Poisson's equation

$$ikE_k = 4\pi\rho_k$$

and then invert the transform.

The final step is to map the field from the grid to the particles. Generally one assigns the force to the particles using the same scheme chosen to assign the charges to the grid. For example:

1. *NGP*

$$F = qE(i)$$

2. *PIC*

$$F = qE(i)(1 - \Delta x) + qE(i+1)\Delta x$$

The particle positions and velocities are then updated by a leap-frog algorithm that defines position and velocity one-half time step apart to achieve a second-order accuracy in the differencing scheme. That is, the equations of motion for a particle

$$\dot{v} = F(x)$$
$$\dot{x} = v$$

are differenced as

$$v^{n+1/2} = v^{n-1/2} + F^n \Delta t$$
$$x^{n+1} = x^n + v^{n+1/2} \Delta t$$

This algorithm represents the basic sequence of calculation steps in each time cycle of the code.

Plasma particle simulation codes are considerably more complex when the full set of Maxwell's equations are used (such as would be necessary in the analysis of electromagnetic wave interactions in plasmas). Frequently multipole expansions of the radiation field are used to simplify the calculation.

How good are plasma simulation results? One interesting comparison of the results given by an electrostatic simulation code with a direct solution of the Vlasov equation is given by Kruer.[64] If we recall that the Vlasov model

$$\frac{\partial f}{\partial t} + v\frac{\partial f}{\partial x} - \frac{e}{m}E\frac{\partial f}{\partial v} = 0$$

$$\frac{\partial E}{\partial x} = -4\pi e\left[\int dv f(x) - n_0\right]$$

corresponds to the $n\lambda_D^3 \to \infty$ limit of the microscopic particle dynamics, then it is apparent that this equation represents the exact behavior of the plasma in the limit in which collective processes dominate. Kruer has compared the nonlinear behavior of large amplitude electron plasma waves with those described by particle codes. The agreement between the Vlasov description and the plasma simulation are found to be quite good.

The state of the art of plasma simulation codes is quite impressive. A two-dimensional electromagnetic, relativistic simulation code can involve as many as 40,000 grid points with a single cell of several Debye lengths in width. This would allow one to simulate the behavior of a plasma of size $400 \times 400 \, \lambda_D$ or $6 \times 6 \, \lambda_0$ (where λ_0 is the wavelength of the electromagnetic radiation). If there are 5 to 10 particles per Debye square, this would imply roughly 10^6 particles. Running the code for 2500 time steps would require roughly 20 hours of CPU time on a CDC-7600 (Cyber 176).

Typically, such plasma simulation codes are used to determine how microscopic phenomena such as turbulence affect the local plasma properties such as energy absorption rates and transport coefficients. This information is then used to adjust ("patch up") the macroscopic description of the plasma provided by hydrodynamics codes.

6.4. CONCLUDING REMARKS

Inertial confinement fusion target simulation codes are extremely complex and expensive to use. Typical one-dimensional calculations require 100 spatial zones and several thousand time cycles for a single simulation. This translates into 20 to 100 min of central processor unit (CPU) time on computers such as the CDC-7600 (Cyber 176). The complexity of the codes arises from the many different physical processes that must be simultaneously described: plasma hydrodynamics, particle transport, driver energy deposition, radiative transfer, equation of state and opacity table look-ups, thermonuclear reaction rate equations, and so on. Each part of the calculation is quite extensive in its own right. Two dimensional calculations are considerably more expensive, and only those laboratories with dedicated computer facilities for such calculations can afford to perform these simulations.

The complexity of the codes arises from the many separate components of program rather than from the sophistication of the physical models used in each component. Generally the models are kept quite simple to avoid troublesome numerical instabilities and the finicky nature of many elaborate numerical algorithms. Most codes strive to maintain conservation of energy and momentum during the course of the calculation. This is difficult to achieve when so many different nonlinear processes are coupled together. However, careful bookkeeping of the various forms of energy in the calculation (e.g., internal, kinetic, radiative, particle) can lead to good conservation properties.

The troublesome details of understanding and running such complex computer codes require a great amount of time to master. Typical operational design codes have had over 25 man-years of effort devoted to their development. This is complicated further by the fact that many of these codes are being changed on almost a weekly basis, with new bits and pieces of physics being patched into them. In fact, there will generally be many different versions of a given code floating around a typical laboratory (sometimes several versions for each user).

The computer simulation of ICF hydrodynamics and transport is a relatively new field that has room for much more development. This conclusion seems apparent from the numerous codes in existence today. The most notable and complete code is the LASNEX code, with two-dimensional Lagrangian hydrodynamics and flux-limited diffusive particle transport. Each of the principal laboratories involved in ICF research has developed its own set of plasma hydrodynamics/transport codes for the simulation of target dynamics.

As more experiments are performed, it has become increasingly apparent that the so-called anomalous effects observed in the experiments are not just a first-order correction to the classical hydrodynamics behavior of the targets, but are, in fact, dominating the results. Modeling these processes requires the treatment of particle distribution functions that may be far from equilibrium. Flux-limited multigroup diffusion theory is a first attempt to include these effects in a hydrodynamics description; however, the necessity for coefficient adjustments to obtain good agreement with experiment and the sensitivity of results to numerical parameters such as zoning, time steps, and differencing techniques raises doubts about the ability of these simple models to provide good insight into inertially confined plasma dynamics and target-driver interaction. Detailed treatment of particle distribution functions and numerical schemes designed for accuracy rather than expediency imply a great increase in computer time for codes that already use large amounts of it. In actual fact, the solution of the hydrodynamics equations in these codes represents only a small fraction of the computing cost. Complicated prescriptions for transport must be recomputed many times for almost all coefficients are both nonlinear and spatially and time dependent. Particle transport calculations are even more time consuming.

The inclusion of transport calculations within hydrodynamics calculations is extremely important, but it will no doubt stretch the capacity of even advanced computers such as the Cray-1S and Cyber 205. Hence the present "modular" approach is likely to continue in which a variety of codes are developed and used to focus on different aspects of the relevant physics, with the hope that these processes decouple sufficiently to allow a separate analysis.

Despite these many drawbacks, plasma hydrodynamics computer codes will remain the mainstay of ICF target design and experimental analysis. The bewildering variety and complexity of the many processes arising in the implosion of an ICF target defy more conventional methods of analysis.

REFERENCES

1. G. A. Moses, "Laser Fusion Hydrodynamics Calculations," *Nucl. Sci. Eng.* **64**, 49 (1977).

2. G. B. Zimmerman and W. L. Kruer, "Numerical Simulation of Laser Initiated Fusion," *Comments on Plasma Physics and Controlled Fusion*, Vol. II, 51 (1975).

3. R. E. Kidder and W. Barnes, "WAZER: A One-Dimensional, Two Temperature Hydrodynamics Code," Lawrence Livermore Laboratory Report UCRL-50583 (1971).

4. G. Fraley, E. Linnebur, R. Mason, and R. Morse, "Thermonuclear Burn Characteristics of Compressed Deuterium-Tritium Microspheres," *Phys. Fluids* **17**, 474 (1974).

5. E. B. Goldman, "Numerical Modeling of Laser Produced Plasmas: Theory and Documentation for SUPER," University of Rochester Laboratory for Laser Energetics Report 16 (1973).

6. J. P. Christiansen, D. E. T. F. Ashby, and K. V. Roberts, "MEDUSA - A One-dimensional Laser Fusion Code," *Comp. Phys. Comm.* **7**, 271 (1974).

7. S. Thomson, "Improvements in the CHART-D Energy Flow Hydrodynamic Code V: 1972/73 Modifications," Sandia Laboratory Report SLA-73-0477 (1973).

8. G. A. Moses and J. J. Duderstadt, "PHD, A Plasma Hydrodynamics Code for Laser Fusion Simulation Studies," *Trans. Am. Nucl. Soc.* **23**, 50 (1976).

9. G. A. Moses and G. R. Magelssen, "PHD-IV, A Plasma Hydrodynamics-Thermonuclear Burn-Radiative Transfer Computer Code," University of Wisconsin Fusion Project Report UWFDM-194 (1978).

10. D. Colombant, K. Whitney, D. Tidman, N. Winsor, and J. Davis, "Laser Target Model," *Phys. Fluids* **18**, 1687 (1975).

11. "LILAC, A 2-D Eulerian Code," University of Rochester Laboratory for Laser Energetics, 1976.

12. J. Friedberg, R. Mitchell, R. Morse, and R. Rudsinski, "Resonant Absorption of Laser Light by Plasma Targets," *Phys. Rev. Lett.* **28**, 795 (1972).

13. K. G. Estabrook, E. J. Valeo, and W. L. Kruer, "Two-Dimensional Relativistic Simulations of Resonance Absorption," *Phys. Fluids* **18**, 1151 (1975).

14. J. Howard, "Classical Energy Deposition and Refraction in Spherical and Planar Laser Fusion Targets," University of Wisconsin Fusion Project Report UWFDM-188 (1976).

15. "Laser Program Annual Report—1975," Lawrence Livermore Laboratory Report UCRL-50021-75 (1975), p. 292.

16. Reference 15, p. 287.

17. D. W. Forslund, J. M. Kindel, and K. Lee, "Theory of Hot Electron Spectra at High Laser Intensity," *Phys. Rev. Lett.* **39**, 284 (1977).

18. K. A. Brueckner, "Fast Electron Production in Laser Heated Plasmas," *Nucl. Fusion* **17**, 1257 (1977).

19. "Laser Program Annual Report—1976," Lawrence Livermore Laboratory Report UCRL-50021-1976, p. 483.

20. K. A. Brueckner and R. S. Janda, "Production of Fast Ions in Laser Heated Plasmas," *Nucl. Fusion* **17**, 1265 (1977).

21. G. Charatis, et. al., "Experimental Study of Laser-Driven Compression of Spherical Glass Shells," in *Plasma Physics and Controlled Nuclear Fusion Research, Tokyo, 1974* (Vienna, International Atomic Energy Agency, 1975), p. 317.

22. J. Lindl, "Effect of a Suprathermal Electron Tail on the Yield Ratio Obtained from DT Targets Illuminated with a Shaped Laser Pulse," *Nucl. Fusion* **14**, 511 (1974).

23. E. L. Lindman, "Convective Instabilities Driven by Electromagnetic Waves," *Phys. Fluids* **10**, 396 (1967).

24. D. Forslund, J. Kindel, K. Lee, and E. Lindman, "Absorption of Laser Light on Self-Consistent Plasma Density Profiles," *Phys. Rev. Lett.* **36**, 35 (1976).

25. G. A. Moses and J. J. Duderstadt, "An Improved Treatment of Electron Thermal Conduction in Plasma Hydrodynamics Calculations," *Phys. Fluids* **20**, 762 (1976).

26. R. Malone, R. McCrory, and R. Morse, "Indications of Strongly Flux-Limited Electron Thermal Conduction in Laser Target Experiments," *Phys. Rev. Lett.* **34**, 721 (1975).

27. J. Stamper and J. Dawson, "Spontaneous Magnetic Fields in Laser-Produced Plasmas," *Phys. Rev. Lett.* **26**, 1012 (1971).

28. Laser Program Annual Report-1977, Lawrence Livermore Laboratory Report UCRL-50021-77 (1977) p. 4-67.

29. S. Braginskii, *Review of Plasma Physics* 1 (New York, Consultants Bureau, 1965), p. 205.

30. J. Shiau, E. Goldman, and C. Weng, "Linear Stability Analysis of Laser-Driven Spherical Implosions," *Phys. Rev. Lett.* **34**, 1273 (1975).

31. J. Lindl and W. Mead, "Two-Dimensional Simulation of Fluid Instability in Laser Fusion Pellets," *Phys. Rev. Lett.* **34**, 1273 (1975).

32. P. Campbell, "A Numerical Method for Discrete Ordinate and Momentum Equations in Radiative Transfer," *Int. J. Heat Mass Transfer* **12**, 497 (1969).

33. K. Brueckner, "Semi-Empirical Estimates of Neutron Production in Shell Implosions," *Nucl. Fusion* 16, 387 (1976).

34. P. Haldy and J. Ligou, "A Moment Method for Calculating the Transport of Energetic Charged Particles in Hot Plasmas," *Nucl. Fusion* **17**, 1225 (1977).

35. G. Pomraning, *Radiation Hydrodynamics* (New York, Plenum, 1972).

36. L. D. Landau and E. M. Lifshitz, *Fluid Mechanics* (New York, Academic 1959).

37. J. von Neumann and R. Richtmyer, "A Method for the Numerical Calculation of Hydrodynamics Shocks," *J. Appl. Phys.* **21**, 232 (1950).

38. J. P. Boris, "Flux-corrected Transport III: Minimal Error FCT Algorithms," *J. Comp. Phys.* **20** (1976).

39. R. Richtmyer and K. Morton, *Difference Methods for Initial Value Problems*, 2nd ed. (New York, Interscience, 1967).

40. Laser Program Annual Report-1975, Lawrence Livermore Laboratory Report UCRL-50021-75 (1975), p. 290.

41. Laser Fusion Program Semiannual Report January-June, 1972, Lawrence Livermore Laboratory Report UCRL-50021-72-1 (1973), p. 107.

42. Laser Fusion Program Semiannual Report, July-December, 1973, Lawrence Livermore Laboratory Report UCRL-50021-73-2 (1974), p. 93.

43. Laser Program Annual Report-1974, Lawrence Livermore Laboratory Report UCRL-50021-74 (1975) p. 368.

44. Ref. 40, p. 289.

45. Ref. 19, p. 4-51.

46. Ref. 28, p. 4-29.

47. W. Schultz, in *Methods in Computational Physics*, Vol. 3 (New York, Academic, 1964), p. 1.

48. Laser Program Annual Report-1978, Lawrence Livermore Laboratory Report UCRL-50021-78 (1979), p. 3-61.

49. D. S. Kershaw, "The Incomplete Cholesky-Conjugate Gradient Method for the Iterative Solution of Linear Systems," Lawrence Livermore Laboratory UCRL-78333 Rev 1 (1977).

50. M. L. Wilkins, "Calculation of Elastic Plastic Flow," Lawrence Livermore Laboratory Report UCRL-7322 Rev (1969).

51. J. J. Duderstadt and W. R. Martin, *Transport Theory* (New York, Wiley-Interscience, 1979).

52. T. Mehlhorn and J. J. Duderstadt, "Discrete Ordinates Solution of the Fokker-Planck Equation," *J. Comp. Phys.* **38**, 86 (1980).

53. F. Beranek and R. Conn, "Neutron Moderation in Inertial Confinement Fusion Pellets and Effects on Damage and Radioactive Inventory," *Nucl. Tech.* **47**, 406 (1980).

54. K. D. Lathrop, "Discrete Ordinates Methods for the Numerical Solution of the Transport Equation," *React. Technol.* **15**, 107 (1972).

55. J. Fleck and J. Cummings, "An Implicit Monte Carlo Scheme for Calculating Time and Frequency Dependent Nonlinear Radiation Transport," *J. Comp. Phys.* **8**, 313 (1971)

56. G. B. Zimmerman, "Numerical Simulation of the High Density Approach to Laser Fusion," Lawrence Livermore Laboratory Report UCRL-74811 (1973)

57. E. Corman, W. Loewe, and G. Cooper, "Multigroup Diffusion of Energetic Charged Particles," *Nucl. Fusion* **15**, 377 (1975).

58. D. Kershaw, "Differencing of the Diffusion Equation in LASNEX," Lawrence Livermore Laboratory Report UCID-17424 (1977).

59. N. Krall and A. Trivelpiece, *Principles of Plasma Physics* (New York, McGraw-Hill, 1973), p. 287.

60. N. K. Winsor, "Velocity Space Methods for Fusion Reactor Plasmas," *Nucl. Sci. Eng.* **64**, 33 (1977).

61. H. Brysk, "Reaction Production Transport in a Fusion Pellet," KMS Fusion Report KMSF-U275 (1975)

62. M. N. Rosenbluth, W. M. MacDonald, and D. L. Judd, "Fokker-Planck Equation for an Inverse Square Force," *Phys. Rev.* **107**, 1 (1957)

63. A. Mirin, "Hybrid-II, A Two-Dimensional Multi-Species Fokker Planck Code," Lawrence Livermore Laboratory Report UCRL-51615 Rev. 1 (1975).

64. W. Kruer, "Lectures on the Interaction of Laser Light with Plasmas," Scottish Summer School, 1979.

65. G. B. Zimmerman and W. L. Kruer, "Numerical Simulation of Laser Initiated Fusion," *Comments on Plasma Physics and Controlled Fusion*, Vol. II, 51 (1975)

SEVEN

Driver Development I: Lasers

In inertial confinement fusion, intense laser or charged particle beams are used to implode fuel pellets to densities and temperatures sufficient for efficient thermonuclear burn. More precisely, the energy deposited by the driver beams incident upon the target produces surface ablation that compresses the pellet core to very high densities ($\rho R > 1$ g/cm^2). The requirements on driver beam intensities to achieve such conditions are severe and correspond to specific energy deposition of several megajoules per gram of target ablation material. Hence a premium is placed on developing ICF drivers capable of yielding the required focused beam intensity ($> 10^{14}$ W/cm^2) and energy densities (> 20 MJ/g).

Pulsed lasers can be used to produce just such focused beam intensities. Laser drivers can convert electrical (or chemical or gas dynamic) energy into an intense beam of coherent light capable of being focused in space and time to achieve the power intensity and energy densities required for ICF applications.

All present ICF laser drivers produce light in the infrared (although there are techniques available to shift this light to frequencies in the visible part of the spectrum). The most extensively used laser system for ICF applications has been the neodymium-glass laser that emits light at a wavelength of 1.06 μm. These solid state lasers are pumped by flashlamps (powered, in turn, by capacitor banks). To date, these lasers have been restricted in energy to less than 1000 J/beam and in efficiency (conversion of electrical energy into light energy) of less than several tenths of a percent. Several laboratories in the United States and abroad have large, multibeam Nd-glass lasers operating in the 10-TW range.

To achieve the high efficiencies and pulse repetition rates required for ICF applications, it will probably be necessary to use gas lasers. Of most interest to

Table 7.1.   **Laser Requirements for ICF Drivers.**

Energy	1 to 5 MJ
Pulse length	1 to 10 ns
Efficiency	>1%
Wavelength	0.2 to ? μm
Repetition rate	1 to 10 Hz

date has been the carbon dioxide laser that emits light at a wavelength of 10.6 μm. Several large CO_2 laser installations are now operating at the 1- to 10-TW level. There has also been considerable interest in the iodine laser (at 1.315 μm) and the hydrogen-fluorine chemical laser (at 2.7 μm). However none of these laser systems presently fulfills the requirements necessary for ICF applications (see Table 7.1).

We begin this chapter by summarizing the essential physics of laser operation. We continue on to discuss the primary laser types used in ICF research programs today (Nd, CO_2, and I) and conclude with a discussion of advanced laser development (including excimer and chemical lasers).

7.1.   LASER PHYSICS

To introduce the basic concepts involved in laser operation,[1-3] consider the interaction of light with a very simple system: a single atom that can exist in one of only two possible energy states, E_1 and E_2. Incident light photons with a frequency ν_{12} such that $h\nu_{12}=E_2-E_1$ can be absorbed by atoms in the ground state E_1. However, photons incident on atoms in the excited state E_2 can stimulate the emission of a second photon of frequency ν_{12} which appears in phase (in coherence) with the incident photon. If there are more atoms in the excited state E_2, then photons incident upon a medium containing such atoms can stimulate a growing cascade of photons of frequency ν_{12} via the stimulated emission process. That is, the medium could be used to achieve *l*ight *a*mplification by *s*timulated *e*mission of *r*adiation; hence the acronym "laser." This process is shown schematically in Figure 7.1.

Under normal circumstances there will be many more atoms in the lower energy state (usually the ratio of population densities goes as the Boltzmann factor, $\exp[-(E_2-E_1)/kT]$). Hence to achieve the laser process, we must somehow achieve a "population inversion" in which more atoms are in the upper than lower state, so that an incident photon will stimulate the emission of other photons rather than being absorbed. A variety of excitation mechanisms can be used, including irradiating the medium with intense light at another frequency, using electrical discharges, or chemical reactions.

A key feature in laser operation involves the fact that the photons emitted in the stimulated process appear in phase with one another. This is in sharp contrast to the light emitted from conventional sources of intense light in which the photons are not only distributed over a relatively broad spectrum,

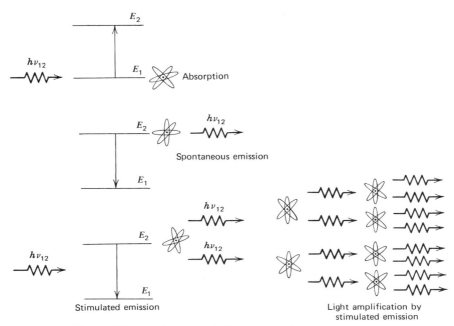

Figure 7.1. A schematic of the stimulated emission process.

but furthermore are incoherent in the sense that there is no correlation between the phases of radiation emitted from two different points of the source. An important theorem of classical optics states that it is impossible to devise an optical system that could focus an incoherent beam of light to an intensity higher than that characterizing the source from which the light originates. That is, the incoherent radiation from an extended source of light cannot be imaged with an increase in brightness.

This is in sharp contrast to the coherent beam of light emitted by a laser. It is possible to concentrate or focus this beam in such a way as to increase the light intensity or brightness to a level greater than that of the original source. In fact, the coherence in many laser systems is sufficient to allow the beam to be focussed down to spot sizes on the order of the wavelength of the light (so-called diffraction limited). Furthermore, in sharp contrast with incoherent light sources, the coherent light in a laser beam can be propagated large distances without appreciable losses from geometric spreading or diffraction effects.

Since laser action essentially involves the collective deexcitation of a number of excited atoms or molecules contained in a resonant cavity, it should be apparent that a study of the various excited states (or energy levels) available to atoms and molecules is of central concern in the understanding of laser operation. Hence we will begin our study of laser physics with a brief review of several concepts from atomic and molecular spectroscopy.

7.1.1. ATOMIC AND MOLECULAR SPECTROSCOPY

It is a well-known consequence of quantum mechanics that atomic systems (atoms, ions, molecules, nuclei) can only be found in certain time-independent states, each of which corresponds to a definite value of energy. These discrete states or energy levels can be characterized by a set of quantum numbers. It is common in atomic physics for several states to be characterized by the same energy level; these states are said to be degenerate, and the corresponding number of such states is called the multiplicity of the level. The lowest energy level available to an atomic system is known as its ground state, while all higher levels are referred to as excited states.

An atomic system can change from one state to another, with the accompanying emission or absorption of the energy difference between the two levels as electromagnetic radiation or energy exchange via an atomic collision process.

Lasers that produce light at optical wavelengths are based on electronic transitions in atomic systems. While the spectroscopy of one-electron atoms such as hydrogen is relatively simple, that characterizing the many electron atoms of most interest in laser applications can become quite complex. We avoid a discussion of atomic spectroscopy and spectroscopic notation at this point and refer the interested reader to several comprehensive references on this subject.

Example: The first gas laser to be demonstrated experimentally (1961) involved transition between various excited states in neutral neon (an "atomic" laser). Helium gas was mixed with the neon to facilitate the achievement of a population inversion using an electrical discharge in the gas. While many lasing transitions have been observed, the three most important transitions in the He-Ne laser are $^3s_2 \rightarrow {}^2p_4$(6328 Å), $^2s_2 \rightarrow {}^2p_2$(1.15 μm), and $^3s_2 \rightarrow {}^2p_4$(3.39 μm).

Literally thousands of lasing transitions have been established for a wide range of atomic and ionic species. Indeed, it has occasionally been suggested that any material can be made to lase—if one is sufficiently clever (and able to invest enough money and effort).

The second class of lasers of interest as ICF drivers are molecular gas lasers that are based on transitions in molecules such as carbon dioxide producing light at infrared wavelengths. The energy level spectra of molecular gases are considerably more complex than those of atomic gases (such as neon). In addition to the electronic energy levels characterizing atoms, a molecule can also have energy levels arising from the vibrational and rotational motion of the atoms in the molecule. Whereas the spacings of the electronic energy levels for molecules are comparable to those for atoms, the vibrational and rotational levels add a "fine structure" whose level spacing is typically smaller by factors of 20 and 500, respectively. A comparison of typical atomic versus molecular spectra is shown in Figure 7.2.

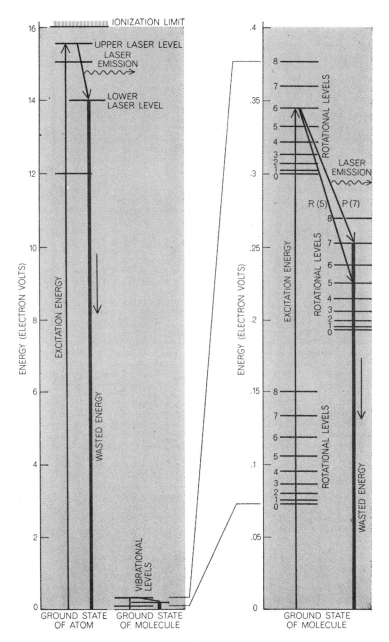

Figure 7.2. A comparison of atomic versus molecular energy levels. (Courtesy of Scientific American.)

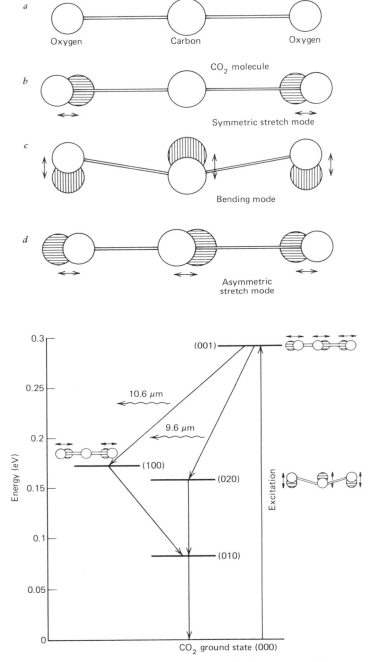

Figure 7.3. The vibrational degrees of freedom and the corresponding energy levels for the CO_2 molecule.

Example: The most important molecular gas laser for ICF applications is the carbon dioxide laser. CO_2 is a linear, symmetric molecule possessing three vibrational degrees of freedom as shown schematically in Figure 7.3. Associated with these vibrational modes are three quantum numbers, v_1, v_2, and v_3 that represent the number of quanta in each mode. The description of a given vibrational level is given by (v_1, v_2, v_3). The vibrational levels of most interest are shown in Figure 7.3. (The rotational levels have been omitted to keep the energy level diagram as simple as possible.) Note that the $001 \rightarrow 100$ transition corresponding to 10.6-μm radiation is of most interest in CO_2 laser applications.

7.1.2. THE INTERACTION OF ELECTROMAGNETIC RADIATION WITH ATOMIC SYSTEMS

Stimulated and Spontaneous Emission. Let us begin by considering the interaction of radiation with a single atom (or molecule) with two levels as shown in Figure 7.1. If the atom is initially in the ground state E_1, then it can be excited into the upper state E_2 by the absorption of a photon of energy $h\nu_{12} = E_2 - E_1$. Similarly if the atom is initially in the excited state E_2, it can spontaneously decay by emitting a photon of frequency ν_{12}. We will denote the transition probability per unit time (i.e., the transition rate) for such spontaneous emission of light as A_{21}.

There is yet a second process for emission, however. An atom in an excited state can be stimulated to emit a photon if one subjects it to electromagnetic radiation of the same frequency as the energy level spacing. That is, there is a probability that an incident photon of energy $h\nu_{12}$ will stimulate the emission of yet another photon. These two photons emerge from the atom in phase. In this sense, stimulated emission can lead to a coherent amplification of light. This is in contrast to the spontaneous emission process in which the photons are emitted randomly in direction and phase.

The transition rate for stimulated emission must be proportional to the photon density or radiant energy density u_{12} at frequency ν_{12}. Let us write the corresponding transition rate as $u_{12} B_{21}$. Then the total transition rate from E_2 to E_1 is given by

$$W_{2 \rightarrow 1} = A_{21} + u_{12} B_{21}$$

The coefficient B_{21} is closely related to a similar coefficient B_{12} that appears in the expression for the absorption transition rate

$$W_{1 \rightarrow 2} = u_{12} B_{12}$$

More generally then, for two energy levels m and n we can write the transition rates

$$W_{n \rightarrow m} = A_{nm} + u_\nu B_{nm}$$

$$W_{m \rightarrow n} = u_\nu B_{mn}$$

The rate coefficients A_{nm} and B_{nm} can be calculated in terms of the electric dipole matrix elements for the atom or molecule. However, there are two very useful relationships, known as the Einstein relations,

$$g_n B_{nm} = g_m B_{mn}$$

$$A_{nm} = \frac{8\pi h\nu^3}{c^3} B_{nm}$$

where g_n is the multiplicity of the energy level. (In a medium with index of refraction η the second relation is modified to

$$A_{nm} = \frac{8\pi h\nu^3 \eta^3}{c^3} B_{nm}$$

Stimulated emission of radiation is the key phenomenon in lasers. If one can prepare an "active" medium in which most of the atoms can be placed in an excited state, then electromagnetic radiation passing through the medium stimulates a cascade of photons, each in phase with one another. In this manner coherent light can be amplified to very high intensities. Of course, the first requirement is to develop a way to excite large numbers of atoms or molecules into the appropriate excited states. We therefore turn our attention to the various mechanisms available for preparing such an "inversion" of excited state populations.

Population Inversion. In most substances the absorption of a beam of incident radiation always dominates the stimulated emission process. To illustrate this, consider a large number N_0 of atoms in thermal equilibrium at a temperature T. Then if E_j is the energy of the jth state, the distribution of atoms among the various states is given by the Boltzmann factor

$$N_j = N_0 \frac{g_j e^{-E_j/kT}}{\sum_i g_i e^{-E_i/kT}}$$

In particular, the populations of any two levels, say E_1 and E_2, are related by

$$\frac{N_2}{g_2} = \frac{N_1}{g_1} e^{-(E_2-E_1)/kT}$$

Hence in thermal equilibrium, the relative population in the upper state will be many times smaller than that of the lower state. If we were to irradiate such a collection of atoms with light of frequency ν_{12}, then obviously absorption would greatly overwhelm any stimulated emission, and light attenuation rather than amplification would result.

More quantitatively, let the incident light be of energy density u_{12} and frequency ν_{12}. Then

$$\text{rate of emission} = (A_{21} + u_{12} B_{21}) N_2$$

$$\text{rate of absorption} = u_{12} B_{12} N_1$$

Hence the net loss in coherent incident beam photons (remembering that spontaneous emission is incoherent) is given by

$$\text{rate of loss} = (N_1 - N_2) u_{12} B_{21}$$

implying attenuation if $N_1 > N_2$.

What we must obviously do is disturb the equilibrium distribution so that we preferentially populate the upper state in such a way that $N_2 > N_1$. Such a population inversion will amplify light:

$$\text{photon rate of gain} = (N_2 - N_1) u_{12} B_{21}$$

Such an amplification or "negative absorption" is the objective of laser design.

Excitation Mechanisms. How does one obtain such a population inversion? Obviously not by trying to directly excite atoms from E_1 to E_2 since this would require absorption from the beam one wishes to amplify. Instead additional levels must be introduced. The simplest scheme is a three-level laser that was first exploited by the development of the ruby laser (see Figure 7.4). The idea is to irradiate the lasing material (in this case, the chromium ions in ruby) with flash lamps to excite them into the upper state 3. Since the light from flash lamps is not monochromatic, only a small fraction of the incident photons will be absorbed to excite the atoms. For this reason, one wants the upper level to have as large a line width as possible, to cover a broad frequency range and hence "catch" as many of the flash lamp photons as possible. The atoms in the upper level 3 then decay very rapidly via fast radiationless transitions into the intermediate level 2, which has a very narrow line width and a relatively long lifetime for spontaneous emission. Hence by using sufficiently intense flash

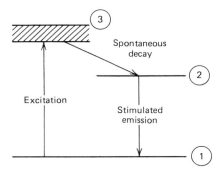

Figure 7.4. A simple three-level laser.

lamp irradiation, one can prepare a population inversion in which the number of atoms in 2 exceeds that in 1. The transition between 2 and 1 is then used as the lasing transition for stimulated emission.

The three level laser is only one of a large variety of level schemes used to excite atoms or molecules into suitable population inversions for light amplification by stimulated emission of radiation. For example, it is easier to establish a population inversion in a material that utilizes a transition to an energy level above ground state, since lasing action can begin as soon as the population of the upper lasing level exceeds that of the lower level rather than the much larger population of the ground state. We will return to discuss such a four-level laser scheme in more detail in the next section.

A variety of excitation mechanisms can be used. Most solid state lasers achieve a population inversion using broad spectral band photon excitation provided by flash lamps. Gas lasers utilize a combination of charged particle collisions and energy transfer via resonant collision processes that occur in electrical discharges in the lasing medium. Other excitation mechanisms including chemical reactions and gas dynamic processes have also been used. These will be discussed in detail in a later section.

It should be apparent that regardless of the specific mechanisms used to excite atomic or molecular systems and achieve population inversions, the relative widths of energy levels and their lifetimes against spontaneous emission play a very significant role in achieving laser light amplification. Hence some discussion of spectral line broadening is of importance.

Spectral Line Broadening. Suppose we consider light of frequency ν propagating in a medium in which there is an excited state at an energy $h\nu_0$ above the ground state. Then if $\nu = \nu_0$ we would expect to find appreciable absorption of the incident light. The attenuation of the light intensity would be described by

$$I_\nu(x) = I_0 e^{-k(\nu_0)x}$$

where $k(\nu)$ is an absorption coefficient for light at frequency ν. In our earlier discussion of atomic energy levels, we implicitly assumed that the levels are of vanishing width so that only light of the proper frequency ν_0 can be absorbed. If this were true, $k(\nu)$ would be nonzero only for $\nu = \nu_0$. In fact, however, an experimental measurement of $k(\nu)$ would reveal that it has the shape as shown in Figure 7.5, being peaked about ν_0. This corresponds to the fact that the excited state E_2 has a finite width, $h\Delta\nu$ (measured at half-maximum). If we recall the Heisenberg uncertainty relation, $\Delta E \Delta t > h$, then a finite width ΔE implies a finite lifetime Δt to the excited state.

There are a number of physical processes responsible for such a broadening of the energy level. Of course, an isolated atom in an excited state will have a finite lifetime against spontaneous decay. But this "natural" line width is extremely small and can usually be ignored. The two major factors contributing to line broadening are the frequency variations resulting from the thermal

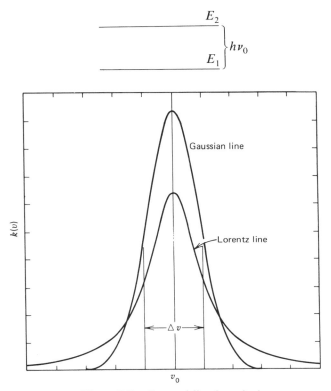

Figure 7.5. Spectral line broadening.

motions of the atoms (Doppler broadening) that give rise to a Gaussian spectral line shape, and a broadening resulting from the disruption of absorption or emission of radiation by atomic collisions (collision broadening) that gives rise to a Lorentzian line shape. A comparison of these two different line shapes is given in Figure 7.5.

7.1.3. THE THEORY OF LASER ACTION

Laser Gain. Suppose we consider a beam of light propagating through a medium in which the atoms are assumed to have only two available energy states, E_1 and E_2. The atomic number densities characterizing the frequency distribution (line shape) of each state will be denoted by $N_1(\nu)$ and $N_2(\nu)$, respectively. Then the equation describing the beam intensity I_ν at any point x can be derived by balancing the change in beam intensity in a frequency interval $d\nu$ against the absorption and stimulated emission over a distance dx:

$$dI_\nu \, d\nu = h\nu \left[B_{21} N_2(\nu) - B_{12} N_1(\nu) \right] I_\nu \frac{dx}{v} \, d\nu$$

If we identify the phase velocity of the light as $v = c/\eta$ we can rewrite this as

$$-\frac{1}{I_\nu}\frac{dI_\nu}{dx}d\nu = \frac{h\nu\eta}{c}\left[B_{12}N_1(\nu) - B_{21}N_2(\nu)\right]d\nu$$

$$\equiv k_\nu\, d\nu$$

We can identify the absorption by integrating over the level widths

$$\int k_\nu\, d\nu = \frac{h\nu_{12}\eta}{c}(B_{12}N_1 - B_{21}N_2)$$

where N_1 and N_2 are the total (frequency-integrated) number densities characterizing each level. If we now use the Einstein relations

$$g_1 B_{12} = g_2 B_{21}$$

$$B_{21} = \frac{c^3}{8\pi h\nu^3\eta^3}A_{21}$$

we can write

$$\int k_\nu\, d\nu = \frac{c^2 A_{21} g_2}{8\pi\nu_{12}^2\eta^2 g_1}\left(N_1 - \frac{g_1}{g_2}N_2\right)$$

$$= \kappa\left(N_1 - \frac{g_1}{g_2}N_2\right)$$

When the material is in thermal equilibrium, then we have seen that

$$\frac{g_1}{g_2}N_2 \ll N_1 \sim N_0$$

Hence we find

$$\int k_\nu\, d\nu = \kappa N_0$$

where κ can then be identified as the integrated absorption cross section per atom for the line of interest, while $\sigma_\nu = k_\nu/N_0$ is the absorption cross section per atom.

When population inversion occurs such that

$$N_1 < \frac{g_1}{g_2}N_2$$

then the absorption coefficient k becomes negative, corresponding to an exponential growth in intensity

$$I_\nu(x) = I_0 e^{\alpha_\nu x}, \qquad \alpha_\nu = -k_\nu$$

We can calculate α_ν as before

$$\int \alpha_\nu \, d\nu = \kappa \left(\frac{g_1}{g_2} N_2 - N_1 \right)$$

If we define the relative population inversion n

$$n = \frac{1}{N_0} \left(\frac{g_1}{g_2} N_2 - N_1 \right)$$

then we find the gain coefficient can be written as

$$\int \alpha_\nu \, d\nu = \kappa N_0 n$$

Threshold Conditions for Laser Oscillation. Consider now an aggregate of atoms in which a population inversion has been achieved (i.e., $n > 0$). Then a beam of light of frequency $\nu \sim \nu_{12}$ will grow in this medium exponentially as $\exp(\alpha_\nu x)$. Of course, in a laser oscillator, the beam must pass back and forth through the active medium many times to be amplified to appreciable intensities. To do this one places the atomic "amplifiers" between two mirrors (one of which is only partially reflecting to allow some of the laser light to escape from the "optical cavity"). For our present analysis, suppose the mirrors have a reflectivity r and the separation distance is L (see Figure 7.6). If we consider only the parallel propagation of light, then upon each reflection, $1 - r$ of the incident light energy is lost. Hence one complete pass yields an amplification of

$$F = r^2 e^{2\alpha_\nu L} = e^{2(\alpha_\nu L - \gamma)}$$

where we have introduced the cavity loss factor, $\gamma = \ln r$. To achieve net amplification, we require $F > 1$. Hence the threshold condition on the gain coefficient α_ν for laser amplification is

$$\alpha_\nu = \alpha_m \equiv \frac{\gamma}{L}$$

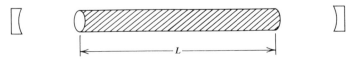

Figure 7.6. A simple schematic of an optical cavity.

Since α_ν will only be greater than or equal to α_m for a narrow range of frequencies, and since only modes with these frequencies will be amplified, the resultant laser output will have a line width much narrower than the atomic linewidth characterizing the lasing transition.

To proceed further, we must analyze the rate equations describing the populations of the various states involved in the lasing transitions. To illustrate, we will consider a four-level laser (the simplest model capable of describing the laser types of interest in ICF applications). We can define

$$W = u_\nu B = \text{stimulated emission rate}$$

$$A = \text{spontaneous emission rate}$$

$$S = \text{radiationless transition rate}$$

so that the rate equations for the state population densities characterizing the four-level laser shown in Figure 7.7 can be written as

$$\frac{dN_4}{dt} = W_{14}N_1 - (W_{41} + A_{41} + S_{43})N_4$$

$$\frac{dN_3}{dt} = W_{23}N_2 - (W_{32} + A_{32})N_3 + S_{43}N_4$$

$$\frac{dN_2}{dt} = W_{12}N_1 - (A_{21} + S_{21})N_2$$

$$N_0 = N_1 + N_2 + N_3 + N_4$$

(More complex models can be analyzed by adding the rate equations characterizing the additional levels involved in the model.) In such a laser, atoms

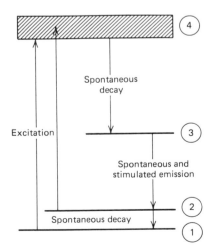

Figure 7.7. A four-level laser.

are excited into the broad band upper state 4 (by flashlamp photons or electron impact). Most of the absorbed energy is transferred by fast, radiation-less transitions into a sharp intermediate level 3 which serves as the upper lasing level. Since the line width of this level is narrow, the lifetime for spontaneous emission is long and the atoms tend to "hang up," accumulating in level 3 until photon emission is triggered by a stimulated emission process. The lasing transition then occurs between levels 3 and 2. Atoms then decay from level 2 to ground state, level 1, via spontaneous decay or collision processes. By using an excited state (level 2) as the terminal lasing level rather than the ground state, the four-level laser can achieve lasing action (population inversion) much earlier since level 2 will be relatively unpopulated (at least compared to the ground state level 1). Most solid state lasers are of the four-level type, including the neodymium laser. (An important exception is the ruby laser which involves a three-level process in which the terminal lasing level is the ground state.)

Although the analysis of the four-level laser rate equations is cumbersome, just before the onset of laser action the populations of states 2 and 4 can be neglected (to first approximation) so that a simpler two-level model applies

$$\frac{dN_3}{dt} = W_{14}N_1 - A_{32}N_3$$

To determine the appropriate threshold condition, we can examine the steady state solution:

$$\frac{N_3}{N_1} = \frac{W_{14}}{A_{32}}$$

$$N_3 = \frac{8\pi\nu^2\eta^3 T_3}{c^3 g T_p}$$

where $T_3 = A_{32}^{-1}$ is the lifetime against spontaneous emission of state 3, T_p is the photon lifetime, and g is the maximum of the line width shape function. If we identify the pump power required to achieve this condition as $P = W_{14}N_1 h\nu_p V$, then we find

$$P = \frac{8\pi\nu^2\eta^3 V h\nu_p}{c^3 g T_p}$$

7.1.4. PUMPING MECHANISMS

A primary goal in the design of a laser is to achieve a sufficient population inversion to overcome losses in the laser cavity and yield light amplification. The general idea behind all pumping mechanisms is to convert low quality energy stored in conventional forms (e.g., electric fields in a capacitor bank or chemical fuels) into high quality energy stored in excited atomic or molecular states that becomes available for extraction as laser light. The traditional mechanisms for achieving such population inversions include:

1. *Optical pumping.* Flash lamp photon sources with a broad frequency distribution can be used to achieve population inversions by using a three- or four-level laser scheme in which an excited state with large line width is used to "catch" a significant number of the flash lamp photons and then transfer this excitation energy to the upper lasing level via radiationless transitions. Such optical pumping is the primary mechanism used in lasers with solid or liquid state materials. Optical pumping is of secondary importance in gas lasers.

2. *Direct excitation by charged particle collisions.* If one maintains an electrical discharge in a gaseous lasing medium, then inelastic collisions between free electrons and atoms or ions can create excited states. Such excited states can then transfer energy to the upper lasing levels by collisions. This combination of electron excitation and collisional energy transfer is the principal mechanism used in gas lasers such as the CO_2 laser.

3. *Excitation through resonant or near resonant energy transfer.* The excitation present in a particular species can be selectively transferred to a particular state (or a narrow band of states) in another species by resonant collisions in which the relative energy between the colliding atoms or molecules is very close to the energy level spacing of their excited states.

4. *Excitation by gas dynamical processes.* Rapid heating or cooling of a molecular gas can generate a population inversion (such as by expanding a gas through a nozzle). Such excitation mechanisms are currently being applied in high power continuous wave (cw) CO_2 lasers.

5. *Excitation by chemical reactions.* It is well known that many chemical reactions yield reaction products in excited states. Hence chemical reactions can be used to create population inversions of molecular gases. In practice, a rather considerable fraction of the available chemical reaction energy can be coupled to the radiation field. Examples of such chemical lasers include the HF and HCl lasers.

Laser pumping mechanisms can be classified into two types: In pumping schemes such as those based upon optical flash lamps or electrical discharges, one essentially begins with a statistical distribution (e.g., a Maxwell-Boltzmann distribution) of states and relies on energy transfer mechanisms to select out only those high energy states in the tail of the distribution that correspond to the upper lasing levels. For example, in an electrical discharge, we start with a statistical distribution of mostly low energy electrons and then rely on resonant energy transfer collisions to produce a nonthermal distribution in the radiating atom or molecule. The trick is to tailor the electron distribution to pump the desired state while avoiding losses into lower states. The combination of the large number of low energy electrons and the much larger cross sections of secondary processes such as electron impact ionization implies that the excited state densities are usually low.

A second class of approaches begins with a narrow distribution of states above the upper lasing level and then relies on energy transfer kinetics to accomplish efficient energy down conversion. By pumping "from the top" one can achieve very high energy densities in the lasing medium. Examples of this approach include electron beam excitation and photolytic processes (in which one lasing transition is used to selectively pump another). Most advanced laser types such as the KrF laser fall into this class of pumping mechanisms, and we consider it further in the last section of this chapter.

7.1.5. PULSED OPERATION

A laser consists essentially of a large number of excited atoms or molecules that act as light "amplifiers" through the stimulated emission process. The active medium of the laser is placed between two mirrors that form an optical cavity. Because of the line width associated with the energy levels involved in the lasing transition, a variety of electromagnetic modes of oscillation are possible in the cavity. Because the length of a laser is typically 10^5 to 10^6 wavelengths, and the amplification occurs over a finite frequency range, usually a number of laser modes will be amplified simultaneously.

One can characterize the ability of the laser to amplify a given mode by the cavity quality factor

$$Q = \frac{2\pi \nu_0 E}{P_d}$$

where E is the energy present in the amplified modes and P_d is the rate at which the mode energy is dissipated in the cavity. The linewidth of the cavity is then given by $\Delta \nu = \nu_0 / Q$. As the laser oscillations build up, those modes falling within the amplification line width with the highest Q will be preferentially excited. The presence of the optical cavity (the mirrors) is essential to the growth of the oscillations, since a given wave must pass back and forth through the amplifying medium many times.

This feature can be used to pulse the laser output. If an electro-optical shutter is placed in front of one of the mirrors, then oscillation growth is inhibited while the shutter is closed. As the pumping continues, a larger and larger population inversion will build up, since little stimulated emission depletion occurs. If the shutter is suddenly opened, the oscillations build up very rapidly, leading to a laser pulse that dies out when the excited state populations have been depleted. Such pulsing amounts to artificially suppressing the Q of the cavity to allow large amounts of energy to be stored in the population inversion and then releasing this energy as laser light by "Q-switching" the optical cavity. Pulse widths as short as 2 to 50 ns can be achieved in this manner.

Gas lasers such as the CO_2 laser can also be operated in a pulsed mode, but Q-switching is unnecessary if one instead pulses the pumping electrical discharge ("gain switching") and uses an atmospheric (or higher) pressure gas in the laser. If the discharge is pulsed in a time less than 1 μs, the laser gain will reach its maximum near the end of the discharge (see Figure 7.8). But it will take some time for the dominant mode of the optical cavity to build up to a power sufficient to deplete the excited levels (typically about 40 round trips in the cavity corresponding to about 300 ns). Hence the gain (i.e., the population inversion) will reach a large value before the laser field becomes sufficiently strong to rapidly depopulate the upper lasing level—without the necessity for Q-switching. This gain switching technique is capable of generating pulses of widths 10 to 300 ns.

Even shorter pulses can be generated by mode-locking the laser. In this technique, a nonlinear optical element (such as a bleachable dye) is used to lock into phase a large number of modes that arise in a laser pumped well above the threshold condition in order to produce a series of sharp pulses. One of these pulses can then be switched out and amplified to achieve subnanosecond pulse widths. Mode-locking requires that the bandwidth of the oscillator cavity be large enough so that a number of modes can be simultaneously amplified. While this condition is easily achieved in solid state lasers (such as the Nd laser), most gas lasers are characterized by very narrow bandwidth (e.g., 3 to 4 GHz in atmospheric pressure CO_2). However by going to higher pressures, collisional broading will cause the lines of the rotational spectrum to broaden and overlap, giving an amplification band considerably larger (in the case of CO_2, a bandwidth of 10^3 GHz can be achieved). This large bandwidth allows many modes to be amplified and become available for mode-locking. Insertion of a bleachable dye into the cavity will then induce the self-locking of the modes into a train of short pulses. One of these pulses can then be switched out. One such pulse switching technique involves irradiating a germanium slab in the optical path with a ruby laser. The germanium is normally transparent to the 10.6-μm CO_2 light, but it becomes reflecting when a cutoff density of charge carriers is optically induced by the incident ruby laser light.

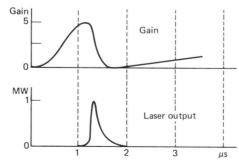

Figure 7.8.   Gain switching in a pulsed gas laser.

7.1.6. LASER SYSTEMS

Thus far we have confined our attention to the essential features of a laser oscillator, that is, an active medium in an optical cavity that can be induced to emit pulses of coherent light. In high-powered laser systems, such oscillators are used to produce low-power pulses of the required pulse widths and shapes. These pulses are then passed through a train of laser amplifier stages to bring the power level of the pulse to that required for ICF driver applications. Most of the complexity and expense of a high-powered laser system is due to the power amplifiers, their power supply, and the laser beam transport system.

The beam transport system includes not only the light beam transmission channels, but also the mirrors, lenses, and windows used to direct the beams and the mounting structures and systems used to align the beams. In most high power laser systems, mirrors are used instead of lenses to focus the beam. The design of suitable windows becomes important in gas laser systems since there must be some interface between the laser gas and the ambient air. Windows are also important in the target chamber, since this usually must be maintained at low pressure.

Another important aspect of high-powered laser systems is the pulsed power system used to drive the laser amplifiers. This system must be capable of storing energy produced by conventional sources (e.g., electrical generators) and then providing this energy in the form of short high-voltage pulses. At present, most high-powered laser systems use a bank of storage capacitors to provide pulsed power. These are typically rated at 120 kV, with 10-kJ energy storage per capacitor. Typical capacitor lifetimes are 10^5 shots, with a pulse repetition rate of up to 1 Hz. It should be noted that both the present lifetime and repetition rate fall considerably short of the goals proposed for reactor applications. However, low efficiency and low repetition rate laser systems are capable of elucidating the physics requirements of ICF.

7.2. PRESENT LASER TYPES

7.2.1. NEODYMIUM LASERS

The most common high-powered laser used in present ICF research is the neodymium laser. Such lasers are based on a solid state lasing material consisting of neodymium ions embedded in a matrix of yttrium aluminum garnet ($Y_3Al_5O_{12}$ or YAG) crystal or glass.[4] Neodymium lasers operate essentially as the four-level scheme described in Section 7.1.3. Such lasers possess a broad absorption band in the upper level 4 so that optical (flash lamp) pumping is possible. The terminal laser level (level 2) is normally sparsely occupied and drains rapidly to the ground state (level 1). Finally, radiationless processes rapidly transfer atoms from the upper excited level 4 to the upper lasing level 3. The interaction with the crystal or glass matrix host splits the degeneracy of the multiple levels of the active Nd^{3+} ion to produce

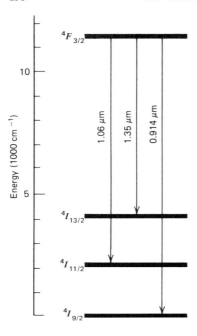

Figure 7.9. The lasing transitions in a neodymium laser.

the broad absorption band (level 4) and facilitate the transition from level 4 to level 3. The lasing transitions of most interest are shown in Figure 7.9.

In the past, high-powered Nd laser system oscillators were usually fabricated from Nd–YAG crystal while the power amplifiers were Nd-glass. Newer systems use phosphate glass amplifiers and Nd–YLF in the oscillator to match wavelengths. The mode-locked Nd–YAG oscillator can produce pulse widths

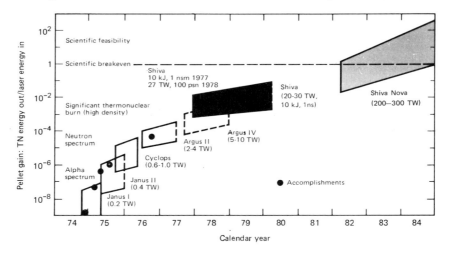

Figure 7.10. Progress in neodymium laser development at the Lawrence Livermore Laboratory.

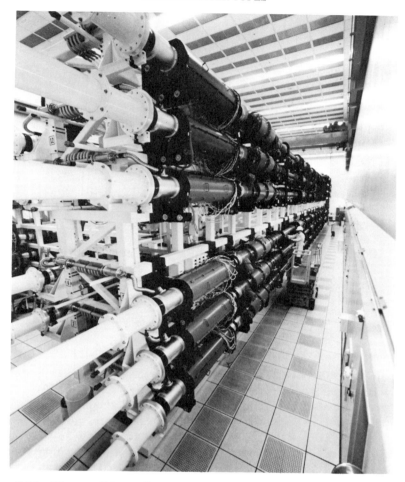

Figure 7.11. The amplifier gallery in the Shiva laser at the Lawrence Livermore Laboratory. (Courtesy of Lawrence Livermore Laboratory.)

of 25 to 1000 ps. Advanced Nd-glass amplifier designs such as the Nova system will utilize phosphate glasses which exhibit superior optical and energy storage properties to silicate glasses.

Glass damage considerations limit the energies produced in Nd-glass lasers to less than 3 J/cm^2. Hence high-powered systems designed for ICF driver applications utilize multiple beams. Several laboratories in the United States and abroad have Nd laser systems operating or under development that approach tens of TW in power level. The timetable for reaching breakeven at the Lawrence Livermore Laboratory along with the laser systems that will accomplish this are shown in Figure 7.10. Amplifiers mounted on the space frame of the Shiva laser system are pictured in Figure 7.11. The Nova system,

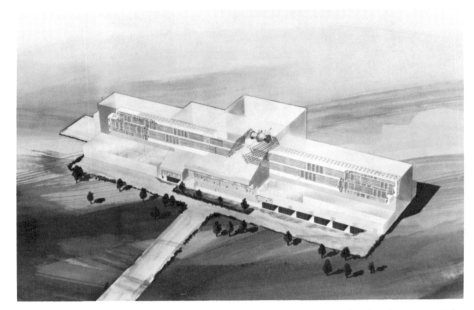

Figure 7.12. An artist's schematic of the Nova laser system under development at the Lawrence Livermore Laboratory. (Courtesy of Lawrence Livermore Laboratory.)

presently under development at the Lawrence Livermore Laboratory, is designed to achieve a power level of 200 to 400 TW (See Figure 7.12).

However since Nd lasers must be pumped using flashlamp sources, they are intrinsically very inefficient (less than 0.2%). Furthermore, the pulse repetition rate is very low (from several minutes to several hours between shots). Hence the Nd laser is not a likely candidate for an ICF reactor application. Nevertheless, the advanced state of Nd laser development has led to the extensive use of this driver type in inertial confinement fusion research.

7.2.2. CO_2 LASERS

To achieve the high efficiencies and pulse repetition rates required by reactor applications, it will probably be necessary to use gas lasers. Of primary interest in present research is the CO_2 gas laser that produces light at 10.6 μm wavelength.[5-7]

The spectroscopy of gas atoms and molecules is much simpler than that of atoms bound in solids, since the former can interact only via collisions. For this reason there are no excited states with broad widths—that is, no broad absorption bands such as one finds in neodymium lasers that are capable of absorbing polychromatic photons. Hence excitation of gas lasers by means of flash lamps is very difficult. Instead one must rely on excitation by electron impact and transfer of excitation by resonant atomic collisions.

As atoms are excited into higher energy states and then cascade down to lower energy levels by nonradiative (collisional) transitions, a nonequilibrium situation will arise in which states of longer lifetimes accumulate a larger number of atoms. Such an accumulation is particularly prevalent for those states for which radiative transitions are forbidden by selection rules (so-called "metastable" states). These metastable states play a key role in influencing the competition between excitation and decay rates of the energy levels necessary for successful population inversion.

The most convenient excitation mechanism is that due to inelastic electron atom collisions in an electrical discharge. In such a discharge, electrons are accelerated and suffer inelastic collisions with atoms, thereby exciting them to higher energy states. Such electron collisions can be used to directly achieve a population inversion, but more frequently the electron excitation is redistributed by means of resonant atomic or molecular collisions. Frequently a second gas with energy levels near to those of the lasing gas is introduced into the discharge to facilitate this energy transfer. This is not essential though, as the Helios laser has no N_2 to suppress parasitic oscillations.

The CO_2 laser utilizes transitions between different vibrational-rotational levels in the CO_2 molecule (see Figure 7.13). The general idea is to use electron inelastic collisions to excite vibrational states of CO_2 and N_2 and then rely on the long lifetime of the CO_2 (001) state and the vibrational excitation exchange occurring in collisions between N_2 and CO_2 to populate the CO_2 (001) state, thereby creating a population inversion with the CO_2 (100) and CO_2 (020)

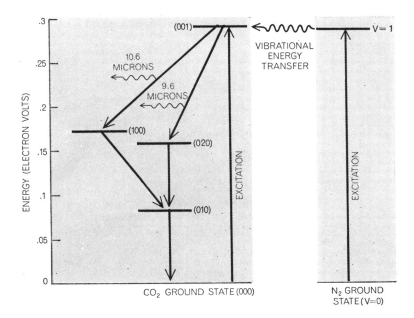

Figure 7.13. The energy levels involved in the CO_2 laser.

states. Lasing action will then occur either from CO_2 (001) to CO_2 (100) (at 10.6 μm) or from CO_2 (001) to CO_2 (020) (at 9.6 μm). Of course, the maintenance of the population inversion relies upon the ability of the lower lasing states to rapidly depopulate in vibrational-translational collisions with either CO_2 (000) or He atoms. In general, the ability of these states to depopulate will diminish as the temperature of the lasing gas rises. Usually, after about a 300°C temperature rise, the population inversion can no longer be maintained, and the lasing action ceases (a "temperature bottleneck").

Low-powered CO_2 lasers are usually designed with the gas mixture placed between two electrodes that produce and sustain an electrical discharge in the gas. The applied electric field is used both to produce the free electrons in the discharge and to accelerate these electrons to energies sufficient to excite the molecular states of the gas. However, using the discharge to perform this dual role limits the control over the electron energy distribution and hence the pumping efficiency.

In high-powered CO_2 lasers one separates the ionization and electron acceleration functions by using a high energy electron beam (*E*-beam) to produce the ionization and then uses an applied drift or "sustainer" field to pump the inelastic collisions. In such *E*-beam lasers, an incident electron beam

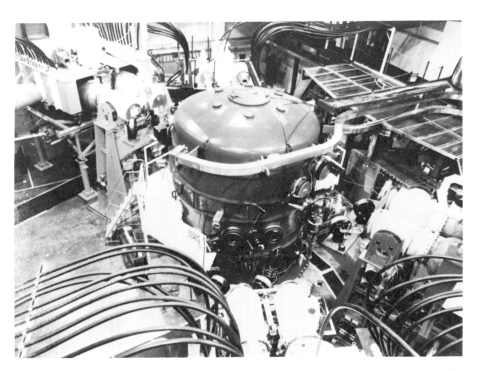

Figure 7.14. The Helios eight-beam CO_2 laser system at Los Alamos Scientific laboratory. (Courtesy of Los Alamos National Laboratory.)

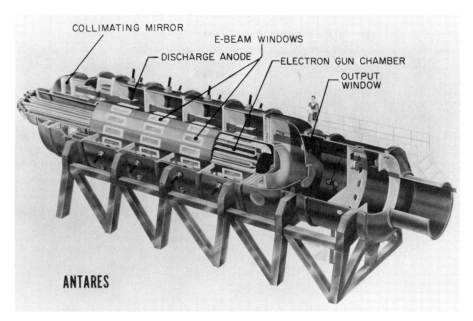

COLLIMATING MIRROR
E-BEAM WINDOWS
DISCHARGE ANODE
ELECTRON GUN CHAMBER
OUTPUT WINDOW
ANTARES

Figure 7.15. A schematic of the amplifier module for the Antares laser system under development at the Los Alamos Scientific Laboratory. (Courtesy of Los Alamos National Laboratory.)

(100 to 300 keV) is used to ionize the laser gas mixture that may be operated at pressures as high as 10 atm. This incident or primary electron beam ionizes the gas mixture, producing a secondary electron density of roughly 10^{13} cm^{-3}. This electron density is primarily limited by electron-ion recombination. The applied field then produces a current density that leads to power dissipation through electron-molecule inelastic collisions.

The Los Alamos Scientific Laboratory has demonstrated single pulse operation at the 10-kJ, 10-TW level with the eight-beam Helios laser system shown in Figure 7.14. The Antares laser system scheduled for completion in 1984 will yield 40 kJ/40 TW (See Figure 7.15). This laser design is modular and should permit scaling to the energies and power levels required for reactor applications. Furthermore, while present CO_2 laser systems designed for ICF applications operate at single pulse efficiencies of 1.5%, there is good reason to expect that this could be improved to 8 to 10% in advanced designs. Since the CO_2 laser can utilize pulse power supplies in the few microsecond range, it does not appear necessary to develop special pulse-forming lines or low inductance pulse charged switches as required by other advanced laser concepts. There also appears to be no fundamental reason why CO_2 lasers could not be operated at the pulse repetition rates (1 to several Hz) envisioned for most ICF reactor designs.

Rather, the primary difficulty faced by the CO_2 laser involves the laser-beam/target plasma interaction at 10.6 μm. We have noted that laser beam energy deposition at long wavelengths leads to energetic electrons that can preheat the fuel, thereby limiting final fuel compression and fusion gain.

7.2.3. THE IODINE LASER

A third laser type under active development and application in ICF experiments is the iodine laser.[8] In this laser photochemical or direct chemical pumping is used to produce excited iodine atoms that lase at a wavelength of 1.315 μm. More precisely, the iodine laser operates on a transition from the $^2P_{1/2}$ state to the ground state $^2P_{3/2}$. This is a forbidden (metastable) transition with a lifetime of the excited level of about 170 ms.

In photochemical iodine lasers,[9] perfluoroalkyliodide gases (C_3F_7I) are used as the parent gas. Upon irradiation with ultraviolet light, these gases produce excited iodine atoms

$$RI + h\nu \rightarrow R + I\left(^2P_{1/2}\right)$$

CF_3I and C_2F_5I are also used. The photon source is usually a xenon flashlamp. By operating at atmospheric pressure or higher, sufficient line width is achieved to allow mode locking and pulse shortening to 1 ns or shorter. The Max Planck Institute in Garching has achieved peak powers of 1 TW in a 280-ps pulse in the Asterix III iodine laser.

An alternative approach is to use a chemical reaction of chlorine, hydrogen peroxide and sodium hydroxide to produce excited O_2. Molecular iodine can then be injected into the excited oxygen mixture and dissociated, thereby producing excited iodine atoms. The chemically pumped iodine laser would eliminate the need for a pulsed power supply, since in principle at least, the pumping energy could be delivered by a continuous flow of chemicals into the laser amplifier cell. Since this laser would operate at a rather low gain, the beam from an optically pumped iodine laser oscillator would be passed through a chemically pumped iodine amplifier many times to achieve sufficient power levels for ICF applications. The overall laser efficiency of a chemically pumped iodine laser, including energy for chemical handling and processing, could be as great as 6%.

7.3. ADVANCED LASER DEVELOPMENT

ICF driver applications place difficult requirements on laser system development. Not only must these systems be capable of high energy (1 to 3 MJ) and power (200 to 500 TW) levels, but they must also operate at efficiencies of 5% or better and pulse repetition rates of several hertz. Furthermore, the laser wavelength must be sufficiently short to allow strong coupling with the target while avoiding the production of suprathermal electrons that could preheat the fuel. None of the laser systems presently in use in ICF research (i.e., neodymium glass, carbon dioxide, and iodine) are close to achieving these goals. Hence some attention has been given to the development of advanced laser systems that appear to offer more potential as ICF drivers.

Advanced laser driver candidates[10] rely on mechanisms that "pump from the top" such as electron beams, photolytic processes, or the conversion of chemical bond energy. In this section we will briefly review several of the more interesting concepts proposed for advanced ICF drivers.

7.3.1. EXCIMER LASERS

One of the most attractive advanced laser concepts is the excimer laser.[11,12] In this laser, a bound molecular state is formed from the association of a ground state atom and an excited state of the same or a similar atom (the "excimer"). This state can then radiate to the unbound ground state formed from the repulsive interaction of the two closed shell atoms.

Of particular interest are excimers formed from rare gases since they are capable of converting the energy from high energy electron beams to specific narrow bands of excited electronic states and then transferring this energy to acceptors. The rare gas excitation process can be outlined as follows:

1. A high energy electron beam is first used to ionize and excite the rare gas atoms (with about 75% of the excited states appearing as ions).

2. Molecular ions R_2^+ are then formed by three-body reactions

$$R^+ + 2R \rightarrow R_2^+ + R$$

3. Dissociative recombination of the molecular ion then occurs

$$R_2^+ + e \rightarrow R^{**} + R$$

4. The highly excited states of the rare gas atom R^{**} are then collisionally deactivated into the lowest metastable state R^* of the rare gas atom.

5. Finally, three-body reactions produce the excited excimers R_2^*

$$R^* + 2R \rightarrow R_2^* + R$$

These excimers can now radiate.

Theoretical calculations and experiments suggest that approximately 50% of the initial electron beam energy deposited in the gas appears in the excited metastable and excimer species. At pressures of one atmosphere this reaction sequence occurs very rapidly, within 10 to 100 ns.

There are four classes of lasers presently being studied as potential ICF advanced drivers. These are listed in Table 7.2. These advanced laser types can be further distinguished by the time scale of their population inversion. In photolytic and rare earth lasers, the energy storing medium in which the

Table 7.2.   Advanced Laser Concepts

Type	Pumping Mechanism	Wavelength (μm)	Efficiency %
Group VI atomic	Optical	0.48	1 to 4
Metal vapor excimers	Discharge	0.33 to 0.47	10 to 15
HF chemical	E-beam	2.6 to 3.4	5 to 10
Rare gas halide	Discharge	0.25 to 0.31	5 to 10
Resonantly excited solid state	Optical	0.28 to 0.45	2 to 6
Optically pumped storage	Optical	0.27 to 0.34	1 to 7

Source. After Stark, Ref. 7.

population inversion occurs is stable for long periods of time in excess of several microseconds. Such lasers store energy for relatively long periods of time compared to energy extraction (lasing) times. Hence these lasers can be pumped for long times, compared to the storage time. However, their effectiveness is limited since the pump species (photons or electrons) can easily deactivate the excited electronic states.

The rare earth and monohalide excimer lasers such as the KrF laser are characterized by highly radiating media with a population inversion decay time of the order of nanoseconds. In order to pump these media for times longer than the required pulse length (10 ns), one must use sophisticated optical extraction techniques (e.g., angular multiplexing) or optical pulse compression techniques.

Photolytic Group VI Lasers. Photolytic lasers utilize an excimer laser such as Xe_2^+ to pump atomic transitions in Group VI elements such as O, S, or Se. For example, one scheme would be to use a xenon excimer pump laser to photolyze OCSe fuel. A diagram of this process is shown in Figure 7.16. These storage lasers utilize the auroral ($^1S \rightarrow {}^1D$) or transauroral ($^1S \rightarrow {}^3P$) transitions in O, S, or Se:

$$S(^1S \rightarrow {}^1D) \qquad 773 \text{ nm}$$

$$S(^1S \rightarrow {}^3P) \qquad 459 \text{ nm}$$

$$Se(^1S \rightarrow {}^1D) \qquad 777 \text{ nm}$$

$$Se(^1S \rightarrow {}^3P) \qquad 489 \text{ nm}$$

The particular photolytic reactions that produce the population inversions are of the form

$$OCS + \gamma \rightarrow S(^1S) + CO \qquad (140 \text{ nm} < \gamma < 160 \text{ nm})$$

$$OCSe + \gamma \rightarrow Se(^1S) + CO \qquad (160 \text{ nm} < \gamma < 180 \text{ nm})$$

A schematic of a xenon excimer pumped OCSe laser is shown in Figure 7.17.

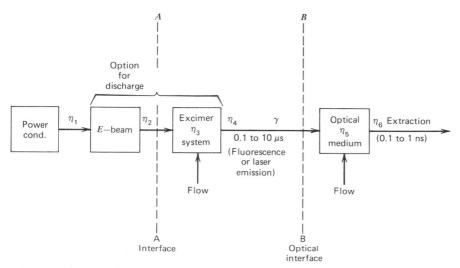

Figure 7.16. A schematic diagram of the operation of a photolytic excimer laser.

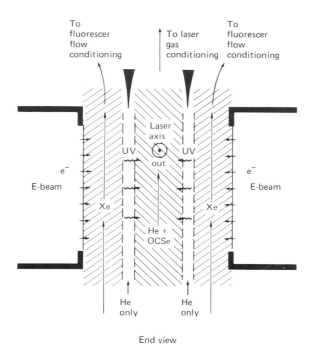

End view

Figure 7.17. A possible design for a xenon excimer pumped O-C-Se laser. (Courtesy of Lawrence Livermore Laboratory.)

Metal Vapor Excimer Lasers. These systems are storage lasers operating on bound-free transitions of excimers formed from lowest metastable states of metal atoms. The principal candidates under investigation are CdHg and Hg_2 lasers. Excitation is provided by an electron beam sustained discharge. Such discharges can efficiently pump metastable atomic states (with possible efficiencies in excess of 10%). In the CdHg laser, the excimer is formed by Cd (^3P) metastable and Hg ground states. The excimer radiates at 470 nm. In the Hg_2 laser, a homonuclear excimer is formed by Hg (^3P) metastable and Hg ground state atoms. This excimer radiates at 335 nm.

Rare Gas Halide Lasers. These are high gain excimer lasers whose upper levels are ion pair states. Examples include ArF (193 nm), KrF (249 nm), XeCl (306 nm), and XeF (353 nm). These lasers are excited by electron beams or electron beam sustained discharges. Since they are highly radiating (short storage times), they must be accompanied by pulse compression or multiplexing to achieve the desired laser pulse lengths.

One can also use rare gas halide lasers to resonantly pump rare earth ions in a solid matrix. The long storage times of the excited ions then permit the rare gas halide pump times to be longer than several microseconds.

7.3.2. THE KrF LASER

Excimer lasers offer the potential of relatively high efficiency and submicron wavelengths to facilitate laser target interaction. Most interest has been in the direct use of an electron-beam excited KrF driver. Since the lasing state has such a short lifetime, KrF cannot be used directly as a storage medium. Therefore the laser power is proportional to the E-beam power, and very large pulse compression is required to produce the high power levels required for ICF applications.

More specifically, KrF amplifies most efficiently pulses of 100 to 200 ns, about 10 to 100 times the pulsewidth considered optimum for driving a target implosion. One scheme for pulse compression utilizes multiple beams passed sequentially through the lasing medium, temporally delayed by different beam paths and appropriate optics, and finally superimposed on the target[13]. A compression factor of 100 would require 100 beams with path length differences of 100 m.

An alternative method of pulse compression would pass the long pulse from a KrF amplifier into a Raman-active medium through which is passed a back-traveling short pulse coupled to the KrF beam by a Raman transition. The short pulse would then be amplified (being pumped by the incident KrF beam). While this process appears capable of high efficiency (60%), it appears limited to a compression factor of about 20 by the onset of secondary processes which deplete the compressed pulse. However additional compression could then be achieved by geometrical methods (described in the previous paragraph).

Although there are many questions concerning the development of high power, short pulse length KrF systems, the attractive laser-target interaction characteristics at the 0.25-μm wavelength produced by this laser type have provided a strong stimulus to proceed with its development. Experiments are being performed using frequency-multiplied Nd light to simulate the KrF wavelength and examine the laser-target interaction phenomena that would be present with this laser driver.

7.3.3.   THE HF LASER

The hydrogen fluorine laser has been studied for a number of years by the Department of Defense. However, we consider it as an advanced laser in the context of ICF because it has not yet been utilized in ICF target experiments.

The HF laser is a chemical laser with a broad band of wavelengths between 2.6 and 3.4 μm.[15] This broad band width characteristic could be very beneficial to laser plasma coupling efficiency, since it might suppress many of the processes that lead to the generation of hot electrons and stimulated back-scattering of the light. This is due in part to the fact that there is no well defined critical density when multiple wavelengths are used. The HF laser uses the chemical reactions

$$F + H_2 \rightarrow HF(v \leqslant 3) + H$$

$$H + F_2 \rightarrow HF(v \leqslant 6) + F$$

$$HF(v, J) \rightarrow HF(v-1, J+1) + h\nu.$$

The lasing energy is supplied by chemical reactions, therefore the electrical efficiency of the HF laser can be over 100%. However, the HF must be reprocessed back into H_2 and F_2 before it can be used again. When this chemical reprocessing is included in the efficiency we find an overall efficiency of 5%. The HF laser amplifier has very high gain and is not a storage medium. Once the gas is excited by electron beam discharge the energy must be extracted or it will be lost. This places severe constraints on the pulsed power equipment that excites the amplifier for it must have very short pulses, on the order of those used in electron or ion beam fusion. The actual laser pulse can be shorted by multiplexing the amplifiers. This means that many shorter pulses traverse the amplifier, one after another, at slightly different angles. In this way, energy is being continuously extracted over 50 ns without resorting to one long laser pulse.

The very high gain of the amplifiers implies that they can be quite compact. It seems possible to build amplifiers one meter in length capable of producing 100-kJ laser pulses.

All of these features make the HF laser a good candidate for laser fusion applications.

7.3.4. SOME FINAL REMARKS

To achieve the ICF target gains necessary for practical applications will require a major advance in laser driver development. It is apparent that none of the laser types presently being used in ICF research approaches the requirements of driver energy and power, efficiency, pulse repetition rate, and driver beam target coupling. Indeed, there is a very real concern that it may be impossible to achieve these goals with these laser systems (i.e., the neodymium glass, carbon dioxide, and iodine lasers).

For that reason there has been a major effort at developing advanced laser systems that exhibit more potential as fusion drivers. Such systems attempt to achieve the high energy storage densities necessary for ICF driver amplifiers by using electron-beam, photolytic, or chemical reaction pumping and to produce beams at the submicron wavelengths felt to be necessary for adequate beam-target coupling using electronic transitions or frequency multiplication of vibrational transitions in molecular lasers. Of most interest has been the class of excimer lasers such as the KrF laser that operate in the 0.2- to 0.4-μm wavelength range and appear to be capable of high energy/power levels and reasonable efficiencies.

The HF laser with its broad bandwidth may also show better laser-plasma coupling. It has the potential for $\sim 5\%$ efficiency and very compact amplifier design that is attractive for reactor applications. The free electron laser also presents a potential for meeting many of the stringent requirements for ICF drivers. However these advanced laser concepts will require a major development effort over the next several years to determine whether they can achieve their potential as ICF drivers.

In the meantime, ICF target experiments will continue with existing laser types. In particular, the Nd-glass laser will continue to be the mainstay of ICF target interaction and implosion research, although more attention may be directed at frequency-multiplying the 1.06-μm light to simulate advanced laser concepts such as the KrF laser. The CO_2 and I laser development programs will also continue, in the hopes that further target interaction experiments at long wavelength will reveal methods for redesigning ICF targets to avoid the interaction phenomena such as suprathermal electron production that plague these laser drivers.

REFERENCES

1. B. A. Lengyel, *Lasers*, 2nd ed. (New York, Wiley-Interscience, 1971).

2. M. Sargent, M. O. Scully, W. E. Lamb, *Laser Physics* (Reading, Mass., Addision-Wesley, 1974).

3. *Lasers and Light*, Readings from Scientific American (San Francisco, W. H. Freeman and Company, 1969).

4. Lawrence Livermore Laboratories Laser Fusion Annual Reports.

5. C. K. N. Patel, "High Power Carbon Dioxide Lasers," *Sci. Am.* **219**, 22 (August, 1968).

6. Los Alamos Scientific Laboratory Laser Fusion Annual Reports.

7. E. E. Stark, Jr., "Lasers and Power Systems for Inertial Confinement Fusion Reactors," Los Alamos Scientific Laboratory Report LA-UR-78-1350 (1978).

8. K. Hohla, "The Iodine Laser: A High Power Gas Laser," in *Third Workshop on Laser Interaction and Related Plasma Phenomena*, Vol 3A (New York, Plenum, 1974), p. 133.

9. "Purely Chemical 1.3μm Iodine Laser Emerges as Fusion-Driver Candidate," *Laser Focus* **15**, 24 (June, 1979)

10. K. A. Brueckner, An Assessment of Drivers and Reactors for Inertial Confinement Fusion, K. A. Brueckner Associates, prepared for the Electric Power Research Institute, EPRI-AP-1371 (1980).

11. P. Hoff, "Laser Fusion Advanced Laser Program," Lectures presented at the AUA-ANL Faculty Workshop on Inertial Confinement Fusion, Argonne National Laboratory, 1978.

12. C. K. Rhodes, Ed., *Excimer Lasers, Topics in Applied Physics*. Vol. **30** (Berlin, Springer-Verlag, 1979).

13. R. M. Hill, D. L. Huestis, and C. K. Rhodes, "Review of High Energy Visible and UV Lasers," *Laser Induced Fusion and X-ray Laser Studies*, (Reading Mass, Addison-Wesley, 1976).

14. "Inertial Confinement Fusion–An LF Meeting Review," *Laser Focus* **16**, 58 (February, 1980).

15. G. Cooper, "HF Laser Design," in *SOLASE-H, A Laser Fusion Hybrid Study*, Univ. of Wisconsin Fusion Engineering Program Report UWFDM-270, May 1979.

EIGHT

Driver Development II: Particle Beams

Present estimates of inertial confinement fusion driver requirements tend to cluster in energy between 1 and 10 MJ and in power from 100 to 1000 TW. Laser drivers are particularly well suited to the task of delivering very high power intensities in short pulses focused on small targets. Unfortunately, however, the task of building laser drivers capable of producing the necessary pulse energies has proven extremely costly. There is a very real concern that the costs of scaling laser pulse energies to the levels of 1 to 10 MJ projected for high gain ICF targets may well eliminate laser drivers as we know them today as a suitable option for reactor applications. In a sense, lasers are power rich and energy poor devices. This feature arises from the low efficiency of most pumping schemes used to convert electrical energy into light energy. Most advanced laser driver designs project efficiencies of the order of 1 to 5%.

The primary source of the inefficiency in laser drivers can be attributed to the laser power amplifiers, rather than the pulsed electrical power systems used to pump the amplifiers. In fact, the efficiency of the pulsed power systems used to excite laser amplifiers is quite high, typically ranging from 80% to 90%. This very high efficiency of pulsed electrical power sources leads one to seek an alternative to the laser for an ICF driver. That is, we seek an alternative type of driver that could eliminate the intermediate (and inefficient) stage of converting the electrical energy produced in a pulsed power source into light to be focused onto an ICF target.

Charged particle beams provide us with a means to couple the energy of pulsed electrical power sources more directly into the ICF target. That is, the electrical energy produced in a pulsed power source can be efficiently converted into charged particle energy by accelerating electrons or ions across a potential difference. These charged particle beams can then be focused on the ICF target.[1-24]

Pulsed power accelerators have been used for many years as intense sources of X-rays. In these accelerators, a capacitor bank is used to store energy at high voltage. The capacitors are then discharged through switches into an insulated pulse-forming line to produce a short pulse of electromagnetic power. This power is propagated through a transmission line to a diode to produce an intense electron beam between a dense plasma that forms on the metal surface of the cathode and the anode. If the anode is formed from a thin foil, the relativistic electrons striking the metal foil anode produce copious quantities of hard bremsstrahlung radiation. Such pulsed power E-beam accelerators have been developed as intense X-ray sources at a number of laboratories, including Sandia Laboratory, Physics International, Maxwell Laboratories, Harry Diamond Research Laboratories, the Naval Research Laboratory, and the Air Force Weapons Laboratory.

In recent years, Sandia Laboratory (Albuquerque) has applied this pulsed power diode accelerator technology to the development of ICF drivers. In particular, in devices such as Proto I and Proto II, intense electron beams have been produced and focused onto ICF targets. More recently pulsed power diode accelerators such as the Particle Beam Fusion Accelerator, PBFA-I, and a follow-on, PBFA-II, have been designed primarily for use as ICF drivers.

At the Naval Research Laboratory pioneering research in diode physics has led to the first production of intense ion beams. These beams have been focused into plasma channels and propagated at high current densities for over a meter in length. This work is fundamental to the eventual application of light ion beam drivers to high gain targets. Important studies of ion beam production have also been underway for many years at Cornell University.

Pulsed power diode accelerators at Sandia have also been modified to accelerate light ions (hydrogen through carbon) as well as electrons. As we noted in Chapter 5, light ions present several significant advantages over electron beams as ICF drivers. For example, ion beams couple energy into the target far more effectively than either laser or electron beams. Ion beams are also easier to propagate from a standoff diode to the target. Conversion from electron to light ion beams has required only minor modification of machines originally designed as electron beam accelerators.

Interest in ion beam drivers has spread to heavy ion accelerators based upon the technology developed for high energy physics research. Heavy ions (xenon through uranium) can be accelerated in bunches to 1 to 10-GeV energies in multiple RF (radio-frequency) cavities and accumulated in charged particle storage rings. Once sufficient particle densities have been accumulated, the beams can be "kicked out" of the storage rings and focussed by magnetic fields onto the ICF target. Once again the potential conversion efficiency of electrical energy to beam energy is very high in such devices. Furthermore, we noted in Chapter 5 that heavy ion beams also exhibit favorable target interaction features. Heavy ion accelerators have the added advantage that high repetition rates are easily obtained.

In this chapter we discuss each of these two quite different approaches to particle beam acceleration. We begin our discussion with pulsed power accelerators, since this technology is already not only highly developed, but is actively being applied to ICF studies in large devices such as the Particle Beam Fusion Accelerator at Sandia Laboratory. While the heavy ion beam accelerator approach is still in its infancy, at least as far as its application to ICF research is concerned, it is based on a highly developed accelerator technology from high energy physics. Furthermore, the beam-target interaction of heavy ion drivers and their repetition rate capabilities have stimulated a strong interest in this technology for ICF applications.

In summary then, while the present estimates of ICF driver power and energy requirements present a serious challenge to high-power laser development, charged particle beam accelerators seem capable of achieving such powers and energies with only a mild extrapolation of existing technology. The ability of charged particle accelerators, whether based on pulsed power diode or RF-accelerator/storage ring devices, to produce beams of the necessary energy and power with existing technology has motivated the serious attention currently being given to particle beams as ICF drivers.

8.1. PULSED POWER DIODE ACCELERATORS

8.1.1. GENERAL FEATURES

Pulsed power diode accelerators are not charged particle accelerators in the usual sense familiar from high energy physics research. Rather they can be regarded as electromagnetic pulse compressors. The general operation of these devices involves the discharge of electrical energy from capacitive storage devices known as Marx generators into a pulse-forming line where a short, high power pulse is formed and then applied to a diode through a transmission line. An intense electromagnetic wave sweeps inward along the transmission line and emerges on a pair of face-to-face particle accelerating electrodes. One electrode, the cathode, is pulse charged negatively with respect to the other electrode, the anode. When millions of volts are applied to the electrodes, the electric fields produced are sufficient to draw electrons out of the cathode material and into the vacuum. Electrons drawn from the cathode dissipate enough energy in both the cathode and anode to vaporize their surface layers and form plasmas. The cathode plasma becomes the electron source and the anode plasma provides a source of positive charge to neutralize the electrostatic field of the beam.

The basic components of a pulsed power diode accelerator include (see Figure 8.1):

1. *Capacitive energy storage system.* This is typically a Marx generator that functions both as a storage device and a voltage multiplier. The Marx

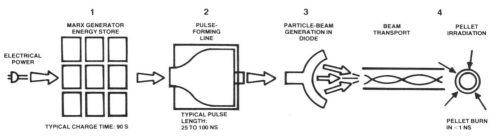

Figure 8.1. Basic components of a pulsed power diode accelerator.

generator charges capacitors in parallel and then discharges them in series to achieve a very high voltage. The rise time of the pulse produced by the Marx generator is too slow for ICF driver applications, so one must next compress the pulse in time.

2. *Pulse-forming lines.* A pulse-forming network is used to compress the electrical energy into a short, fast pulse, roughly doubling the output voltage in the process. This can be either a single pulse line or a more complex transmission device such as a Blumlein network.

3. *Transmission line.*   The energy pulse then travels as an intense electromagnetic wave down a transmission line to the diode. The high-voltage insulation of the transmission line is of particular concern since it determines the maximum voltage pulse allowable.

4. *Diode.*   The diode consists of a cathode and an anode foil, separated by a short gap. When the high voltage pulse arrives at the cathode, its voltage causes intense field emission from the cathode tip which produces a plasma on the surface of the cathode and leads to the production of an intense stream of electrons moving toward the anode. At high energies the electrons can easily penetrate a thin anode foil and continue on to form a relativistic electron beam with beam currents of up to several megaamperes. The diode polarity can be reversed and the electron current suppressed to produce an ion beam in the device.

To illustrate such a pulsed power diode accelerator, let us consider the Particle Beam Fusion Accelerator (PBFA-I) device at Sandia Laboratory (Albuquerque).[25] This device, shown schematically in Figure 8.2, is typical of pulsed power accelerators that can produce either relativistic electron beams (1 to 10 MeV) or light ion beams (H to C at energies of several MeV). The PBFA-I accelerator is modular in design, consisting of 36 pulsed power modules connected in parallel. Each module consists of a series of energy storage devices separated by synchronized switches, as shown in Figure 8.3. The primary energy storage devices used in such pulsed power accelerators are Marx generators. These consist of a set of capacitors that are charged to several hundred kilovolts in parallel and then discharged in series to obtain a 3.2-MV source. The switching is accomplished with triggered gas switches. In

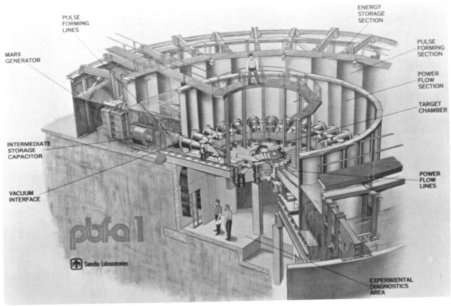

Figure 8.2. The Particle Beam Fusion Accelerator (PBFA-1) at the Sandia National Laboratory.

PBFA-I the energy is transferred from the Marx generators in 650 ns at a peak module power of 2×10^{11} W.

The pulse length of the output of the Marx generators is too long and the power too low for ICF applications. Hence the electromagnetic pulse produced by discharging the Marx generators must be shortened in a pulse-forming line (PFL). However the transfer time is also too long to charge the pulse-forming line directly. Instead, a water-filled dielectric intermediate storage capacitor is charged by the Marx generator. This energy is then transferred to the PFL in 250 ns by a single triggered gas switch for each of the 36 modules. These switches provide the timing for synchronizing all of the modules, being characterized by a RMS uncertainty in switching time of 1.6 ns. Each module has two parallel pulse-forming lines. The lines are switched into a wave mixer with multichannel water dielectric self-triggered switches. The output pulse has a duration of 40 ns. This surge flows through a pulse conditioner, then through a water-vacuum interface, and finally along a magnetically self-insulated transmission line.

It is interesting to note the difference in dielectric media at each stage of the pulse formation and compression process. The Marx generators are submerged in transformer oil. The intermediate storage capacitors, triggered gas switches, pulse-forming lines, output switches, and pulse conditioners are in water. The power flows through plastic insulators into vacuum at a distance of 6 m from the target. Power is channeled through long self-magnetically insulated transmission lines to the diodes. The diodes convert the electromagnetic energy in

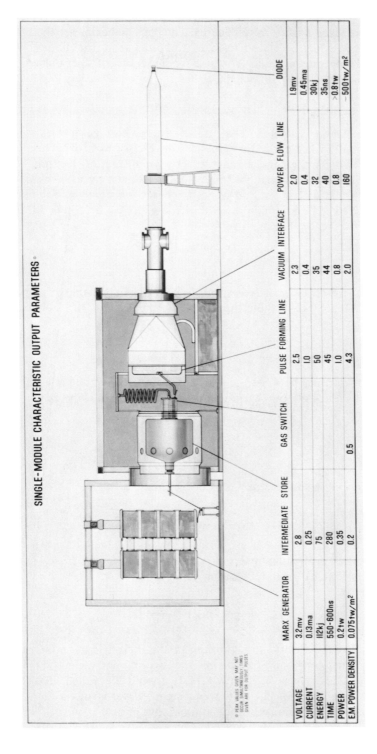

SINGLE-MODULE CHARACTERISTIC OUTPUT PARAMETERS*

* PEAK VALUES GIVEN MAY NOT
OCCUR SIMULTANEOUSLY TIMES
GIVEN ARE FOR OUTPUT PULSES

	MARX GENERATOR	INTERMEDIATE STORE	GAS SWITCH	PULSE FORMING LINE	VACUUM INTERFACE	POWER FLOW LINE	DIODE
VOLTAGE	3.2mv	2.8		2.5	2.3	2.0	1.9mv
CURRENT	0.13ma	0.25		1.0	0.4	0.4	0.45ma
ENERGY	112kj	75		50	35	32	30kj
TIME	550-600ns	280		45	44	40	35ns
POWER	0.2tw	0.35	0.5	1.0	0.8	0.8	>0.8tw
E.M. POWER DENSITY	0.075tw/m²	0.2		4.3	2.0	160	~500tw/m²

Figure 8.3. Pulsed power modules used in PBFA-1. (Courtesy of Sandia National Laboratory.)

the pulse to charged particle beam energy. The particle beams are then focused onto the target, which is placed from 20 to 50 cm from the diodes. We will consider in more detail each of these components of the pulsed power amplification and compression chain.

8.1.2. COMPONENTS OF PULSED POWER ACCELERATORS

Capacitive Energy Storage.   The primary energy storage devices in pulsed power accelerators are capacitors. In its simplest form, a capacitor can be pictured as two parallel plates separated by an insulating medium. Let us suppose for the moment that the plates are separated by a vacuum. If the plates are charged with equal and opposite charge density $\pm\sigma$ coulombs/cm^2 and are assumed to be infinite in extent, then the electric field between them is given by

$$E = \frac{\sigma}{\varepsilon_0}$$

where ε_0 is the emissivity of free space. We can calculate the corresponding potential difference between the capacitor plates as

$$V = Ed = \frac{\sigma d}{\varepsilon_0} = \frac{dQ}{\varepsilon_0 A} = \frac{Q}{C}$$

where d is the distance between the plates, A is the area of the plates, and $\pm Q$ is the charge on each plate. We can identify the coefficient of the charge as just the capacitance C

$$C = \frac{\varepsilon_0 A}{d}$$

The energy stored in the electric field between the plates is

$$\text{E} = \frac{1}{2}\frac{Q^2}{C} = \tfrac{1}{2}CV^2$$

If a dielectric (insulating) material is placed between the plates, then the capacitance is changed from its vacuum value to

$$C = \frac{\varepsilon_0 A}{d}(1+\chi) = \frac{\kappa\varepsilon_0 A}{d}$$

Here χ is called the electric susceptibility of the dielectric material, and κ is called its dielectric constant. Properties of the two common dielectric materials, oil and water, are given in Table 8.1. It should be noted that on the microsecond time scales of pulsed power systems, water is an excellent insulator (with a dielectric constant roughly 30 times that of oil).

Table 8.1 Properties of Oil and Water as Dielectrics

	Oil	Water
Dielectric constant	2.3	80
Coaxial impedance	$40 \ln r_2/r_1$	$6.7 \ln r_2/r_1$
Useful field strength (positive electrode)	200 to 300 kV/cm	100 to 150 kV/cm
Energy density (J/l)	4 to 9	35 to 80
Current density (kA/m)	80 to 120	240 to 360
Polarity effect	Variable $\approx 1.5:1$	$2:1$

Common properties
Breakdown is:
1. Self-healing
2. Time dependent $(t^{-1/3})$
3. Electrode initiated
4. Electrode-surface dependent
5. Variable, hence area dependent

Marx generators consist of a bank of capacitors that can be charged in parallel to low voltages (100 kV) and discharged in series to provide high voltages (1 to 10 MV). Hence the capacitors comprising a Marx generator will experience substantial stress from high voltages only during their short discharge phase. In this way very high storage densities can be achieved. A Marx generator is shown in Figure 8.3.

Intermediate storage capacitors (shown in Figure 8.3) are often coaxial rather than flat plates. They most often use water as a dielectric medium so that they can store more energy than an equivalent sized oil dielectric capacitor.

Switches. Once the Marx generators or intermediate storage capacitors have been charged, the energy must be rapidly switched out to achieve short pulses. Switches are therefore a critical element in the design of pulsed power accelerators. Switches are generally spark gaps designed to discharge between two electrodes when some programmed initiation mechanism is activated. For example, a short voltage pulse might be used to initiate the discharge, thereby closing the switch. In this case the voltage across the switch would be less than the breakdown voltage until the external voltage is applied. Switches use various dielectric materials to insulate the electrodes from one another. High pressure gas such as air or SF_6 are commonly used in so-called gas switches.

Switches are also very important in affecting the shape of the current pulse. When extremely short pulses are required, the switching time becomes critical to the rise time of the pulse. The rise time is determined by the inductance of the switch, since inductance limits the time rate of change of the current. The inductance can be decreased by increasing the number of breakdown channels in the switch.

Self-triggered switches discharge when the voltage drop across the electrodes exceeds their breakdown value. These switches are most useful for transferring energy from the rapidly charged final pulse forming lines that feed the load. This self-triggered breakdown is common for water or oil switches. These switches are much smaller than gas switches. This can pose a critical problem at the pulse-forming line stage of the pulse compression.

Pulse-Forming Lines. Pulse-forming lines (PFL) are yet another form of energy storage device that serve to shorten the pulse length and thus increase its power.[26-29] In a simple coaxial transmission line the output voltage is one-half of the input voltage, but the output current is equal to the switch current. In the Blumlein triaxial transmission line, this situation is reversed. In practice, the simple coaxial PFL has been associated with low impedance systems and water dielectrics, and the Blumlein with high impedances and oil dielectrics.

The pulse-forming line length is determined by the desired electrical pulse length. This is twice the length of the line divided by the speed of the electrical signal in the dielectric. This can be understood by recognizing that once the output switch is closed, L/v_s seconds are required to propagate this informa- tion to the far end of the transmission line, and L/v_s seconds are required for this charge to flow to the switch.

The staged pulse compression that is produced by the synchronized switch- ing of power between the Marx generator, intermediate storage capacitor, pulse-forming line, and finally the transmission line in a single PBFA-I module is shown in Figure 8.3.

Self-Magnetically Insulated Transmission Lines. The sequence of power flow through oil and water dielectrics into the vacuum interface before the diode is called the power flow chain. At each stage of this chain the medium has an electrical breakdown strength that is determined by the medium and the pulse length, among other factors. The weakest link in this chain is the vacuum interface between the PFL and the diode. To prevent breakdown at the very high power levels required for ICF driver applications, many square meters of insulator surface area would be required with conventional dielectrics (oil or water). Simple geometric considerations imply that this vacuum interface must be many meters from the target, and hence power must be transported over this distance. This problem can be solved by the use of self-magnetically insulated transmission lines.[30-33]

If a voltage is applied to a vacuum dielectric transmission line, only the displacement current flows if the electric field is less than 25 MV/m. This line behaves as a classical transmission line. If the voltage increases, and the electric field exceeds 25 MV/m, a conduction current flows across the vacuum gap. If the voltage source has sufficiently low impedance, the current increases with the loss current until it reaches a critical value at which electrons are deflected in the self-generated magnetic field of the current and are thereby prevented from reaching the anode surface (see Figure 8.4). Once the initial loss current is established, the power flow in this self-magnetically insulated transmission line is very efficient ($\sim 100\%$).

In PBFA-I, the same power that would require 4000 cm^2 of insulator surface area in a traditional transmission line can be carried by 50 cm^2 of magnetically insulated transmission line at an electrical stress of 2 MV/cm and a power density of 16 GW/cm^2. After the lossy front has propagated down the line at half the speed of light, the power transport is 100% efficient. This increased power density is the key to scaling pulsed power machines to the power levels required by ICF driver applications.

Diodes.   The electromagnetic pulse energy is converted into charged particle beam energy in a diode.[15–17,34–40] The simplest example is provided by the electron emitting diode shown in Figure 8.5. When the high voltage pulse is applied across the anode-cathode gap, a breakdown process begins. Microscopic whiskers or imperfections that typically cover the cathode explode, forming a dense plasma. Because of the electric field enhancement on these

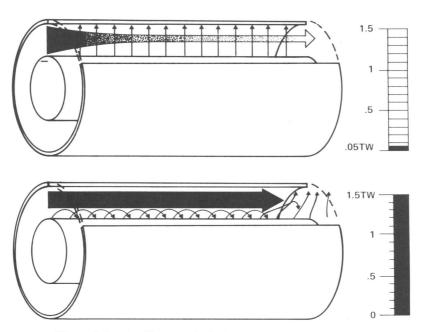

Figure 8.4.   A self-magnetically insulated transmission line.

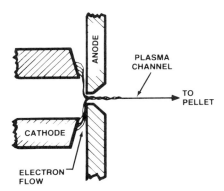

Figure 8.5. An electron emitting diode.

protrusions, the initial field emission current density is of the order of 10^9 A/cm^2. This leads to rapid resistive heating and vaporization of these whiskers. The vapor is easily ionized and heated, and explodes toward the anode at a velocity of 2 to 3×10^6 cm/s.

If the current density from regions of field enhancement on the cathode is sufficiently high, then anode material can be vaporized and move toward the cathode. The anode and cathode plasmas act as virtual electrodes, thereby decreasing the effective region of electron acceleration and the subsequent diode impedance. The cathode emission is from the dense expanding plasma.

For low currents, electrons stream across the gap reaching the anode with small angles of incidence. As the current is increased, the angle of intersection with the anode increases because of the self-magnetic field of the beam. This critical pinch current can be expressed as

$$I_c = 8.5 \times 10^3 \gamma \beta R / d \qquad \text{(amperes)}$$

where $\gamma = E/mc^2 + 1$, R is the cathode radius, d is the gap width, and $\beta = (1 - \gamma^2)^{1/2}$.

In early electron beam fusion experiments, the target was mounted directly onto the anode, where it was bathed in relativistic electrons swarming from the cathode. A great deal of work has been devoted to the design of these electron diodes to achieve efficient conversion of electromagnetic pulse energy into relativistic electron energy. This transfer can be accomplished with efficiencies approaching 100%.

Such pulsed power diode electron beam accelerators can be applied to the production of hard X rays quite readily. These machines are relatively inexpensive, wall-plug to electron power conversion is high, and the required power levels are attainable. However, for fusion applications many problems remain. There appear to be limitations on the electron current that can be focused onto a target due to magnetic field and space charge effects near the target. The electron beam–target interaction also raises serious questions. The classical energy deposition range of relativistic electrons is quite large. Furthermore, these electrons produce appreciable bremsstrahlung radiation during energy deposition, which can penetrate into the target, preheating the fuel. These beam propagation and energy deposition problems cast serious doubt about the suitability of electron beam drivers to inertial confinement fusion.

These problems have been mitigated to some extent with the possibility of accelerating light ions rather than electrons in pulsed power accelerators. This is made possible through ion diode designs and a reversal of the machine polarity. Both of these modifications can be implemented rather easily, so that pulsed power devices can serve a dual purpose of accelerating electrons for X-ray production and ions for fusion target irradiation.

It has been known for some time that pulsed power diode accelerators could be used to produce ion beams. In fact, in many instances high energy ions have been found traveling along with the electron beams even when such devices are operated in the electron beam mode. This occurs because if the electron current density is high enough (greater than several kA/cm^2), then the anode material and anode surface contaminants are strongly heated and turned into a plasma sheet before the end of the pulse. The diode's electric field then pulls positively charged ions from this anode plasma and accelerates them to the cathode, thereby producing a very powerful ion beam.

Hence to produce ion beams for use as fusion drivers, one must first reverse the polarity of the diode. Then the electron flow must be suppressed, since in normal diode operation, the small mass of the electrons compared to the ions causes the majority of the current to be carried by this species. For example, if the ions were protons, then the proton current (and therefore the proton beam power) would only be about 2% that of the electron beam.

One scheme for suppressing the electron flow from the cathode is to impose an external magnetic field on the anode-cathode gap that is strong enough to impede the electron flow across the gap. This is shown schematically in Figure 8.6. With the electrons orbiting the field lines, ions introduced in the vicinity of the anode from a plasma layer or an external injector are accelerated toward the negative electron cloud. Because of the large ion mass, the ion current is not affected by the magnetic field. By shaping the ion source region properly, the ions can be ballistically focused through the anode onto a target or into the end of a plasma transport channel. In a variation of this idea, the magnetic field can be generated by currents in the diode itself.

Ion diode research currently centers on improving the power density brightness factor, JV/θ^2, where J is the ion source current density, V is the accelerating voltage, and θ is the divergence of the beam. To improve this figure of merit, we see that an increase in voltage is desirable. This will allow larger anode-cathode gaps while maintaining a high value of J. If J is kept the same and V is increased then the stiffer beam will be less susceptible to defocusing effects due to structure in the diode. For these reasons, the power brightness is expected to scale as $V^{2.5}$. With these high voltages (4 to 10 MV rather than 1 to 2 MV), it is also desirable to accelerate more massive ions such as helium or carbon because they are less affected by magnetic fields in the

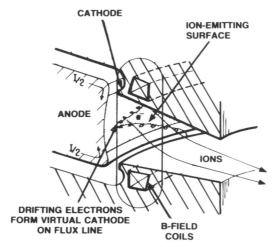

Figure 8.6. An ion beam diode.

diode. This improves focussability. Furthermore, to maintain a constant ion range in the target ablator for high voltage (kinetic energy) ions, the Z value of the ion must increase. (Recall the discussion in Chapter 5.)

Particle Beam Propagation. A major concern arising in the application of pulsed power accelerators (either electron or ion beam) as ICF drivers involves the propagation of the particle beam from standoff diodes to the target.[19,41-47] The focusing of the beams onto the target involves a competition between the momenta of the particles as they are aimed at the target and the repulsive Coulomb or space charge forces that arise between the particles as they are squeezed together near the target. These competing effects limit the focal spot size and the distance over which the beams can be propagated.

Space charge effects rapidly cause beam spreading and defocusing for electron beam propagation into a vacuum. However, by allowing the beam to propagate through a low pressure gas (air), ion production neutralizes the space charge of the beam and permits it to propagate. The electron beam propagation also generates a strong magnetic field that can pinch down the beam radius. If the beam current is too high, however, instabilities can develop and the beam does not propagate. However, by allowing the beam to propagate through a plasma channel, the return current will allow currents in excess of this critical limit (although the net current is still below the limit).

Even with such plasma channels, it is now felt that space charge effects may prevent electron beam propagation over the distances or focusing to the intensities required by ICF applications. The situation appears more hopeful for light ions. Here the mass and hence the momentum of the particles is large so there is hope that they can be propagated ballistically over distances up to 50 cm in the megaampere current range. However this is still insufficient for reactor applications since a diode at a standoff distance of only 50 cm would suffer major damage from each ICF shot. A possible solution to this problem is to use very many ion beams so that the current in each beam is small, and the space charge effects are not so pronounced. Such schemes have been proposed, but they are not compatible with the high current pulsed power approach to particle acceleration. Rather they appear more suitable for the particle accelerator approach of high energy physics that is discussed later in this chapter.

A method that is more suited to pulsed power technology involves the propagation of the ion beam in a preformed, ionized plasma channel as shown in Figure 8.7. In this scheme the plasma channel is initiated by preionizing a fine line of gas with an exploding wire or laser beam. A capacitor bank is then discharged through this ionized path, creating a hot plasma channel with a current of roughly 50 kA. This current establishes a magnetic field around the channel. The hydrodynamics of the discharge pushes gas from the channel, forming a high density layer at the outer channel radius. Ions (or electrons) are then injected into the end of the channel where they are trapped by the magnetic field as they propagate toward the target at the other end. The relatively high density of the channel compared to the beam density neutralizes the beam space charge while the current in the channel neutralizes the beam current. These effects negate the space charge problems encountered in vacuum propagation.

Ion orbits in the channel can be estimated if we assume that the beam is entirely space and current neutralized.[19] In this case the single ion betatron orbits are computed as follows: If we model the channel current by a uniform profile, the azimuthal magnetic field is given by

$$B = B_0 \frac{r}{r_c} \qquad r < r_c$$

$$= B_0 \frac{r_c}{r} \qquad r > r_c$$

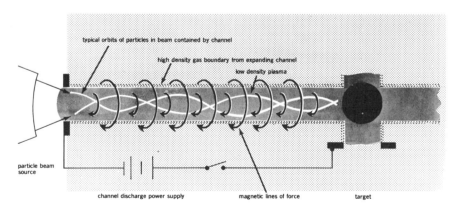

Figure 8.7. Propagation of an ion beam in an ionized plasma channel. (Courtesy of Sandia National Laboratory.)

where r_c is the channel radius. The equations of motion for an ion confined in such a channel are

$$\ddot{r} = -\omega_{cb}\dot{z}\frac{r}{r_c}$$

$$\ddot{z} = \omega_{cb}\frac{r\dot{r}}{r_c}$$

where $\omega_{cb} = eB_0/m_ic$, the cyclotron frequency of the beam ion. For the case of $\dot{r}/\dot{z} \ll 1$, an expansion can be used to solve these equations, giving

$$z = \left(V_0\cos\alpha_0 - \frac{\omega_{cb}\bar{r}^2}{4r_c}\cos 2\phi\right)t$$

$$+ \frac{\bar{r}^2}{8r_c}\left(\frac{r_c\omega_{cb}}{V_0\cos\alpha_0}\right)^{1/2}\left[\sin 2\phi(\omega_\beta t + \phi) - \sin 2\phi\right]$$

$$r = \bar{r}\cos(\omega_\beta t + \phi)$$

where

$$\omega_\beta = \Omega\left(1 - \frac{\omega_{cb}r^2}{16r_cV_0\cos\alpha_0} + \frac{\tan^2\alpha_0}{4}\right)$$

$$\tan\phi = -\left(\frac{r_cV_0\sin^2\alpha_0}{\omega_{cb}r_0^2}\right)^{1/2}\tan\alpha_0$$

$$\bar{r} = \left(r_0^2 + \frac{r_cV_0\sin^2\alpha_0}{\omega_{cb}\cos\alpha_0}\right)^{1/2}$$

and $r(0) = r_0$, $z(0) = z_0$, $\dot{r}(0) = V_0\sin\alpha_0$, and $\dot{z}(0) = V_0\cos\alpha_0$. V_0 is the injection velocity, α_0 the injection angle, and $\Omega^2 = \omega_{cb}V_0/r_c$. Since ions should be confined in the channel, a relationship between the maximum angle of injection α_m and the current in the channel is

$$I_0(A) \geq \left(1.57\times 10^7\frac{\mu\alpha_m^2V_0}{c}\right)\left(1 - \frac{r_s^2}{r_c^2}\right)^{-1}$$

where μ is the ratio of the ion mass to the proton mass, and r_s is the beam spot size at injection. For a 2 MeV proton beam with $\alpha_m = 0.2$ rad, $r_s = 0.4$ cm, and $r_c = 0.6$ cm, we find $I_0 \geq 180$ kA.

The z motion of the ions has both a streaming and an oscillatory component. Hence ions injected at the same time will arrive at the target at different times.

**Table 8.2   Energy Loss of Protons in a 4-m Long Plasma Channel
of Density $= 10/760$ atm
(Corresponds to 50 torr Chamber Prefill)**

Gas	Energy Loss for 2-MeV Proton (MeV)	Energy Loss for 4-MeV Proton (MeV)	Energy Loss for 10-MeV Proton (MeV)
He	0.16	0.09	0.04
Li	0.24	0.14	0.07
N_2	0.90	0.55	0.26
Ne	0.55	0.36	0.18
Ar	0.90	0.55	0.28
Xe	1.80	1.20	0.65

This spread in arrival time is given by

$$(\Delta t_a)_s = \left(\frac{Z_t}{4V_0} \right) (\alpha_m^2 + \omega_{cb} r_s^2 / V_0 r_c)$$

where Z_t is the channel length. This time spreading can be very important for beam-bunching considerations. If the diode voltage is ramped so that the end of the ion pulse has a higher velocity than the beginning of the pulse, then the later ions can overtake the earlier ions on route to the target. This axial beam compression or bunching can shorten the pulse and hence increase the power by as much as a factor of 3 to 5. In this case the actual pulse length is determined by the accuracy of the voltage ramp and the spread in arrival time.

There are many other effects that determine the efficiency of the beam propagation. These include beam rippling, bumpy channels, electric fields that slow the beam, and possible instabilities arising from nonperfect charge and current neutralization. Table 8.2 indicates the loss of beam energy in channels of differing ionized gases. Channels have been created in the laboratory using both fine wires, laser beams and wall confined discharges, and ions have been propagated through these channels, thus demonstrating the feasibility of this approach.

8.1.3.   SUMMARY OF PULSED POWER ACCELERATORS

We have noted that while pulsed power diode accelerators have been operated since the mid-1960s as intense X-ray sources, they have only recently been applied to inertial confinement fusion research. A list of the most notable machines, past, present, and future, along with their operating characteristics is given in Table 8.3. Many of these machines are serving double duty as X-ray generators and test-stands for fusion related experiments. The major programmatic effort in pulsed power accelerators is centered at Sandia Laboratory (Albuquerque) where actual ICF target implosions have been driven with both

Table 8.3   A Brief List of Pulsed Diode Accelerators

Machine	Location	P (TW)	E (kJ)	V (MV)	I (MA)	T (ns)
Proto I	SNL	1.1	22	2.0	0.55	24
Proto II	SNL	8	100	1.6	5	22
PBFA I	SNL	30	1000	2.0	15	42
PBFA II	SNL	100	4000	4.0	25	35
HYDRA	SNL	0.4	35	1.0	0.4	80
BLACKJACK 5	Maxwell	10	1000	2.0	5.0	100
AURORA	Harry Diamond	20	2500	15.0	1.6	125
GAMBLE II	NRL	2.5	150	2.0	1.2	60
PITHON	PIC	5	500	2.0	2.5	100

electron and light ion beams. These have been performed on spherical targets using electron beam "in diode" configurations on the Proto I machine. Cylindrical and conical targets have been irradiated with light ions using the Proto II accelerator. The first major ion beam experiments on spherical D-T targets are scheduled for the PBFA-I machine. It is anticipated that 250 kJ of the 1000 kJ of energy available at the diode will be focused onto the target. This should be sufficient to verify the light ion beam approach to ICF target implosion.

The PBFA-II accelerator will be built as a follow-on addition to PBFA-I in 1984 (see Figure 8.8). This facility has been designed to accommodate another full set of Marx generators, intermediate storage capacitors, and transmission lines to bring the total to 72 beams. With enhanced energy densities in the Marx generators, it is anticipated that 3.7 MJ will be delivered to the diodes and 1 MJ of ion energy focused onto the target. Fusion energy breakeven should result from these experiments. Other pulsed power accelerators at other

PBFA-I   **PBFA-II**

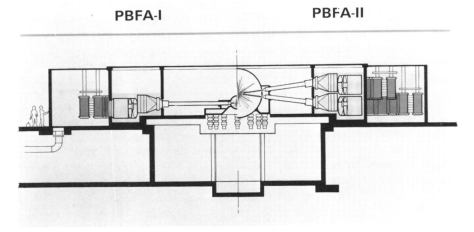

Figure 8.8.   A comparison of the PBFA-I and PBFA-II accelerators.

laboratories will allow research groups to test ideas about ion diode design and ion beam transport, thereby contributing to the light ion beam ICF program.

It is useful to briefly summarize the relative advantages and disadvantages of pulsed power accelerators (primarily using light ion beams) as ICF drivers when compared to other driver types. To begin on a positive note, the coupling of light ion beams (although not necessarily with electron beams) is expected to be far superior to that found with laser drivers with infrared wavelengths (i.e., Nd and CO_2). Apparently, light ions will deposit their energy in the target via classical collision mechanisms, while it is known that a significant fraction of the energy deposited by long wavelength laser light appears as suprathermal electrons.

Furthermore pulsed power machines are much more efficient than lasers. The conversion efficiency of wall-plug electricity to ions at the target is expected to be as high as 30 to 35%. For lasers, the best that can be achieved is 5 to 10%. This implies that the minimum economical target gain requirement is relaxed considerably for light ion drivers.

Pulsed power accelerators can deliver very large amounts of energy. The PBFA-I facility is rated at 1 MJ at the diode, while the largest present laser facility, Shiva (at the Lawrence Livermore Laboratory) can deliver only about 20 kJ. Furthermore, pulsed power machines are relatively inexpensive compared to lasers. The cost of PBFA-I is only \$7/J while lasers generally cost several times this amount.

Pulsed power accelerators share an advantage with lasers over high current heavy ion accelerators in that the technology developed for these drivers has other applications aside from ICF research.

But there are also several drawbacks with pulsed power accelerators. Although pulsed power accelerators are energy rich, they are also power and intensity poor. Certainly the 100 TW projected for large machines such as PBFA-I is a significant power level, but it may still not be sufficient for ICF. Intensities of 50 TW/cm^2 are expected from PBFA-II, but again this may not be sufficient. All of the other advantages of pulsed power accelerators are for nought if targets cannot be designed to yield adequate performance at these relatively low intensities.

The pulse lengths characterizing pulsed power accelerators are limited to greater than about 10 ns. If shorter pulses at higher powers are required to drive ICF targets, these drivers may prove unsatisfactory.

Both pulsed power and laser drivers face problems for repetitive operation. For reactor applications, devices must fire 1 to 10 times per second, while current experimental devices operate at 1 to 10 shots per day.

8.2. HEAVY ION BEAM FUSION

The use of particle beam accelerators as ICF drivers assumes two quite different forms. In the previous section we discussed the application of pulsed power diode accelerators to produce beams of light ions in the MeV range for use as ICF drivers. Of more recent vintage are proposals that high energy physics accelerator technology be applied to accelerate heavy ions (e.g., Xe to U) to GeV energies and then focus these heavy ions beams on ICF targets.[48–53] Not only do such beams couple quite effectively to drive ICF targets, but the technology developed for high energy physics research appears capable of extrapolation to the high current pulses required by ICF applications. A rather substantial theoretical effort has been directed toward the use of heavy ion accelerators for ICF, and workshops have been held annually for a number of years to summarize the state-of-the-art in the design of heavy ion beam drivers.[48–51]

However almost the entire heavy ion beam (HIB) fusion effort to date has been of a theoretical nature, with only a very modest experimental component. HIB accelerator concepts are generally very expensive (ranging in cost from $100 million to $1 billion for an ICF facility) and do not appear to be capable of being studied on a small experimental scale. Hence the "entry fee" into this approach to ICF is very high, and limited funding has kept the HIB program at the level of paper studies. It is expected that the feasibility demonstration for ICF will come with a large laser or light ion beam driver (such as PBFA-II). Once high gain targets have been verified by one of these facilities, then serious programs directed at engineering development with the eventual goal of commercialization can be undertaken. At this stage, the HIB approach might step in as the ultimate commercial ICF driver. Of course, this is highly speculative, but it does serve to place the HIB driver in perspective aside laser and pulsed power ion beam drivers.

In the following sections we review the principal types of HIB drivers being proposed, including their major components and several of the relevant physical principles underlying the operation of these accelerators.

8.2.1. HEAVY ION BEAM DRIVER TYPES

There are at least three approaches to the acceleration of heavy ions.[52] These include (1) the RF linac (radio frequency linear accelerator), (2) the induction linac, and (3) the synchrotron based accelerator. An example of the RF linac system is shown in Figure 8.9. This accelerator differs dramatically from the pulsed power machines discussed in Section 8.1. Starting with an ion source at low energy and current, the ions are accelerated through several stages of the system, to then be stored or accumulated in storage rings where the pulse length can be compressed. Bunches of ions are then switched out and guided to the target through magnetic focusing elements. Further pulse compression is achieved as the ions drift from the storage rings to the target.

The size of these heavy ion accelerators is immense, with typical lengths being measured in kilometers. One similarity to the pulsed power or laser drivers is the fact that heavy ion drivers are composed of many different elements. For example, HIB fusion approaches deal with conventional accelerator components such as ion sources, injectors, Wideroe linacs, Alvarez linacs,

Type of machine	Schematic Assembly	Beam Output
8 Cockcroft–Walton injectors		40mA each of U^{+1} at 500keV
8 2 – MHz Wideroe linacs		20mA each of U^{+1} at 8MeV
Electron stripper		20mA each of U^{+2} at 6MeV
4 4–MHz Wideroe linacs		40mA each of U^{+2} at 13MeV
2 8–MHz Wideroe linacs		80mA each of U^{+2} at 30MeV
48–MHz Alvarez linac		160mA of U^{+2} at 120MeV
96–MHz Alvarez linac		160mA of U^{+2} at 480MeV
192–MHz Alvarez linac		160mA of U^{+2} at 20GeV
Multiplier ring–1 km radius		1.6A of U^{+2} at 20GeV
Multiplier ring–100 m radius		16A of U^{+2} at 20GeV
8 Accumulator rings–100m radius with beam compression factor of 30		500A each of U^{+2} at 20GeV
8 Beam compressors with factor of 5		2500 A each of U^{+2} at 20 GeV, 10 MJ, 200 TW
Pellet assembly boilers etc		Fusion energy

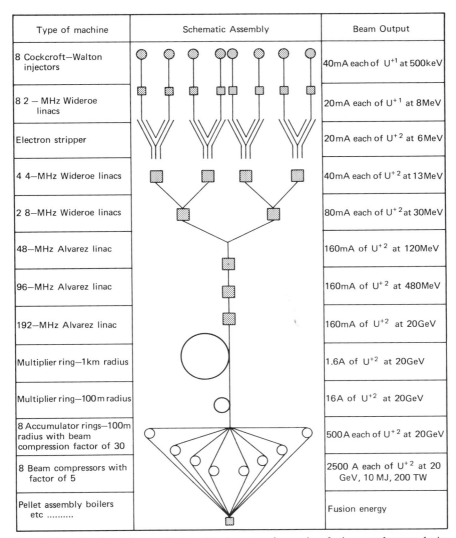

Figure 8.9. The Brookhaven National Laboratory heavy-ion fusion accelerator design. (Courtesy of **Fusion** magazine.)

multiplier rings, accumulator rings, synchrotron rings, and so on. Part of the intent of this section is to introduce the reader to the concepts and jargon used in the discussion of such accelerators.

8.2.2. COMPONENTS OF HEAVY ION ACCELERATORS[53]

Ion Sources. In contrast to electron sources that can be based on conventional cathode or field emission processes, most high current ion sources involve applying electric fields to gaseous discharges or plasmas to extract ion beams. Advanced ion sources could employ pulsed power diode technology similar to that discussed in Section 8.1. Typical ion sources operate with extraction voltages of tens of kilovolts. The velocity of keV ions is too low and their space charge repulsion too high to allow direct injection into a main accelerator stage. Hence it is common to house the ion source in a terminal maintained at a high DC voltage (typically 750 kV) by conventional Cockroft-Walton voltage multiplying circuits.

Low Beta Accelerators. DC acceleration to energies of 1 MeV is adequate for injection of protons into conventional RF linacs, even at high currents (several hundred milliamperes). However heavy ions of that energy are moving so slowly that another element must be added into the accelerating chain. Such "low beta" accelerators (where "beta" refers to the ratio of ion speed to the speed of light) represent a new development in accelerator technology. Oscillating electric fields are difficult to use since such low frequencies would be required by the slow ion speeds. Resonant cavities would require excessive size and power consumption to handle such high currents. Of present interest is the Wideroe accelerator, developed several decades ago at the low frequencies required for heavy ion acceleration. Present designs still operate at low currents and frequencies still somewhat high for ICF applications. Drift tubes are used to facilitate the operation of the accelerator chain as a single pass device, that is, to allow a single bunch of ions to pass from ion source to target.

Main Accelerators. The energies and currents required for HIB fusion eliminate conventional heavy ion accelerators such as cyclotrons or synchrocyclotrons as possible options. Instead, HIB fusion concepts tend to favor RF linacs, induction linacs, or synchrotron accelerators. The type of RF linac best suited to heavy ion beam fusion was developed by Alvarez and consists of a succession of cylindrical cavities, resonant at frequencies of 50 to 100 MHz, in which there is a uniform axial electric field. In each cavity a succession of smaller cylinders or "drift tubes" is suspended to shield the ions from the electric field in its decelerating phase as they pass through the cavity. Focusing magnets or electrodes are contained in each of the drift tubes to provide transverse containment for the beam. The RF power necessary to maintain the accelerating field and to supply energy to the ions is coupled into the sides of the cavities from power amplifiers. Such Alvarez linac devices are a common component in all high energy proton accelerators.

A second type of accelerator with potential for ICF driver application is the induction linac developed by Christofilos in the late 1950s. The accelerating action in this type of device is analogous to the operation of a transformer. It consists of ferromagnetic rings with a one-turn primary. The beam itself acts as a one-turn secondary winding. Induction linacs are capable of high current operation. They accelerate the ions in a single pass, eliminating the current multiplying operations required in more conventional accelerators. Transverse focusing is provided by magnets placed between the accelerating modules. In contrast to the RF linac, in which beam intensity is limited by the rate at which the cavity stored energy can be replaced, the induction linac works best when the pulse length is short and the current is high. The ability of the focusing fields to contain high currents is a dominant factor in the design of an induction linac.

The third type of accelerator of interest in ICF applications is the alternating gradient synchrotron. Such machines consist of an array of bending and quadrapole magnets closing on themselves around an approximately circular path. Accelerating cavities are distributed around the ring. Ions are injected into the ring from a linac or smaller synchrotron, after which the magnetic fields are increased in strength. Bunches of ions are locked in step with the RF frequency and gain energy from the RF cavities. The frequency is gradually increased to correspond to the rotation frequency demanded by the increasing magnetic field. At peak magnetic field and ion kinetic energy, the ions can be extracted from the ring, usually by pulsing special magnets. Although this accelerator provides the simplest and most economic means for achieving high ion energies, peak currents are severely limited. Hence the beam must be extracted in many small bunches, perhaps a hundred or more, greatly complicating delivery to the ICF target. Furthermore, the energy efficiency of a synchrotron is inherently low due to the power required by the magnets.

Accumulator (Storage) Rings. Conventional ion sources and low-beta accelerators produce currents on the order of 25 mA. It is proposed to use several such units funneling together in pairs to form a treelike collection of RF linacs yielding a current of 500 mA. But ICF applications will require beam currents on the kiloampere level. Hence it is proposed to feed the RF linac beams into one or more accumulator or storage rings. These rings would consist of arrays of bending and focusing magnets (similar to a synchrotron), probably based on superconducting magnets. The linac beams could be injected into an accumulator ring for as many as a hundred turns. The resulting circulating current could then be built up to 50 A.

Linear Compressors. To compress the ion bunch length and increase the beam current to the levels required for ICF drivers, induction accelerator modules can be used in which the voltage waveform is such that the early-arriving ions are decelerated slightly and the late ones strongly accelerated. The ions would then drift freely, constrained transversely by quadrapole magnets, until the faster ions catch up with the slower ions to compress the bunch. The induction accelerator modules could be located in the accumulator rings or in the transport lines following beam extraction.

Beam Transport Lines. Between components of the accelerator chain, the beams are guided by a succession of focusing magnets (probably superconducting quadrapoles). Such beam transport lines are standard equipment in conventional accelerators. However, they will represent a significant contribution to the capital cost of an HIB driver facility.

Final Focusing.   The last set of two or three quadrupoles in the transport lines will be used to focus the beam on the target. The demands placed on these final focusing magnets are severe, since they must focus the beam on a spot a few millimeters in diameter located 5 to 10 m from the end of the transport line. Beam optics require an expansion to as much as a meter in diameter in the final focusing lenses. Fortunately, at this beam size, space charge effects are relatively unimportant, and will remain so until the beam is within a short distance of the target. The design of the final focusing lens system will play an important role in the achievement of beam intensities sufficient to drive ICF targets.

8.2.3.   BEAM PHYSICS CONSIDERATIONS

Beam Loss Mechanisms.   Physical structures such as electrodes or current carrying sheets must often be located in places where they may intercept some fraction of the ion beam during its formation, acceleration, and focusing. By careful design such beam losses can be kept to a few percent. Other important loss mechanisms include atomic collision processes. If an ion collides with a residual gas molecule and gains or loses an electron, the change in its radius of curvature in bending magnets or lenses will force it into the wall of the vacuum chamber. Hence a premium is placed on achieving a good vacuum, typically in the range of 10^{-10} torr in accumulator rings and 10^{-7} torr in linacs. A second collision loss mechanism is intra-beam charge exchange. Within a given bunch, ions may collide with one another as they oscillate about. Any such collisions resulting in a change in charge state can result in ion loss. Such loss processes can limit the storage time in accumulator rings.

Beam Focusing Constraints[54]   Throughout the acceleration process, the forces applied to the ions to constrain and accelerate them or exerted by the ions on one another are electromagnetic in nature. If the individual particle encounters are neglected compared to long-range collective forces, then Liouville's Theorem implies that the volume occupied by the required number of particles in the six-dimensional phase space (position and momentum) is a constant of the motion. The area of the projection of this volume onto a plane defined by one coordinate and its corresponding momentum is called the "emittance" of the beam in that degree of freedom. If the three degrees of freedom are uncoupled, then each of the three emittances is a constant of the motion. If they are coupled the situation is more complicated. In any case the product of the three emittances corresponding to horizontal, vertical, or longitudinal motion, is the six-dimensional volume and is a constant of the motion.

Liouville's Theorem provides a necessary condition on the accelerator design. If the transverse emittances at the final lens are limited by geometric aberrations, and the longitudinal emittance is limited by pulse duration and the longitudinal momentum spread, and the maximum number of beams is limited by practical considerations, then the ion source and low-beta accelerator must supply the required number of ions in a phase space volume less than the product of the final three emittances and the number of final beams. If this is not achieved, then no degree of complexity or ingenuity in the intervening hardware can produce the desired result. In practice it is impossible to manipulate the beam through the various stages of the accelerator without "stirring some air" into the phase space volume (much as the volume of an egg is increased by beating it). These dilution effects must be taken into account in accelerator design.[55-58]

Space Charge Limits. Space charge effects tend to defocus the beam. This can be particularly important during circular motions such as in a synchrotron or storage ring. To compensate for this effect, one can lower the charge state of the ions, increase the beam emittance, or inject ions at a higher kinetic energy. Such changes require more expensive injectors, larger synchrotron aperatures, and more elaborate manipulation of the beam to reduce the final emittance per beam on target to an acceptable level.

Beam Transport Limits. There is a cost premium on keeping the ion bunch length and instantaneous current as high as possible. A question arises as to what level of current can be transported for long distances in a quadrupole beam line without serious degradation in longitudinal or transverse emittance. Both theoretical and computer-based studies have been directed at this question in recent years. These investigations have revealed the possibility of instabilities arising in the transverse motion of the beam leading to emittance growth factors of two or three. However a more definitive conclusion awaits experimental studies.

8.2.4. FINAL BEAM TRANSPORT IN THE REACTION CHAMBER

In the heavy ion beam approach to ICF, many beams are focused by magnets at a distance of 5 to 20 m from the target. After final focusing, the beam propagation to the target can be via several different mechanisms depending on the gas pressure in the cavity. This is shown schematically in Figure 8.10. At pressures below 10^{-4} torr the beam propagates as it would in a vacuum. Here the major problem arises from space charge effects due to the nonneutralized beam. At pressures greater than 10^{-4} torr, the high value of the cross section for ionization of the beam ions ($\sigma \sim 10^{-16} - 10^{-17}$ cm^2) implies that the beam ion charge state grows while the background gas is ionized. As a result, the beam may be partially or fully charge and current neutralized; however a two-stream instability may be excited, thereby deflecting the beam. As the pressure is increased to about 1 torr, collisions with gas atoms inhibit the growth of the two-stream instability, however current filamentation can now occur. This again will defocus the beam. There may be a "window" at about 1 torr where instabilities are suppressed, and the beam will propagate. This is a

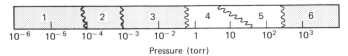

Pressure (torr)

Figure 8.10.   Background gas pressure regimes for heavy ion beam transport.

very speculative subject at this time, and final verification will only come with high current experiments. At the higher pressures, the neutralized current decays because of plasma resistivity and significant magnetic field production. These fields confine the beam enroute to the target in a fashion somewhat analogous to the pinched mode of beam propagation used in the pulsed power diode, light ion approach.

REFERENCES

1. G. Yonas, "Fusion Power with Particle Beams," *Sci. Am.* **239**, 50 (November, 1978).

2. G. Yonas, "Particle Beam Fusion Program: Publications and Related Reports," Sandia National Laboratory Report SAND-80-0809 (1980).

3. G. Kuswa, "Progress Toward Fusion with Light Ions," *Eighth International Conference on Plasma Phys. and Controlled Nuclear Fusion Research*, Brussels, 1980 (Vienna, International Atomic Energy Agency, 1979).

4. S. L. Bogolyubskij et. al., "Demonstration of the Possibility of Using Electron Beams for Heating Thermonuclear Targets," 6th Int. Conf. on Plasma Physics and Controlled Thermonuclear Fusion Research (Vienna, 1977).

5. A. J. Toepfer, "Particle Beam Fusion," lectures presented at the ANL-AUA Faculty Institute on Inertial Confinement Fusion, Argonne National Laboratory, 1978.

6. S. Nakai, K. Imasaki, and C. Yamanaka, "Relativistic-Electron Beam Induced Fusion," lectures presented at the ANL-AUA Faculty Institute on Inertial Confinement Fusion, Argonne National Laboratory, 1978.

7. J. W. Poukey et. al, "Focused Intense Ion Beams Using Self-Pinched Relativistic Electron Beams," *Phys. Rev. Lett.* **35**, 1806 (1975).

8. J. T. Verdeyen et. al., "The Use of Electronic Space Charge to Accelerate, Focus, and Bunch Ions for Pellet Fusion," *Appl. Phys. Lett.* **27**, 380 (1975).

9. W. Bostick, V. Nardi, and O. Zucker, Eds., *Energy Storage, Compression, and Switching* (New York, Plenum, 1976).

10. G. Yonas, Ed., *Proc. Int. Top. Conf. Electron Beam Res. Technol.*, Vol. I and II, November, 1975 (Albuquerque, NM).

11. P. A. Miller, et. al., "Light Ion and Electron Beams for Inertial Fusion," *Comments on Plasma Physics* **5**, 95 (1979).

12. Electron-Beam Fusion Progress Report, 1975, Sandia National Laboratory Report SAND-76-0148, June–July, 1976, SAND-76-0410, July–September, 1976, SAND-76-0711, October, 1976–March, 1977, SAND-77-1414, April–September, 1977, SAND-78-0080. Particle Beam Fusion Progress Report, October, 1977–March, 1978, SAND-79-0002, April–December, 1978, SAND-79-1011, January–June, 1979, SAND-79-1944.

13. G. Cooperstein, D. Mosher, J. R. Boller, D. G. Colombant, W. F. Oliphant, S. J. Stephanakis, F. C. Young, S. A. Goldstein, R. J. Barker, R. A. Meger, P. F. Ottinger, F. L. Sandel, and A. Drobot, "NRL Light Ion Beam Research for Inertial Confinement Fusion," NRL Memorandum Report 4387 (November 20, 1980).

14. D. J. Johnson, S. A. Goldstein, R. Lee, and W. F. Oliphant, "Time-Dependent Impedance Behavior of Low-Impedance REB Diodes During Self-Pinching," *J. Appl. Phys.* **49**, 4634–4643 (1978).

15. A. E. Blaugrund and G. Cooperstein, "Intense Focusing of Relativistic Electrons by Collapsing Hollow Beams," *Phys. Rev. Lett.* **34**, 461–464 (1975).

16. S. A. Goldstein and R. Lee, "Ion-Induced Pinch and The Enhancement of Ion Current by Pinched Electron Flow in Relativistic Diodes," *Phys. Rev. Lett.* **35**, 1079–1082 (1975).

17. S. J. Stephanakis, D. Mosher, G. Cooperstein, J. R. Boller, J. Golden, and S. A. Goldstein, "Production of Intense Proton Beams in Pinched-Electron-Beam Diodes," *Phys. Rev. Lett.* **37**, 1543–1546 (1976).

18. S. A. Goldstein, G. Cooperstein, R. Lee, D. Mosher, and S. J. Stephanakis, "Focusing of Intense Ion Beams from Pinched-Beam Diodes," *Phys. Rev. Lett.* **40**, 1504–1507 (1978).

19. P. F. Ottinger, D. Mosher, and S. A. Goldstein, "Propagation of Intense Ion Beams in Straint and Tapered Z-Discharge Plasma Channels," *Phys. Fluids* **23**, 909–920 (1980).

20. P. F. Ottinger, D. Mosher and S. A. Goldstein, "Electromagnetic Instabilities in a Focused Ion Beam Propagating Through a Z-Discharge Plasma," *Phys. Fluids* **24**, 164–170 (1981).

21. P. F. Ottinger and D. Mosher, "Microstability of a Focused Ion Beam Propagating Through a Z-Pinch Plasma," *Phys. Fluids* **22**, 332–337 (1979).

22. P. F. Ottinger, D. Mosher, and S. A. Goldstein, "Stability Considerations for Light-Ion Beam Transport in Z-Discharge Channels," NRL Memorandum Report, (March 1981).

23. D. G. Colombant, S. A. Goldstein, and D. Mosher, "Hydrodynamic Response of Plasma Channels to Propagating Ion Beams," *Phys. Rev. Lett.* **45**, 1253–1256 (1980).

24. D. Mosher and D. G. Colombant, "Beam Requirements for Light-Ion-Driven Inertial-Confinement Fusion," NRL Memorandum Report 4397 (November 27, 1980).

25. J. Van Devender, "Light Ion Fusion Driver Technology: Pulsed Power Technology," Fusion Energy Technology, NE 712, In-Hours Technical Courses, Sandia Laboratory, August, 1980.

26. T. H. Martin and K. R. Prestwich, "EBFA, a 20 TH Electron Beam Accelerator," in *Energy Storage, Compression, and Switching*, edited by W. H. Bostick et. al. (New York, Plenum, 1976).

27. C. W. Mendel, Jr., and S. A. Goldstein, "A Fast-Opening Switch for Use in REB Diode Experiments," *J. Appl. Phys.* **48**, 1004 (1977).

28. K. R. Prestwich, "Harp, a Short Pulse, High Current Electron Beam Accelerator," *IEEE Trans. Nucl. Sci.* NS-22, 975 (1975).

29. I. Smith "Liquid Dielectric Pulse Line Technology," *Energy Storage, Compression, and Switching*, W. Bostick, V. Nardi, and O. Zucker, Eds. (New York, Plenum, 1976).

30. K. D. Bergeron, "Equivalent Circuit Approach to Long Magnetically Insulated Transmission Lines," *J. Appl. Phys.* **48**, 3065 (1977).

31. K. D. Bergeron, "One and Two Species Equilibria for Magnetic Insulation in Coaxial Geometry," *Phys. Fluids* **20**, 688 (1977).

32. K. D. Bergeron, "Relativistic Space-Charge Flow in a Magnetic Field," *Appl. Phys. Lett.* **27**, 58 (1977).

33. K. D. Bergeron, "Theory of the Secondary Electron Avalence at Electrically Stressed Insulator-Vacuum Interfaces," *J. Appl. Phys.* **48**, 3073 (1977).

34. A. E. Blaugrund, G. Cooperstein, and S. Goldstein, "Relativistic Electron Beam Pinch Formation Processes in Low Impedance Diodes," *Phys. Fluids* **20**, 1185 (1977).

35. P. A. Miller, J. W. Poukey, and T. P. Wright, "Electron Beam Generation in Plasma-Filled Diodes," *Phys. Rev. Lett.* **35**, 940 (1975).

36. J. A. Pasour et. al., "Reflex Tetrode with Unidirectional Ion Flow," *Phys. Rev. Lett.* **40**, 448 (1978).

37. J. W. Poukey, "Ion Effects in Relativistic Diodes," *Appl. Phys. Lett.* **25**, 145 (1975).

38. J. W. Poukey, "Two-Dimensional Ion Effects in Relativistic Diodes," *J. Vac. Sci. Technol.* **12**, 1214 (1975).

39. J. P. Quintenz and J. W. Poukey, "Ion Current Reduction in Pinched Electron Beam Diodes," *J. Appl. Phys.* **48**, 2287 (1977).

40. D. W. Swain et al., "Measurements of Large Ion Currents in a Pinched Relativistic Electron Beam Diode," *J. Appl Phys.* **48**, 118 (1977).

41. Yu. L Bakshaev and E. I. Baranchikov, "Transfer and Focusing of High-Current Relativistic Electron Beams onto a Target," (IAEA-200, 1976) p.25

42. J. R. Greig, "Electrical Discharges Guided by Pulsed CO-2 Laser Radiation," *Phys. Rev. Lett.* **41**, 174 (1978).

43. W. L. Johnson, G. B. Johnson, and J. T. Verdeyen, "Ion Bunching in Electronic Space-Charge Regions," *J. Appl. Phys.* **47**, 4442 (1976).

44. P. A. Miller, J. Chang, and G. W. Kuswa, "Electron Beam Concentration Enhanced by a Laser-Produced Plasma," *Appl Phys. Lett.* 23, 423 (1973).

45. P. A. Miller et al, "Propagation of Pinched Electron Beams for Pellet Fusion," *Phys. Rev. Lett.* **39**, 92 (1977).

46. P. A. Miller and J. B. Gerardo, "Relativistic Electron Beam Propagation in High-Pressure Gases," *J. Appl. Phys.* **43**, 3008 (1972).

47. G. Yonas et al., "Electron Beam Focusing Using Current-Carrying Plasmas in High nu/gamma Diodes," *Phys. Rev. Lett.* **30**, 164 (1973).

48. Proc. ERDA Summer Study of Heavy Ion Fusion. Oakland/Berkeley, Lawrence Berkeley Laboratory Report LBL-5543, July, 1976.

49. Proceedings of Brookhaven National Laboratory Heavy Ion Fusion Workshop, BNL-50769, October, 1977.

50. Proceedings of Argonne National Laboratory Heavy Ion Fusion Workshop, ANL-79-41, September, 1978.

51. Proceedings of Lawrence Berkeley Laboratory/Stanford Linear Accelerator Heavy Ion Fusion Workshop, LBL-10301, SLAC-PUB-2575, UC-28, CONF-7910122.

52. J. Schoonover and M. Levitt, "Heavy Ion Fusion," *Fusion* **24** (February, 1979).

53. Discussion of the principal components is taken largely from W. Hermannsfeldt, "The Development of Heavy-Ion Accelerators as Drivers for Inertially Confined Fusion," Lawrence Berkeley Laboratory Report LBL-9332, June, 1979.

54. This discussion of phase space considerations is taken from lectures by K. Symon, Department of Physics, University of Wisconsin, to the University of Wisconsin Fusion Engineering Program Heavy Ion Beam Reactor Design Group, February, 1980.

55. J. Lawson, *Particle Beam Acceleration* (London, Oxford University Press, 1972).

56. P. Arnold, "Heavy Ion Beam Inertial Confinement Fusion," *Nature* **276**, 19 (1978).

57. J. Rosenblatt, *Particle Acceleration* (London, Mehtuen, 1968).

58. E. Persico, E. Ferrari, and S. Segre, *Principles of Particle Accelerators* (New York, W. A. Benjamin, 1968).

59. K. A. Brueckner, "An Assessment of Inertial Confinement Fusion Drivers," K. A. Brueckner and Assoc., EPRI Report AP-1371 Feb. 1980.

NINE

Target Design, Fabrication, and Diagnostics

The design of inertial confinement fusion (ICF) targets is a rather transient and uncertain endeavor since much of the physics underlying inertial confinement fusion is not yet well established. The behavior of matter under such extreme pressure and density conditions, the production and transport of hot (suprathermal) electrons, and the effects of plasma instabilities on driver energy deposition are all examples of unresolved physics problems that strongly influence the design of ICF targets. Furthermore, some aspects of ICF target design are classified because of their presumed relation to nuclear weapons physics. With these physical and political constraints, any discussion of target design can provide only a limited picture. However, in the spirit of completeness, we attempt a brief review of those aspects of target design that have been made available in the open literature.

Target designs can be classified according to driver type: lasers,[1-31] electron beams,[32-39] light ion beams,[40-45] and heavy ion beams. They can also be distinguished by intended application: power production, physics experiments, and military applications. Most effort to date has been directed at target designs for physics experiments.[8, 9, 16, 17, 28, 31, 38, 44, 46-50] Early designs (so-called slab targets) were intended for the study of the interaction of intense laser or charged particle beams with matter. More recent designs have emphasized spherical targets designed for implosion experiments. Some reactor targets have also been designed to study the viability of inertial confinement fusion as a possible source of electric power.[1, 7, 10, 18, 41, 46-50] Targets for military applications are used to study either nuclear weapons related physics or to simulate blast and radiation effects. We will not dwell on this last application.

The target design procedure is strongly dependent on the intended application. For example, in the design of targets for physics experiments, the designer takes as input the specific driver characteristics (e.g., energy, wavelength, and temporal and spatial pulse distribution) and attempts to design a target with the highest yield (or neutron production or other experimental characteristic) subject to these driver constraints. The design of high gain targets for reactor applications takes the desired target performance characteristics as given and attempts to design targets to achieve these goals subject to the constraints of anticipated driver development.

The primary tools in target design studies are hydrodynamic computer codes that simulate the dynamics of an inertial confinement fusion target. We have considered such codes in some detail in Chapter 6. However, for purposes of reference here, we have provided a simple schematic of the essential physics described by ICF hydrodynamics codes in Figure 9.1 (in this particular case, for the LASNEX code[51] developed at the Lawrence Livermore Laboratory).

A variety of other more specialized computer codes are used in target design. These include:

1. Plasma simulation codes to study the microscopic behavior of driver energy deposition and energy transport mechanisms in target plasmas.

2. Fluid instability codes to study Rayleigh-Taylor instability growth.

3. Particle transport codes to describe the transport of suprathermal particles.

The complexity of target design calculations should not be underestimated. For example, most high gain pellet designs are composed of multilayered shells of different materials and densities. Furthermore there is strong dependence on

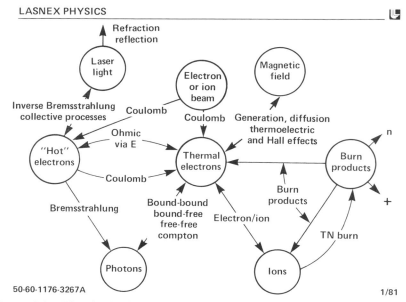

Figure 9.1. The physics included in a target design code (in this case, LASNEX).

the incident beam pulse shape, driver type (photons, electrons, or ions), number and location of beams, focal spot size and spatial distribution of the beam, and wavelength, polarization, or energy spread. By way of example, a single one-dimensional LASNEX run may require from 30 to 120 min of CDC-7600 time. Two-dimensional calculations require from one to many hours of computer time. Hence it is essential for the target designer to begin with simple estimates and back-of-the-envelope calculations whenever possible, to learn as much about the design as possible before commiting to the expense of a full-blown hydrodynamics simulation. The hydrodynamics simulation is then used to "fine tune" the design. The number of simulations required depends strongly on the proximity of the original guess to the final design.

ICF target design requires a high degree of artistry, intuition, and imagination. Designs are often as dependent on their designer as buildings are on their architect.

9.1. GENERAL GUIDELINES FOR TARGET DESIGN

Despite the complex nature of actual target design, there are some simple "rules of thumb" that provide the designer with useful guidelines. We consider several such guidelines most suited for the design of high-gain reactor targets.

Fuel Loading.   The complete combustion of deuterium-tritium fuel leads to a specific yield of 340 MJ/mg. In a practical target design one can expect 30 to 50% of the fuel to burn, depending on its ρR value and other features. Hence a yield of 100 MJ requires a pellet fuel loading of about 1 mg of D-T.

ρR Value.   A ρR value in the fuel of about 1 to 3 g/cm^2 is required for efficient thermonuclear burn. For 1 mg of D-T these ρR values correspond to compressions of 300 to 1600 times liquid density. The larger the ρR value, the more efficient the ICF burn—that is, the higher the burn fraction (see Figure 9.2). [Recall our earlier estimate of the burn fraction, $f_b = \rho R / (\rho R + 6.3)$.]

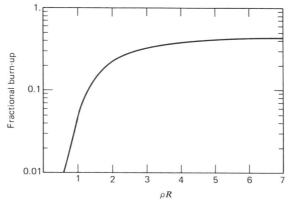

Figure 9.2.   Fractional burnup of D-T fuel versus ρR.

Larger values of ρR can be achieved by surrounding the fuel with a tamper material.

Central Ignitors.   The hot spot in the center of the fuel that ignites and burns into the surrounding cold compressed D-T must have a ρR value of at least 0.5 g/cm² and a temperature of 3 to 10 keV. For high gain, this hot spot should correspond to a small fraction (<10%) of the fuel mass. The ignition temperature may be reduced from 5–10 keV to about 3–4 keV by separating the hot-spot fuel from the bulk fuel using a central ignitor or "spark-plug" concept.[52] In this design a small part of the fuel is contained in a high-Z shell (see Figure 9.3). The main fuel layer and pusher are driven inward and implode the central ignitor as well. All of the ignitor fuel is heated to ignition conditions. The required ignition conditions can be reduced by surrounding the ignitor with a high-Z tamper.[7,52] To understand the function of this high-Z layer, recall that the ignition condition is defined as that situation where the fuel is being heated more rapidly than it can lose energy so that its temperature increases. Reducing the energy loss rate causes the fuel to ignite at a lower temperature. This can be done by surrounding the hot fuel with a high-Z tamper that absorbs radiation that would otherwise be lost from the fuel. Reradiation of this energy back into the fuel effectively traps this energy that would have been lost and lowers the ignition temperature. This is very important, since the ignition temperature is directly related to the final collapse velocity of the fuel and thereby to the required driver energy.

Shock Sequencing.   The final collapse velocity of the shock that raises the hot spot temperature to ignition conditions must be about 2 to 3×10^7 cm/s. This must be achieved by timing the shock waves created in the fuel during the implosion so that they all converge to the center near the instant of maximum compression. This timing process is very difficult to achieve. However, proper shock sequencing is essential for high gain, for it is through this process that the energy required to ignite a small amount of fuel is minimized.

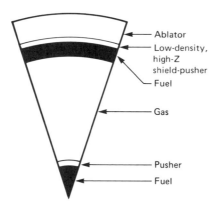

Figure 9.3.   An ICF target design having a central ignitor region and an outer fuel layer.

Implosion Velocity.   The efficiency with which the fuel can be accelerated to the required implosion velocity can be estimated using a simple analogy which models the imploding fuel as a rocket.[53] Let

$$m = \text{rocket mass (fuel mass)}$$

$$m_0 = \text{initial mass}$$

$$v = \text{rocket speed (implosion speed)}$$

$$v_0 = \text{exhaust speed (ablation speed)}$$

We can now apply Newton's law to this process

$$F = v_0 \frac{dm}{dt}$$

to solve for the speed

$$v = v_0 \ln \frac{m_0}{m}$$

Using this result, we can solve for the implosion efficiency as

$$\text{Eff} = \frac{x \ln^2 x}{1 - x}, \qquad x \equiv \frac{m}{m_0}$$

If we include thermal energy loss, then the efficiency is reduced to

$$\text{Eff} = \frac{x \ln^2 x}{2.5 \, (1 - x)}$$

For good efficiency, $x \sim 0.1$, we need $v \sim 2v_0$. But using our earlier estimate of $v \sim 3 \times 10^7$ cm/s, we find that the ablation velocity should be $v_0 \sim 1.5 \times 10^7$ cm/s. This implosion efficiency indicates that only about 10% of the driver energy can ultimately end up in the fuel. One mg of mass traveling at a velocity of 3×10^7 cm/s has a kinetic energy of 45 kJ. Therefore, without accounting for any other loss mechanisms such as the acceleration of nonfuel material, we require 450 kJ of driver input energy. Of course this is not strictly correct because not all of the fuel mass need be accelerated to 3×10^7 cm/s to achieve final shock velocities of 3×10^7 cm/s. This again demonstrates the desire to have proper shock sequencing to achieve ignition.

Velocity Multiplication with Multiple-Shell Targets.   Should the required implosion velocity not be consistent with the power characteristics of the driver, or if the driver pulse cannot be adequately shaped, then one can multiply the implosion velocity by using multishell targets.[7,15,32,35] This works just like the collision of two billiard balls. That is, if the outer shell has a mass μ times greater than the inner shell, then the maximum velocity multiplication of the inner shell which occurs when the outer shell collides with it is

$$\text{velocity multiplication} = \frac{2\mu}{1+\mu}$$

The implosion efficiency then becomes

$$\text{Eff} = \frac{4\mu^2}{(1+\mu)^2}$$

Such designs could be advantageous for drivers that are capable of delivering large amounts of energy but at low power levels (such as ion beams). In multishell targets care must be taken to ensure an elastic collision between the shells. This is usually accomplished by introducing a gas between the shells. This gas is heated to high temperature as the shells collide and smooths the collision process. Furthermore, the collision process between the two shells can be programmed to achieve a pressure profile on the inner shell that matches the one needed for isentropic compression.[54] This can be accomplished with an unshaped pulse of energy applied to the outer shell by the driver.

Driver Energy Deposition.   To generate the desired implosion velocity of 3×10^7 cm/s, we require about 20 MJ/g of specific energy deposition in the target. This is roughly the specific kinetic energy of matter at a velocity of 3×10^7 cm/s. Hence we assume that the ablator must be blown off at a velocity that is roughly the same as the implosion velocity. This is certainly the case, for the rocket efficiency of the implosion is drastically reduced for ablation velocities that are much smaller than the implosion velocity. Therefore the range of the driver particles must be such that this specific energy can be produced for reasonable driver energies. We require a driver energy E_D

$$E_D = (20 \text{ MJ/g})\,(4\pi R^2)\lambda\rho$$

where λ is the driver particle range and ρ is the mass density of the absorption zone. We can solve for

$$\lambda\rho = \frac{PT}{4\pi R^2 T\,(20 \text{ MJ/g})} = \frac{IT}{(20 \text{ MJ/g})}$$

where P is the average power of the driver pulse, T is the pulse length, and I is the intensity. If we require about 10^{14} W/cm^2 in intensity to effectively implode the target, and the pulse length is about 20 ns, then the driver energy deposition range must be about $\lambda\rho \sim 100$ mg/cm^2. This is roughly the range of light and heavy ions in cold matter. The situation is a bit more complex for laser drivers. Here the range of the hot electrons generated by the laser-target interaction is a better estimate than the range of the laser photons. From this argument, we can see the possible advantages of short wavelength lasers over long wavelength lasers, even if the absorption efficiencies are the same. Although the hot electron temperature is a weak function of the laser wavelength

$$T_{\text{hot}} \sim \lambda_L^{0.6-0.8}$$

the electron range scales essentially as T_{hot}^2 so that

$$\lambda_e \sim \lambda_L^{1.6-1.8}$$

This increase in range with laser wavelength makes the 20 MJ/g criterion more difficult to achieve for longer wavelength lasers.

Driver Pulse Shape. A final very important aspect of target design is the driver pulse shape. Simple theory shows that the "ideal" power profile for the achievement of isentropic compression of a hollow shell is given by [54]

$$P(t) = \frac{P_0}{\left[1 - (t/t_c)^2\right]^{5/2}}$$

where t_c is the collapse time. In more sophisticated target designs with tampers and multiple shells, this exact form is not applicable. However, in general we desire a pulse that starts at low power and increases to high power at the time of collapse. The initial low power phase of the pulse helps to avoid driving shock waves into the cold target and preheating the fuel while the high power finale brings the final collapse velocity up to 3×10^7 cm/s for ignition.

   With these general rules of thumb, the target designer can arrive at some idea of the relations between target size, fuel mass, and driver input energy. But this does not complete the story of target design. Added to these general considerations are a number of very important and very troublesome constraints.

9.2. MAJOR CONSTRAINTS IN TARGET DESIGN

9.2.1. RAYLEIGH-TAYLOR INSTABILITY

When we try to compress a low density fuel (such as D-T) with a high density tamper shell, we can encounter a fluid instability known as the Rayleigh-Taylor instability. (See Chapter 3 for a more detailed discussion.)[55-62] In the pellet implosion process, the acceleration force can cause the boundary between the heavier tamper shell and the lighter fuel to become unstable. This is most important when the fuel and tamper begin to decelerate as the fuel reaches its final stages of compression just before ignition. At this point the larger inertial force of the heavy tamper material can result in jets of high-Z material streaming into the fuel. This can destroy the ignition process much in the same way that high-Z impurities are detrimental to magnetic fusion plasmas. Therefore a rule of thumb can be established that states that the target should be designed to ignite before the 'free-fall" line of the tamper-fuel interface reaches the hot spot radius. Hence ignition must occur before this time. Even if ignition does occur, this mixing of impurities with the fuel can degrade the efficiency of the thermonuclear burn (i.e., the fractional burnup).

The Rayleigh-Taylor instability can limit the allowable aspect ratio $R/\Delta R$ of the target shells. That is, very thin-walled targets tend to be more susceptible to the instabilities. The instability grows most rapidly for large density differences at shell interfaces, for large accelerations, and for short wavelength disturbances. However, very short wavelength disturbances quickly grow out of the linear instability regime. It is found that the most serious instabilities are those that occur at wavelengths comparable to the shell thickness, for these do not saturate before becoming disruptive. Detailed analysis leads to the conclusion that aspect ratios of no greater than $R/\Delta R \sim 10$ should be used for compressed shells. This in itself does not guarantee that a stable implosion is possible, but it does generally help to mitigate instability problems by creating density gradients that are longer than the instability wavelengths. Another helpful design feature is to avoid placing high and low density shells adjacent to one another. So-called low-Z pushers are useful for this purpose where a high-Z–impregnated plastic (e.g., TaCOH) with a density of about 1 g/cm³ is used as a pusher next to the fuel rather than a high-Z material with a density greater than 15 g/cm³.

Fluid instabilities are an extremely important concern in the implosion process and must not be underestimated by the target designer. Their presence (or absence) remains a crucial unanswered question regarding the viability of inertial confinement fusion.

9.2.2. FUEL PREHEAT

A second important design constraint is fuel preheat.[27] We saw in Chapter 3 that the work that is needed to compress the D-T fuel depends on the isentrope along which the compression occurs. If the fuel is initially preheated by some mechanism (such as hot electrons or photons generated by the driver beam deposition in the outer layers of the target or premature shocking by an unshaped driver pulse), then more work is required to reach the same final pressure. To avoid this problem, the fuel must be isolated from these hot particles. This often determines the thickness of the fuel tamper or the use of a double shell target to further reduce the flux of hot particles incident on the fuel. For high gain targets, only a few eV temperature in the fuel is allowed in the initial stages of the implosion.[52] This corresponds to a specific energy of about 10^4 J/g for D-T. For 1 mg of D-T this represents only 10 J or about one part in 10^5 of the input energy. Therefore we can easily see the sensitivity of the target performance to preheat.

9.2.3. TARGET FABRICABILITY

Another practical constraint on target design is the ability to fabricate the target. It is easy for a computer code to mathematically levitate concentric shells within one another, but it is quite another matter to build such a target. For reactor applications this fabrication process must be automated to the point where tens of targets per second can be produced at a cost of a few cents apiece. Levitation may be achieved in practice by mounting the inner shell on a very thin film or spokes. The designer must be sensitive to these facts when designing targets for future applications.

9.3. SPECIFIC TARGET DESIGNS

Much of our discussion in earlier chapters has been confined to a theoretical analysis of the implosion of either solid or shell D-T targets. Many early studies of laser fusion assumed solid spheres of D-T.[1,3,27] Although such pellets would be straightforward to fabricate, they would require very high power levels in excess of 1000 TW for high gain. Furthermore, it is now clear that such targets are unrealistic from the point of view of beam coupling and symmetry requirements. In most experimental work to date, the targets have been glass shells containing D-T gas at high pressure. As time goes on, targets are becoming more and more complex in order to achieve the desired performance.

9.3.1. GLASS MICROBALLOON TARGETS

The glass microballoon target, consisting of a single thin walled glass shell containing about 30 atm of D-T gas, has been the primary workhorse of the laser fusion experimental program since the first successful neutron bearing shots at KMS Fusion in 1974. This target is probably the best understood and characterized design in inertial confinement fusion research. A schematic of the dynamics of a typical glass microballoon target is shown in Figure 9.4. These targets behave in an exploding pusher mode.[63] That is, the incident laser light deposits energy so rapidly in the glass shell that it explodes, half inward, half outward. The imploding shell half acts as a piston, driving a shock ahead of it into the D-T fuel that compresses and heats the fuel.

For the highest neutron yields the glass microballoon target is irradiated with a high power, very short laser pulse. The target dynamics are very different than the ideal isentropic compression described as the ultimate goal of inertial confinement fusion in earlier chapters. In fact the fuel is shock heated to a high isentrope and compressed to rather low densities. However, this combination of high ion temperature and low density corresponded to the optimum neutron yield conditions for early laser systems.

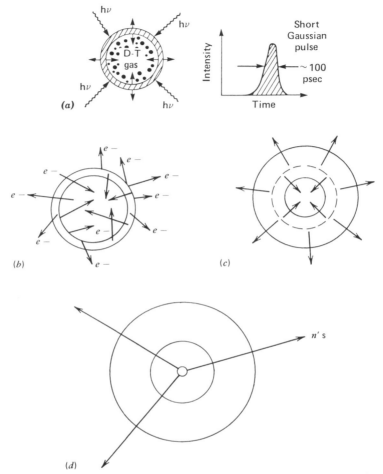

Figure 9.4. A schematic of the dynamics of a glass microballoon target behaving in the "exploding pusher" mode. (*a*) Short pulse of laser light incident on the target. (*b*) Hot electron generation and transport throughout the target. (*c*) "Explosion" of the glass shell, half moving outward and half moving inward. (*d*) Compression and heating of the DT fuel to TN temperatures.

We noted in Chapter 5 that the absorption of short, high power laser pulses is dominated by resonance absorption over the classical inverse bremsstrahlung absorption. This occurs because of the rapid heating of the underdense plasma corona and density profile modification by the incident laser light pressure. The fractional absorption for inverse bremsstrahlung scales as

$$f_{IB} \sim \frac{ZL}{\lambda_L T_e^{3/2}}$$

where L is the scale height of the density gradient and T_e is the electron temperature. Hence it is apparent that the efficiency of inverse bremsstrahlung drops markedly with increasing temperature and shorter scale heights. By way of contrast, the fractional absorption characterizing the resonance absorption process scales as

$$f_{RA} \sim (L/\lambda_L)^{2/3} \exp\left(-\frac{4}{3}\frac{L}{\lambda_L}\right)$$

Hence a small value of L ($L \sim 1$ μm) is required for good resonance absorption efficiency, and the ponderomotive forces at the critical density surface provide this kind of profile. The large fraction of absorbed energy by resonance absorption leads to the formation of suprathermal electrons characterized by a temperature

$$T_{hot} \sim T_{ave} + 3 \times 10^{-6}\left(I_L \lambda_L^2\right)^{0.425}$$

These energetic electrons have very long collisional mean free paths in comparison to target dimensions. In essence, they swarm about and through the target like bees in a hive. They are confined from escaping the target by the electrostatic potential that is built up once a few have escaped. These hot electrons lose energy in the glass shell surrounding the fuel, very rapidly heating it almost isothermally. This causes the glass shell to explode (i.e., the "exploding" pusher target) with roughly half of the mass traveling inward and the remainder traveling outward. Exploding pusher targets designed for optimum neutron yield require that the laser pulse rise to its peak power in a time that is short compared to the explosion of the shell. This generates a tremendous acceleration of the shell-fuel interface and drives a strong shock wave into the fuel ahead of the converging interface. This process places the fuel on a high isentrope. The subsequent compression of the fuel by the inward directed glass shell debris is nearly isentropic. The initial shock and the following pdV work preferentially heat the ions.

The coupling and partitioning of the laser light energy into the glass microballoon target is quite poor. Only about 20 to 40% of the incident laser light is absorbed, and most of this appears in the form of hot electrons. About 50% of the absorbed energy appears eventually as "fast" ions and escapes the target on the time scale of the laser pulse. This represents a pure loss term that must be added to the reflected laser light. The remainder of the energy in hot electrons couples to the pusher and fuel at an efficiency of about 50% so that only 250 J of the absorbed 1000 J is available for useful work. Of the total absorbed energy, about 4 to 5% appears as thermal fuel ion energy. This is not a bad value for the hydrodynamic coupling efficiency. However, the very poor effective absorption efficiency and coupling to hydrodynamic motion means that such targets are ill-suited for high gain applications. Furthermore, they do not address either of the two critical design issues for high gain targets, fuel preheat and hydrodynamic instabilities. The fuel is severely preheated, by design, and the implosion is stable because the pusher is actually decompressing rather than compressing.

On the positive side, exploding pusher targets offer several advantages at this early stage in the experimental program. They are relatively easy to fabricate and are amenable to analysis using existing theoretical models. Their implosion characteristics can be thoroughly studied using X-ray, ion, and neutron diagnostics. Therefore glass microballoon targets offer a good opportunity to compare computer code predictions with experimental results. This has been vital to the understanding of laser light interaction with matter, thermal electron transport inhibition, and fast electron generation and transport. These targets are simple enough that relatively straightforward theories can be used to accurately predict the neutron yield.[64,65]

While the glass microballoon does not really meet the requirements of a design suitable for high gain, it does provide a good example of how target design has had to conform to the available driver and target fabrication capabilities. The present generations of laser and ion beam drivers in the multikilojoule energy range are large enough to implode more complex targets that have many of the characteristics of high-gain reactor targets. Ablator layers of plastic can be added to the glass microballoon to study ablatively driven implosions. It is this type of target that has led to the high compressions achieved with current laser systems.

9.3.2. HIGH GAIN LASER FUSION TARGETS

High gain targets must be designed to implode via the adiabatic ablation process described in Chapters 2 and 3. Several high gain target designs are compared with the simple glass microballoon target in Figure 9.5.

Of particular interest are high gain target designs with multiple fuel regions. In Figure 9.6 we show a target consisting of an outer shell with a LiH ablator, a TaCOH pusher region, and D-T fuel, levitated at an aspect ratio of about 3 to 5.[66] The densities of the ablator and pusher are matched to help avoid fluid instabilities. Although the densities are matched, these materials have different properties. The high-Z impregnated plastic (TaCOH) has a higher opacity and electron stopping power than a low Z material with similar density. This tailoring of material properties is an important part of target design. These materials also meet the requirements of fabricability and surface finish quality. Inside of this outer shell is levitated a smaller gold capsule containing D-T fuel. This is the central ignitor or "spark-plug" region referred to in our discussion of Section 9.1. The implosion dynamics of this target can be summarized as follows: Laser light is absorbed in the LiH ablator, thus imploding the pusher and D-T fuel down onto the central ignitor. The ignitor then implodes to ignition conditions. These ignition conditions are minimized by the gold layer which traps the X rays generated in the hot D-T fuel, thus lowering the ignition temperature to 3 to 4 keV. The ignitor explodes and ignites the inner surface of

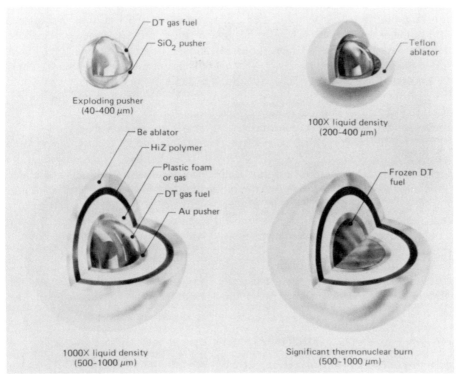

Figure 9.5. Various laser fusion target designs. (Courtesy of Lawrence Livermore Laboratory.)

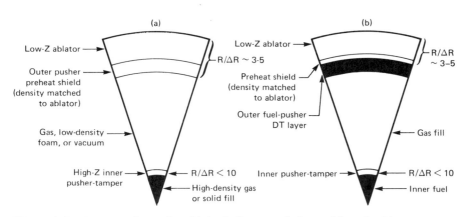

Figure 9.6. A comparison of multiple shell target designs with and without an outer fuel region.

the D-T fuel that has been crushed down around it. This double shell design allows lower final shock velocities because of the central ignitor. Furthermore, velocity multiplication between the shells allows the implosion velocity of the outer shell to be reduced to 1.4×10^7 cm/s. Therefore, twice as much fuel may be imploded to one-third the density for the same amount of energy as would be required to implode the fuel at 2×10^7 cm/s. Through velocity multiplication, the inner fuel capsule is accelerated to 2×10^7 cm/s. The ρR value of the fuel is roughly 3 g/cm^2 at ignition.

This highly efficient utilization of the input laser energy results in very high gains. For input energy of 4 MJ of 0.2 μm laser light in a shaped pulse, the target gain is calculated to be about 1000. Two significant problems associated with this target are the mixing of the outer fuel with the gold ignitor shell and the fabrication of the target. The mixing would make the ignition of the outer fuel more difficult. Of course, if ignition did not occur, then the target gain would be minimal. Fabrication difficulties will be important for such double-shelled targets. The degree of difficulty and hence the cost of the target manufacturing will depend on the surface finish required and the allowable methods of levitating the ignitor capsule. Any support structure such as thin film or spokes between the shells must meet the test of fluid instability analysis.

9.3.3. HIGH GAIN ION BEAM TARGETS

A high gain ion beam driven target is shown in Figure 9.7.[39, 41, 67] This is a simple single shell target with an aspect ratio of 10. The shells consist of a Pb tamper, a TaCOH ablator/pusher and a frozen D-T fuel layer. The driver ions

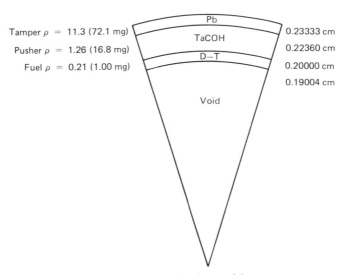

Figure 9.7. A high gain ion beam driven target.

are stopped in the Pb tamper and TaCOH pusher. Note that the range of the 6.5 MeV protons assumed as driver in this design is about 120 mg/cm^2, and that they are preferentially stopped in the low density pusher due to the Bragg peak in the stopping formula (see Chapter 5). The pulse duration is about 20 ns with a low initial power to avoid shocking the cold fuel onto a high isentrope. It rises to a maximum power of 2.4×10^{14} W or an intensity of approximately 3.5×10^{14} W/cm^2. The pulse energy is 1.3 MJ. The Pb shell serves as an inertial tamper to improve the implosion efficiency. Hence the implosion in this case is more analogous to a bullet fired from a gun than to a rocket. These somewhat different implosion dynamics are shown as a radius versus time plot in Figure 9.8. Such plots are generally produced from target design codes using computer generated graphics to help the designer visualize the implosion process.

In the implosion of this target, the pusher-fuel interface starts inward very slowly for the first 15 ns of the pulse and then accelerates to high velocity in the final 8 ns. The maximum compression occurs at about 23 ns when the interface is driven inward to a radius of 0.015 cm. This corresponds to a D-T density of 68 g/cm^3 and a fuel ρR value of about 1 g/cm^2. To this is added a pusher ρR value of about 1 g/cm^2 for a total inertial confinement ρR parameter of 2 g/cm^2, enough to allow efficient thermonuclear burning. The target yield is 88 MJ in this particular design, representing a D-T fractional burnup of 25%.

In this target, the bulk of the fuel is compressed along an isentrope that comes very close to following the Fermi degeneracy line. However, the inner

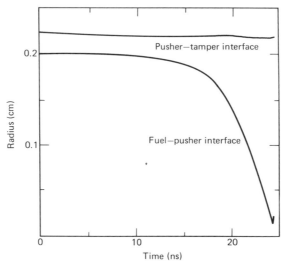

Figure 9.8. Radius as a function of time for pusher-tamper and fuel-pusher interfaces for ion beam target.

**Table 9.1.   Ion Energy Corresponding to a
Range of 100 mg/cm^2**

Ion Type	Ion Energy (MeV)
H	6
He	40
Ne	400
Fe	2000
U	8000

edge of the fuel is initially decompressed into the central void until it reaches the center. It then reflects until it collides with the still solid fuel and reflects back inward. These multiple reflections heat the fuel to a higher isentrope. This inner fuel (a small fraction of the total) is then compressed isentropically, but starting at high temperature and lower density. All of the fuel is then compressed until the central part ignites.

Target designs of this type have several significant advantages. The single shell design should be more easy to fabricate than multishell designs. The low pusher density ameliorates the problems of pusher-fuel mixing due to fluid instabilities. The gain of roughly 100 for a few megajoules of input energy should be quite acceptable for high efficiency (25%) ion beam drivers. Although this specific design is for 6.5 MeV protons, the generic design concept is applicable to all types of ions. The ion energy must, of course, be determined by the range of the ions. Heavier ions must be incident at higher energy to achieve the same range. Table 9.1 indicates that 6-MeV protons correspond roughly to the same energy deposition characteristics as 8-GeV uranium ions.

We should note a significant problem with the design of almost all such ion beam targets. These designs are quite sensitive to the uniformity of the beam illumination. This occurs because the ions deposit their energy quite deeply in the target (compared to laser drivers). Hence there is little room for lateral thermal conduction to smooth the energy transfer between the energy deposition region and the ablation front.

9.3.4.   TARGET GAIN CURVES

Target designs are usually quite unique and depend strongly on the particular driver used (and the target designer). However, once a design has been completed, it can be scaled in principle to arbitrary size while holding certain intrinsic quantities constant. These quantities are the specific energy deposition ε and the focused intensity I_{max}. The rules of thumb discussed earlier in this chapter gave these quantities as roughly 20 MJ/g and 10^{14} W/cm^2, respectively. If targets were only described by hydrodynamic phenomena (and not other processes such as suprathermal electron transport), then they could be scaled to arbitrary size according to the proportionalities

$$\text{energy:} \quad E \sim r^3$$

$$\text{power:} \quad P \sim r^2$$

$$\text{range:} \quad R \sim r$$

$$\text{time:} \quad t \sim r$$

However, this scaling is not strictly true because of transport processes during the implosion and the rate of thermonuclear reactions as a function of temperature and density. The target gain is therefore not independent of input energy. For this reason several point designs of the same generic target design must be performed to establish so-called gain curves. Such gain curves are show in Figure 9.9. Two different types of targets are displayed in this figure. The high performance double-shell design is similar to the high gain laser-driven target described earlier. The single shell design is similar to the ion beam target. Note that at low enough energies, below 400 kJ, the single shell design out-performs the double shell design, while at high input energies the double shell design is superior.

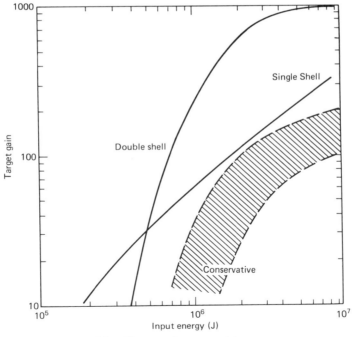

Figure 9.9. Target gain versus driver energy.

Table 9.2.   Assumptions in Target Gain Calculations

Parameter	Theoretical Limit	Conservative
Absorption	100%	80%
Ablation efficiency	~10% (LASNEX)	~10% (LASNEX)
Transport inhibition	None	None
Fast ion losses	Negligible	Negligible
Entropy of compressed matter	Fermi limit	Fermi limit
Preheat	Not significant	Not significant
Pulse shaping	Optimum	Optimum
Asymmetries/fluid instabilities	No effect	No effect
Ignition efficiency	100% (LASNEX)	50% (LASNEX)
Propagation/burn efficiency	LASNEX	70% (LASNEX)

The band labeled "conservative" is derived from the double shell results. The difference between the double shell line and the band is due to uncertainty in the calculation. These uncertainties are listed in Table 9.2. In effect, the amount of energy required to reach ignition conditions is assumed to be twice the value predicted by the target design code. Furthermore, once the target is ignited, it is assumed that the yield is only about 70% of that predicted by the codes. These two "fudge factors" are an attempt to bring the code predictions into closer agreement with reality. In fact, target implosions are not one-dimensional as modeled in these calculations—or even two-dimensional—but rather three-dimensional. Perfectly isentropic compression is also only an ideal, and there will surely be some energy loss through fast ions or electrons in the absorption region. Once the fuel has ignited, it will not burn with the theoretical efficiency because of asymmetries and mixing of the fuel and the surrounding pusher material. Therefore the best guess on target gain is generally taken to be these conservative curves. These curves are also specific to short wavelength laser and ion beam drivers. Each of these drivers is expected to efficiently couple energy into the target (assuming 80% absorption efficiency). For drivers with lower coupling efficiency, such as long wavelength lasers, one must hope that the "allowances" for the conservative gain band are not so pessimistic, and that these high efficiency energy sources will actually perform nearer to their theoretical limits. This would then imply that there is sufficient room for degradation to allow drivers with lower coupling efficiency to provide interesting results.

9.4.   TARGET DIAGNOSTICS

The physical processes occurring during the implosion of an inertial confinement fusion target are characterized by length scales of 10 to 100 μm and time scales of 10 ps to 10 ns. ICF target implosions are also characterized by exceptionally high energy densities (10 MJ/g) and driver power intensities (10^{15} W/cm^2). Hence it should be apparent that the development of diagnostic methods suitable for monitoring the dynamical behavior of inertial confinement fusion targets presents a very significant challenge.[68,69]

The length and time scales characterizing the dynamics of ICF targets eliminate the possibility of using many of the diagnostic methods employed in magnetic fusion research, such as placing physical probes or detectors into the target. In inertial confinement fusion experiments, most attention is placed on the emission of radiation or reaction products from the target, although limited use can be made of probe beams in some situations. The primary emission signatures of interest include electromagnetic radiation at all wavelengths, from the infrared to the visible to the ultraviolet and X-ray regions; charged particles arising from the blowoff plasma (fast ions), fusion reaction products, and energetic electrons produced by the driver-target interaction; and neutrons produced in fusion reactions (primarily D-T reactions) in the fuel core of the target. Light beams can also be used as probes to infer information about the driver-target interaction process, for example, via holographic or interferometric techniques.

In a general sense, ICF experiments seek to determine the spatial and temporal behavior of the target during the implosion process, with particular

Figure 9.10. A schematic of the instrumentation of the SHIVA target chamber. (Courtesy of Lawrence Livermore Laboratory.)

attention given to energy transfer and transformation processes (e.g., driver-target interaction, hydrodynamic compression, and fusion yield). Such experiments play the key role in determining and studying the various physical processes occurring during the implosion process. They are also invaluable in providing data to "calibrate" computer code models that are used in target design. (See Figure 9.10).

9.4.1. GENERAL REQUIREMENTS ON ICF TARGET DIAGNOSTICS

We have noted that the implosion dynamics of an ICF target can be analyzed by considering three different regions of the target. In the outer layers of the target, driver energy deposition and plasma blowoff are of primary interest. In the region between the energy deposition and the ablation surface, energy transport via thermal conduction or radiation transport are the dominant processes. And, finally, in the central fuel core of the target, hydrodynamic compression and thermonuclear burn are the processes of major concern.

Each of these various energy transport and transformation processes must be studied experimentally. The particular diagnostic method used will differ from region to region in the target (driver deposition corona, ablation zone, implosion core) because of the large variations in conditions such as density and temperature. (See Figure 9.11.)

Since the plasma corona is underdense, it can be probed with optical methods. One of the more successful methods for studying the corona regions involves classical optical techniques such as interferometry and polarimetry.[70-75] The corona densities and gradients demand a probe wavelength in the visible or ultraviolet. However, this can be obtained by frequency multiplication of driver laser beams such as Nd laser light at 1.06 μm. As the probe beam is passed transversely through the corona plasma, its phase and polarization are modified by the refractive index of the plasma. Interferometry can then be

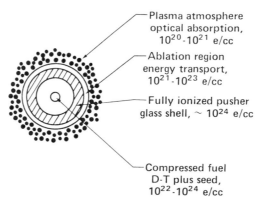

Plasma atmosphere optical absorption, 10^{20}-10^{21} e/cc

Ablation region energy transport, 10^{21}-10^{23} e/cc

Fully ionized pusher glass shell, $\sim 10^{24}$ e/cc

Compressed fuel D-T plus seed, 10^{22}-10^{24} e/cc

Figure 9.11. Various regions of interest in target diagnostics.

used to infer the plasma density profile. Considerable information concerning the corona region can also be obtained by analyzing the transmitted, back-scattered, or refracted components of the incident driver beam. X-ray and charged particle emission measurements[76-80] can also be used to infer the presence of suprathermal electron generation in the corona region.

The ablation zone between the critical surface and the ablation surface is overdense, with electron densities in the 10^{23} to 10^{24} e/cm^3 range. Such densities require X-ray wavelengths to be above cutoff. Furthermore, in many target designs (such as glass microballoons), photoelectric absorption can be strong. Hence X rays in the 1- to 10-keV range (1 to 10 Å) are most useful for studying this region. In exploding pusher targets, the pusher material is heated to sufficient temperatures (several keV) to produce strong X-ray emission. These X rays can then be used for diagnostic purposes. For example, time-integrated measurements can be made with X-ray pinhole cameras to de-termine the spatial distribution of the emission. X-ray streak cameras can measure the temporal dependence of the emission. Both techniques can be combined to obtain space-time information on the implosion process.

While X-ray emission is a valuable diagnostic in exploding pusher targets, the lower temperatures occurring in ablative (isentropic implosion) targets significantly reduces the emission of keV X rays. Hence diagnostic X-rays must be provided by an external or secondary source. One scheme involves using a secondary target that can be driven to higher temperatures to produce an intense X-ray source to illuminate ("backlight") the ablative target. These X rays then pass through the primary target, serving as an incoherent X-ray probe. Transmission can then serve to characterize the target configuration during implosion.

The dynamics of the very high density, compressed fuel core of the target occurs on the time scale of 10 to 100 ps. Of most use in providing data on the core dynamics have been the measurement of reaction products (alpha par-ticles and neutrons) and the measurements of the X-ray line widths from high-Z seed atoms in the fuel region. Neutron yield can provide information on the degree of thermonuclear burn. Zone plate coded imaging using alpha particles can provide density measurements.[76] Both density and density-radius information can be provided by spatially resolving Stark broadened X-ray line emission from high-Z seed atoms such as argon or neon. As experiments with high yield target designs are performed, additional diagnostic tools such as neutron activation and thermonuclear induced K-shell vacancy techniques provide valuable information.

9.4.2. DIAGNOSTICS BASED ON THE EMISSION OF ELECTROMAGNETIC RADIATION

The electromagnetic radiation emitted (or reflected) from an ICF target can range in wavelength from the infrared to the hard X-ray region. Such emission provides important information on the target dynamics, the energy balance and transformation within the target, and the driver-target interaction phenomena.

Of particular interest in laser fusion experiments is the amount of incident light absorbed or reflected by the target. Incident or reflected beam intensities can be measured by calorimeters. For example, one can divert a portion of the incident beam into calorimeters to measure the incident beam intensity. Calorimeters can also be placed about the target to measure the reflected or scattered light. Of particular use are spheroidal mirror geometries that can collect both forward scattered and backscattered light over large solid angles and focus this light into calorimeters. Filters can be placed in front of the calorimeters to restrict measurements to the wavelength of the incident beam, or to examine harmonic emission of incident light (for example, at 2ω or $3/2\omega$).

Another valuable diagnostic tool involves measurements of soft X rays (100 eV to 5 keV) emitted by the target. To study the target implosion process, various methods are used to image the X rays. X-ray pinhole cameras provide time-integrated measurements of the spatial distribution of the X-ray source (and hence direct information on compression and implosion symmetry). However, pinhole cameras are restricted to small solid angles. Hence for high resolution, the cameras must be placed close to the target, limiting the lifetime of the pinhole to several shots. Fresnel zone plate imaging allows larger solid angles and better resolution, but the analysis of the FZP data is complex, and once again the measurements must be performed close to the target. X-ray telescopes and microscopies (such as the Wolter or Kirkpatrick-Baez types) can be placed far from the target. However, these are difficult to fabricate and align.

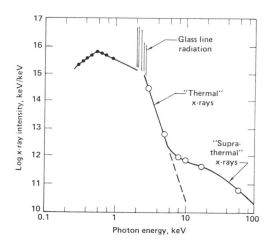

Figure 9.12. A schematic of X-ray spectra from a glass microballoon target. (Courtesy of Lawrence Livermore Laboratory.)

The spectral resolution of X-ray emission provides information on target temperatures and the presence of suprathermal electrons. (See Figure 9.12.) The most common technique is to use filters based on using the absorption edges of various elements as bandpass filters. By using several filters, one can cover the spectrum of interest. The filters can be placed in front of various sensing devices, including metal diodes, scintillator-photomultiplier tubes, or other imaging devices. More detailed spectral resolution can be obtained with X-ray spectrometers based on crystal diffraction methods. Slits can be used to obtain spatial resolution, and streak camera methods can be used to provide both spectral and temporal resolution.

The measurement of hard X rays provides information concerning the presence of suprathermal electrons. Filters can be placed in front of fluorescent foils to provide signals for a scintillator-PM tube. Magnetic fields are used to reject charged particle debris from the target.

9.4.3.   DIAGNOSTIC METHODS BASED ON PARTICLE EMISSION

ICF targets will emit a variety of energetic particles during the implosion and thermonuclear burn process. During the driver beam absorption and implosion process, ions and electrons will be accelerated in the blowoff plasma. The burning thermonuclear fuel will produce reaction products such as alpha particles and fast neutrons. These particles can provide valuable information about the target dynamics.

The detection of fast ions can provide information concerning the driver-target interaction and the presence of thermal conduction inhibition in the ablation region. A variety of methods can be used to measure the ion spectrum including Thomson Parabola methods, electrostatic deflection, magnetic analyzers, or time of flight measurements. Fresnel zone plate methods can be used to provide spatial resolution of the ion source.

A host of methods are available for the detection and analysis of fast neutrons emitted by the burning thermonuclear fuel of the target. Time of flight methods and fast scintillators can be used to measure the neutron energy spectrum. The neutron yield can be measured by activation techniques, such as Cu/Ag activation or Si activation within the target.

REFERENCES

1. Yu-Li Pan and D. S. Bailey, "Super Liquid Density Target Designs," Lawrence Livermore Laboratory Report UCRL-78472 (1976); *Bull. Am. Phys. Soc.* **21**, 1134 (1976).

2. J. H. Nuckolls, "Inertial Confinement Fusion Targets," Lawrence Livermore Laboratory Report UCRL-80166 (1978).

3. R. C. Kirkpatrick, C. C. Cremer, L. C. Madsen, H. H. Rogers, and R. S. Cooper, "Structured Fusion Target Designs," *Nucl. Fusion* **15**, 333 (1975).

4. Yu V. Afanas'ev, N. G. Basov, P. P. Volosevich, E. G. GamLii, O. N. Krokhin, S. P. Kurdyumov, E. I. Levanov, V. B. Rosanov, A. A. Samarskii, and A. N. Tikhonov, "Laser

Initiation of Thermonuclear Reactions in Inhomogeneous Spherical Targets," *JETP Lett.* **21**, 68 (1975).

5. S. I. Anisimov, M. P. Ivanov, P. P. Pashinin, and A. M. Prokhorov, "Gas Shell Target for Laser Initiation of Thermonuclear Reactions," *JETP Lett.* **22**, 161 (1975).

6. J. Daiber, A. Hertzberg, and C. E. Wittliff, "Laser-Generated Implosions," *Phys. Fluids* **9**, 617 (1966).

7. J. Lindl, "Low Aspect Ratio Double Shell Targets for High Density and High Gain and a Comparison with Ultrathin Shells," Lawrence Livermore Laboratory Report UCRL-79735 (1977).

8. W. C. Mead, J. D. Lindl, J. H. Nuckolls, J. T. Larsen, D. S. Bailey, and Y. L. Pan, "Simulations of Intermediate Density Laser Fusion Targets: Recent Progress in Design and Analysis," Lawrence Livermore Laboratory Report UCRL-80005 (1977).

9. J. M. Kindl and M. A. Stroscio, "Double Shell Target Designs for the Los Alamos Eight-Beam System," Los Alamos Scientific Laboratory Report LA-7167-MS (1977).

10. J. H. Nuckolls, R. O. Bangerter, J. D. Lindl, W. C. Mead, and Y. L. Pan, "High Performance Inertial Fusion Targets," Lawrence Livermore Laboratory Report UCRL-79373 (1977).

11. M. A. Stroscio, "Structured Target Design for Laser Fusion: A Numerical Determination of the Optimum Mass Ratio," Los Alamos Scientific Laboratory Report LA-UR-77-737 (1977) and LA-6986-MS (1977).

12. R. J. Mason, "The Calculated Performance of Structured Laser Fusion Pellets," *Nucl. Fus.* **15**, 1031 (1975).

13. G. S. Fraley, "Implosion Characteristics of Deuterium-Tritium Pellets Surrounded by High Density Shells," Los Alamos Scientific Laboratory Report LA-6378-MS (1976).

14. J. Nuckolls, J. Lindl, W. Mead, A. Thiessen, L. Wood, and G. Zimmerman, "Laser Driven Implosion of Hollow Pellets," Plasma Physics and Controlled Fusion Research, Vol. II (Vienna, International Atomic Energy Agency, 1974).

15. G. S. Fraley, W. P. Gula, D. B. Henderson, R. L. McCrory, R. C. Malone, R. J. Mason, and R. L. Morse, "Implosion, Stability, and Burn of Multi-Shell Fusion Pellets," Los Alamos Scientific Laboratory Report LA-5783-MS (1975).

16. B. Yaakobi and L. M. Goldman, "Laser Compression Studies with Neon-Filled Glass Microballoons," *Phys. Rev. Lett.* **37**, 899 (1976).

17. I. Pelah, E. B. Goldman, and B. Yaakobi, "Hydrodynamic Efficiency Measurements in Laser-Imploded Targets," *Phys. Rev. Lett.* **37**, 829 (1976).

18. K. A. Brueckner, "Laser Driven Implosion of Spherical Shells," *Nucl. Fusion* **15**, 417 (1975).

19. R. J. Mason, D. V. Brockway, and E. L. Lindman, "2-D Implosion of Structured Pellets for Laser Fusion," Los Alamos Scientific Laboratory Report LA-UR-76-2319 (1976).

20. R. J. Mason and R. L. Morse, "Tamped Thermonuclear Burn of DT Microspheres," *Nucl. Fusion* **15**, 935 (1975).

21. D. E. Ashby, "Illumination Asymmetries in Laser Induced Compression," *Nucl. Fusion* **15**, 933 (1975).

22. R. E. Kidder, "Laser Driven Isentropic Hollow Shell Implosions: The Problem of Ignition," *Nucl. Fusion* **19**, 223 (1979).

23. R. E. Kidder, "Energy Gain of Laser Compressed Pellets: A Simple Model Calculation," *Nucl. Fusion* **16**, 403 (1976).

24. R. C. Kirkpatrick, "An Overview of Design Space for Small Fusion Targets," *Nucl. Fusion* **19**, 69 (1979).

25. R. E. Kidder, "Theory of Homogeneous Isentropic Compression and Its Application to Laser Fusion," *Nucl. Fusion* **14**, 53 (1974).

26. R. E. Kidder, "Laser Compression of Matter: Optical Power and Energy Requirements," *Nucl. Fusion* **14**, 797 (1974).

27. J. D. Lindl, "Effect of a Suprathermal Electron Tail on the Yield Ratio Obtained from DT Targets Illuminated with a Shaped Laser Pulse," *Nucl. Fusion* **14**, 511 (1974).

28. E. B. Goldman, J. A. Delettrez, E. I. Thorsos, "A Theoretical Interpretation of Exploding Pusher Laser Fusion Experiments," *Nucl. Fusion* **19**, 155 (1979).

29. K. Lee, D. W. Forslund, J. M. Kindel, and E. L. Lindman, "Vacuum Insulation as a Way to Stop Hot Electrons," *Nucl. Fusion* **19**, 1447 (1979).

30. S. D. Bertke and E. B. Goldman, "The Dynamics of High Compression of Laser Fusion Targets," *Nucl. Fusion* **18**, 509 (1978).

31. M. A. Stroscio, D. B. Henderson, and A. G. Petshek, "Numerical Simulation of the Density Profile Produced by 10.6 Micron Irradiation of an SiO-2 Microballoon," *Nucl. Fusion* **18**, 1425 (1978).

32. M. A. Sweeney and M. M. Widner, "'Thick-Shell Shock-Focusing Electron Beam Targets," *Nucl. Fusion* **18**, 429 (1978).

33. M. J. Clauser, "Ion Beam Implosion of Fusion Targets," *Phys. Rev. Lett.* **35**, 848 (1975).

34. J. W. Shearer, "Ion Beam Compression of Thermonuclear Pellets," *Nucl. Fusion* **15**, 952 (1975).

35. J. D. Lindl and R. O. Bangerter, "Low Power Multiple Shell Fusion Targets for Use with Electron Beam and Ion Beams," Lawrence Livermore Laboratory Report UCRL-77042 (1975).

36. R. O. Bangerter, J. D. Lindl, C. E. Max, and W. C. Mead, "Stability and Symmetry Requirements of Electron and Ion Beam Fusion Targets," Lawrence Livermore Laboratory Report UCRL-77048 (1975).

37. E. Nardi and Z. Zinamon, Weizmann Institute of Science Report WIS-76/37Ph (1976).

38. D. J. Meeker, J. H. Nuckolls, and R. O. Bangerter, "Fusion Targets Designed to Match Present Relativistic Electron Beam Machine Parameters," Lawrence Livermore Laboratory Report UCRL-77045 (1976).

39. R. O. Bangerter and D. J. Meeker, "Charged Particle Fusion Targets," in *Second International Topical Conference on High Power Electron and Ion Beam Research and Technol.* (Ithaca, 1977), p. 183.

40. R. O. Bangerter, W. B. Hermannsfeldt, D. L. Judd, and L. Smith, *ERDA Summer Study of Heavy Ions for Inertial Fusion*, Lawrence Berkeley Laboratory Report LBL-5543, 1976.

41. R. Bangerter and D. Meeker, "Ion Beam Fusion Target Designs," Lawrence Livermore Laboratory Report UCRL-78474 (1976).

42. E. L. Lindman and J. M. Kindel, "Compression and Burn by Fast Ions," Los Alamos Scientific Laboratory Report LA-UR-76-2333 (1976).

43. J. M. Kindel and E. L. Lindman, "Target Designs for Energetic Ions," *Nucl. Fusion* **19**, 597 (1979).

44. J. D. Lindl and R. O. Bangerter, "Low Power Multiple Shell Fusion Targets for Use with Electron and Ion Beams," *Int. Conf. Electron Beam Res. Technol.* (Albuquerque, 1975), Vol. I, p. 37.

45. W. S. Varnum, "Electrically Imploded Cylindrical Fusion Targets, *Nucl. Fusion* **15**, 1183 (1975).

46. Laser Program Annual Report-1978, Lawrence Livermore Laboratory UCRL-50021-78 (1978), Chap. 3.

47. Laser Program Annual Report-1977, Lawrence Livermore Laboratory UCRL-50021-77 (1977), Chap. 4.

48. Laser Program Annual Report-1976, Lawrence Livermore Laboratory UCRL-50021-76 (1976), Chap. 4.

49. Laser Program Annual Report-1975, Lawrence Livermore Laboratory UCRL-50021-75 (1975), Chap. 5.

50. Laser Program Annual Report-1974, Lawrence Livermore Laboratory UCRL-50021-74 (1974), Chap. 8.

51. G. B. Zimmerman and W. L. Kruer, "Numerical Simulation of Laser Initiated Fusion," *Comments in Plasma Physics and Controlled Fusion* **2**, 51 (1975).

52. J. Nuckolls, "ICF Target Physics Overview," Topical Meeting on Inertial Confinement Fusion, OSA, San Diego (February, 1980).

53. F. Mayer, private communication (1979).

54. R. E. Kidder, "Laser Driven Compression of Hollow Shells: Power Requirements and Stability Limitations," *Nucl. Fusion* **16**, 3 (1976).

55. J. R. Freeman, M. J. Clauser, and S. L. Thompson, "Rayleigh-Taylor Instabilities in Inertial Confinement Fusion Targets," *Nucl. Fusion* **17**, 223 (1977).

56. J. D. Lindl, W. C. Mead, "Two-Dimensional Simulation of Fluid Instability in Laser Fusion Pellets," *Phys. Rev. Lett.* **34**, 1273 (1975).

57. W. C. Mead, J. C. Lindl, *Proc. Orbin Scientiae* II, Coral Gables, Florida (1975).

58. K. A. Taggart, R. L. Morse, R. L. McCrory, R. N. Remund, "Two Dimensional Calculations of Asymmetric Laser Fusion Targets Using IRIS," *Bull. Am. Phys. Soc.* **20**, 1378 (1975).

59. K. A. Taggart, R. L. Morse, R. L. McCrory, and R. N. Remund, "Two-Dimensional Studies of Turbulent Instabilities in High Aspect Ratio Laser Fusion Targets," *IEEE International Conference on Plasma Science*, Austin, Texas (1976).

60. J. P. Boris, "Dynamic Stabilization of the R-T Instability on Laser-Imploded Shells," NRL Memorandum Report 3427 (Dec. 1976).

61. D. L. Book, "Linear Stability of Self-Similar Flow: 3. Compressional Waves in Imploding Spherical Shells," NRL Memorandum Report 3799 (May 1978).

62. D. L. Book and I. B. Bernstein, "Stability of Self-Similar Flow: 6. Uniform Implosion of an Ablatively Driven Shell," NRL Memorandum Report 4132 (Dec. 1979).

63. Reference 46, pp. 3-2 to 3-10.

64. D. V. Giovanielli and C. W. Cranfill, "Simple Model for Exploding Pusher Targets," LA-7218-MS, May 1978.

65. E. K. Storm, J. T. Larsen, J. H. Nuckolls, H. G. Ahlstrom, K. R. Manes, "A Simple Scaling Model for Exploding Pusher Targets," UCRL-79788.

66. Reference 47, pp. 4-15 to 4-19, See also Ref. 7.

67. Reference 48, pp. 4-44 to 4-46, See also Ref. 37.

68. L. W. Coleman, "Fusion Target Diagnostics," Lawrence Livermore Laboratory Report UCRL-81099, presented at the ANL-AUA Faculty Institute on Inertial Confinement Fusion, Argonne National Laboratory (1978).

69. D. T. Attwood, "Diagnostics for the Laser Fusion Program—Plasma Physics on the Scale of Microns and Picoseconds," *IEEE J. Quantum Electron.* **QE-14**, 909 (1978).

70. D. Kania, private communication.

71. V. W. Slivinsky et al., "Measurement of the Ion Temperature in Laser-Driven Fusion," *Phys. Rev. Lett.* **35**, 1083 (1975).

72. N. M. Ceglio and L. W. Coleman, "Spatially Resolved Emission from Laser Fusion Targets," *Phys. Rev. Lett.* **39**, 20 (1977).

73. D. W. Sweeney, D. T. Attwood, and L. W. Coleman, "Interferometric Probing of Laser Produced Plasmas," *Appl. Opt.* **15**, 1126 (1976).

74. J. A. Stamper, E. A. McLean, and B. H. Ripin, "Studies of Spontaneous Magnetic Fields in Laser Produced Plasmas by Faraday Rotation," *Phys. Rev. Lett.* **40**, 1177 (1978).

75. V. W. Slivinsky, H. N. Kornblum, and H. D. Shay, "Determination of Suprathermal Electron Distributions in Laser Produced Plasmas," *J. Appl. Phys.* **46**, 1973 (1975).

76. N. M. Ceglio, D. T. Attwood, and E. V. George, "Zone Plate Coded Imaging of Laser Produced Plasmas," *J. Appl. Phys.* **48**, 1566 (1977).

77. R. A. Lerche et al., "Laser Fusion Ion Temperatures Determined by Neutron Time-of-Flight Techniques," *Appl. Phys. Lett.* **31**, 645 (1977).

78. B. Yaakobi, D. Steel, E. Thoros, A. Hauer, and B. Perry, "Direct Measurement of Compression of Laser-Imploded Targets using X-ray Spectroscopy," *Phys. Rev. Lett.* **39**, 1526 (1977).

79. H. G. Ahlstrom, "Progress of Laser Fusion at Lawrence Livermore Laboratory," Lawrence Livermore Laboratory Report UCRL-82835 Rev. 1 (1979).

80. N. M. Ceglio and L. W. Coleman, "Spatially Resolved Alpha Emission from Laser Fusion Targets," *Phys. Rev. Lett.* **39**, 20 (1977).

TEN

Applications

Inertial confinement fusion research has expanded quite rapidly along several fronts during the past decade. A primary goal has been to demonstrate scientific breakeven for ICF targets, that is, to create that situation in which the fusion energy yield is equal to or greater than the incident driver energy. A summary of experimental programs (both achieved and under development) is provided in Table 10.1. This table suggests that drivers capable of achieving breakeven should become available by the mid to late 1980s. Hence it is appropriate that we look beyond these break-even experiments to identify possible applications of inertial confinement fusion.

We can identify three general classes of applications of inertial confinement fusion: (1) power production, (2) weapons applications, and (3) fundamental physics studies. Certainly the most significant application of ICF will be to the production of energy which can then be used for a variety of purposes such as the generation of electricity,[1,2] the production of process heat or synthetic fuels,[3] or propulsion.[4] Unfortunately, it appears that this will also be the most difficult application to achieve. Nevertheless we concern ourselves almost entirely with the energy production applications of ICF in this chapter.

On a shorter term basis, much of the funding for ICF research has been directed toward military applications.[5-7] The environment created by the implosion and thermonuclear burn of a tiny ICF fuel pellet is similar in many respects to that of a thermonuclear weapon. Hence there has been considerable interest in using ICF targets to simulate on a microscopic scale weapons physics and effects.

Perhaps the most immediate application of ICF will be in basic physics studies. The imploded ICF pellet produces conditions of temperature and pressure which are quite unusual (at least on a terrestrial scale). ICF implosions can be used to study properties of matter under extreme conditions, the interaction of intense radiation with matter, and aspects of low energy nuclear

Table 10.1. U.S. Experimental ICF Programs

Institution	Driver	Significant Results
Lawrence Livermore National Laboratory	SHIVA Nd-glass laser 30 kJ — 30 TW	$100 \times$ liquid density 10^{12} neutrons Resonance absorption phenomena
	NOVA Nd-glass laser 300 kJ — 300 TW	Completed in 1983
Los Alamos National Laboratory	HELIOS CO_2 laser 10 kJ — 10 TW	50–$100 \times$ liquid density 10^9 neutrons Hot electron phenomena
	ANTARES CO_2 laser 40 kJ — 40 TW	Completed in 1983
Sandia National Laboratory	PBFA-I 2-MeV ions 1 MJ — 25 TW	Completed in 1981
	PBFA-II 4-MeV ions 4 MJ — 100 TW	Completed in 1983
KMS Fusion	CHROMA-II Nd-glass laser 1 kJ — 2 TW	First thermonuclear neutrons Comparison of implosions at 0.52 and 1.06 μm
Univ. of Rochester	OMEGA-10 Nd-glass laser 10 kJ — 10 TW	10^{11} neutrons Implosions with 0.35-μm light
Naval Research Laboratory	PHAROS-II Nd-glass laser 1 kJ — 1 TW	Ablative acceleration Laser-plasma interaction phenomena
	GAMBLE-II 2-MeV ions 150 kJ — 2.5 TW	Ion diode development Focussing and propagation of ions in plasma channels

physics. Indeed, ICF presents us with a unique opportunity to study certain aspects of astrophysics such as stellar interiors on a laboratory scale.

But, as we noted earlier, perhaps the most important application of ICF will be to the production of energy, whether directly through the conversion of fusion energy into electricity, or indirectly, by using the fusion energy or reaction products to produce synthetic chemical or nuclear fuels. We will examine each of the major applications of inertial confinement fusion to energy production in this chapter:

1. Electric power generation.
2. Fissile fuel production.
3. Process heat and synthetic fuel production.
4. Propulsion.

10.1. INERTIAL CONFINEMENT FUSION REACTORS

Perhaps the central question concerns how to capture the energy released in an ICF microexplosion and convert it into a useful form such as electrical power. As with all early generation fusion systems based on D-T fuels, this energy will appear primarily as the kinetic energy of fast 14-MeV neutrons. Therefore our goal is to burn the ICF fuel pellets using appropriate drivers (lasers or particle beams), confine the pellet microexplosion in a suitable blast chamber or reactor cavity, capture the kinetic energy of the fusion reaction products (primarily fast neutrons) as heat in a surrounding blanket, and then use this heat to perform useful work (e.g., driving a steam thermal cycle to produce electricity). Since most attention has been directed toward the study and design of laser-driven inertial confinement fusion, we consider here for the most part the design of laser fusion power systems.

A detailed design must address and answer a host of complex questions involving the fuel pellet design and fabrication, the driver, the blast chamber and blanket design, and the thermal cycle. Several of these questions are summarized below:

1. *Laser Pellet Studies.* What range of pellet gain is required? What yield should one choose? What value of laser energy is appropriate? What are the spectra associated with various forms of pellet debris? What degree of target illumination uniformity is required, and how many laser beams are needed? What do targets cost, how are they fabricated, and how are they delivered to a spot inside the chamber?

2. *Laser Studies.* What is the range of viable wavelengths? What is the laser energy? What is the laser pulse shape? How are beam lines designed, and what are the problems with the last focusing mirror? What repetition rate is required, and how does this influence laser design, power supplies, and component lifetimes?

3. *Optical Beam Train.* How many last mirrors are there, and how are they designed? How is the beam train integrated with the laser system? Do we combine beams? What is the shape and location of the last mirrors? How does wavelength influence optics system design?

4. *Cavity and Blanket Design.* Is a protective liner for the first wall required? How does the liner or first wall respond to X rays, ions, and reflected laser light? How does this response vary as the spectra and fraction of energy in each category change? How will we design the first wall? How will the blanket be designed to remove the heat and breed tritium?

5. *Materials and Neutron Radiation Damage.* Are there special rate effects in ICF reactor systems? How does the damage vary with temperature? What structural material should one choose? What should one choose for neutron wall loading? What are the dynamic stress problems and how should they be treated?

6. *Tritium Systems and Power Cycle.* What does the complete tritium cycle look like? How do we breed and recover tritium? How do we minimize leakage effects? How do we integrate the various forms of heat flow into an optimum power cycle? Is an intermediate loop needed?

It has become apparent that the successful application of inertial confinement fusion to the production of electric power will require a number of technological goals to be achieved:

1. A high average power driver with the required efficiency ($>5\%$) and reliability ($>70\%$).

2. The development of high gain fusion targets.

3. The ability to manufacture cheaply fusion targets on a mass production basis.

4. A first wall able to withstand the effects of X rays, debris, and neutrons from ICF microexplosions.

5. Structural materials that can withstand the cumulative damage effects of high energy neutrons and cyclical stresses.

6. Final beam-focusing elements that can be protected or placed sufficiently far from microexplosions or easily replaced to prevent compromising the availability of the power plant.

10.1.1. POWER BALANCE IN INERTIAL CONFINEMENT FUSION REACTORS

All ICF reactors must meet certain power balance and related economic requirements. These constraints essentially reflect the requirement that the reactor must produce substantially more power than it uses in order to be economically competitive with other sources of electricity. The power flow in an ICF reactor is shown schematically in Figure 10.1.

The quantity of most interest for reactor economics studies is the recirculating power fraction (RPF).[8] This is defined (see Figure 10.1) as the ratio of the power needed to drive the driver system to the gross electrical power. The RPF can be expressed in terms of the basic system efficiencies and the target gain Q. If the blanket power multiplication (fusion neutron power in/thermal power

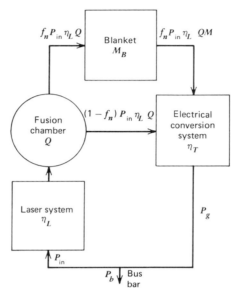

where:

$\eta_L = $ laser system efficiency

$$Q = \text{pellet gain} = \frac{\text{thermonuclear energy}}{\text{laser light energy}}$$

$$f_n = \frac{\text{neutron energy}}{\text{thermonuclear energy}}$$

$M_B = $ blanket energy multiplication

$\eta_\tau = $ electrical conversion efficiency

$P_{in} = $ electrical power input to laser

$P_g = $ gross electrical power

$$\text{RFP} = \frac{P_{in}}{P_g} = \frac{1}{\eta_L Q \eta_i [1 + f_n (M_B - 1)]}$$

Figure 10.1. Power flow in an ICF reactor.

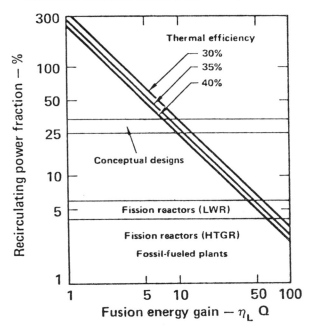

Figure 10.2. Plant recirculating power requirements as a function of fusion energy gain.

out) is taken to be 1.0, then the recirculating power fraction can be plotted as a function of the product $\eta_D Q$ as shown in Figure 10.2. (As an aside, we should note that for hybrid reactors, the blanket multiplication can be substantially greater than 1.0, ranging to as high as 100. Furthermore, pure fusion reactors will typically have blanket multiplications of 1.1 to 1.3, and this can be quite important to system economics.) For reference, the recirculating power fraction of fossil-fueled and fission reactor power plants are also given. It is generally assumed that the recirculating power fraction cannot exceed 25% for an economically viable system. This translates into a value of $\eta_D Q = 10$. Hence the attractiveness of inertial confinement fusion as the basis for a power reactor depends equally upon the driver efficiency and target gain. A 10% efficient driver such as a CO_2 laser must achieve a target gain of 100 to be economically attractive. If the driver efficiency is only 3 to 5% (such as with a KrF laser), then the target gain must be 200 to 300. On the other hand, ion beam accelerators with projected efficiencies of 25 to 30% allow target gains as low as 30 to 40.

10.1.2. TARGET YIELD AND REPETITION RATES IN ICF REACTORS

Once the relationship between the driver efficiency and target gain has been established, then the driver energy that is required to produce the target gain must be determined from gain curves such as those presented in Chapter 9. These gain curves depend on target design and on the coupling efficiency of the driver energy into the target. With this information, the target yield and the pulse repetition frequency are simply determined by the relations

$$Y = QE_D$$

$$P_F = Y\omega$$

where P_F is the fusion power and ω is the pulse repetition frequency. This then determines the type of reactor design that is required for a given driver. The reactor must be designed to withstand the blast of an explosion of Y MJ and must reestablish itself to firing conditions in a time between pulses of ω^{-1}.

With this general background, we will now turn our attention to specific ICF reactor designs.

10.2. REACTOR CAVITY (BLAST CHAMBER)

The design of a chamber to contain the blast of the ICF microexplosion would appear at first to be a challenge. A yield of 100 MJ of energy is equivalent to 48 pounds of TNT, which might be expected to blow a well-constructed reactor cavity to bits with a repetition rate of typically 1 to 10 blasts/s. Fortunately, this is not the case because the force on the walls of the chamber due to such a blast is proportional to the square root of the debris mass. Thus for a pellet mass of 1 mg, the fusion pellet debris produces less than $\frac{1}{1000}$ the force of the debris from a chemical explosion.

However of far more significance is the radiation emitted by the microexplosion. This produces an extremely hostile environment that places severe demands on cavity structures.

10.2.1. RADIATION

In general, four types of radiation will be emitted from an ICF fusion pellet: X rays, charged particles (including alphas and pellet debris), fusion neutrons, and reflected laser light (for laser drivers). Typical release fractions and energies for these products for the high gain ion beam target design discussed

Table 10.2. ICF Target Yield (Ion Beam Drivers)

Total fusion yield	100 MJ
Neutron yield	71 MJ
X-ray yield	20 MJ
Ion yield	7.4 MJ
Endoergic reactions	1.6 MJ
Neutron multiplication	1.046
Average neutron energy	12 MeV
Average gamma energy	1.53 MeV
Tritium breeding ratio	0.01
X-ray Spectrum	~1 keV (black body)

INTEGRATED RADIATION SPECTRUM

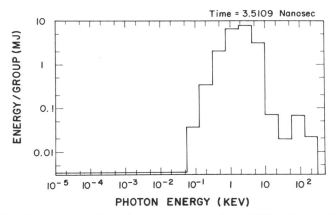

Figure 10.3. Time-integrated radiation spectrum from ICF microexplosion. (LIB target design.)

in Chapter 9 are given in Table 10.2. The time integrated X-ray spectrum is shown in Figure 10.3. Although the 14.1-MeV neutrons initially constitute 80% of the fusion reaction energy, computations have shown that through collisions experienced while leaving the pellet, these neutrons will moderate and lose a substantial fraction of energy to the fuel. (See Figure 10.4). In fact, for a pellet of $\rho R \sim 5$, 60% of the 14.1-MeV neutrons suffer a collision in the pellet, losing as much as 90% of their original energy in a single collision with a deuteron.

The X-ray spectrum for bare pellets is quite hard and nonthermal, but it can be approximated as an 8-keV black body spectrum. In structured pellets, more neutron energy is transferred to the particle debris via increased moderation.

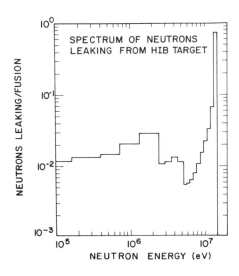

Figure 10.4. Time-integrated neutron spectrum from ICF microexplosion. (LIB target design.)

The energetic charged collision products will quickly radiate this additional energy away in the form of X rays. Thus, up to 20% of the fusion energy can be released in the form of X rays in complex pellets. This X ray spectrum may range from 300 eV to 1 keV (blackbody). In such pellets there is some production of MeV gammas due to the inelastic scattering of neutrons with pellet constituents.

Reflected laser light can also be significant in certain situations. For low yield targets, the reflected light energy may well exceed the energy of the reaction products. For short wavelength lasers with short pulses and high pellet reflectivity, the reflected laser light can cause excessive surface heating of the wall.

Although fast neutrons have large mean free paths in most wall materials (~ 10 cm) and will pass easily through to the reactor blanket, the other types of radiation will be totally absorbed by the first wall.

10.2.2. WALL LOAD MECHANISMS

The sudden deposition of the ICF burn product energy in the first wall will result in a stress due to thermal gradients in the material from nonuniform heating and conduction as well as inertial effects. One can distinguish among several different load mechanisms: (1) pellet debris impact, (2) blast wave reflection, (3) evaporation recoil, (4) blanket expansion, and (5) thermal distortion. The principal source of energy deposition at the first wall will be due to the charged particles and X rays supplemented to a lesser extent by any reflected photons.

The task of calculating the first wall response (i.e., the surface temperature rise and mechanical stress due to a pulse) is quite difficult. Because of the extreme conditions at the first wall surface, phase relations will play a large role. Furthermore, thermal relaxation time constants for first wall materials are often of the same order as the heat deposition times. More specifically, if τ_m is the mechanical relaxation time and τ_T is the thermal relaxation time of the wall, we find that the deposition time spread of the neutrons, X rays, and high energy alphas is $\tau < \tau_m \ll \tau_T$, while that of the pellet debris is typically much longer $\tau_m \ll \tau_T \sim \tau$.

The rapid energy deposition and short deposition range of charged particle debris and X rays can cause severe damage to the surface of first wall materials. In fact, studies of wall lifetime have concluded that an unprotected dry wall cavity design made from any reasonable material will not survive microexplosions at economically reasonable wall loadings ($\gtrsim 1$ MW/m^2) because of excessive thermal ablation and sputtering of the wall surface material.[9] Both graphite and metallic first walls (Mo, Ta, SS) will experience large surface temperature excursions since the ions and soft X-rays deposit their energy in a thin surface layer. Excessive ablation (~ 1 cm/yr) will take place. Sputtering will also be significant (~ 1 cm/yr) since damage occurs at elevated temperatures. (Sputtering yields increase sharply as the surface approaches the melting

Table 10.3. Anticipated Structural Materials Requirements for Fission and Fusion Reactors

Parameter	LMFBR	MCF	ICF
Temperature (°C)	300–600	300–500 (Steel)	300–500 (Steel)
		500–1000 (refractory)	500–1000 (refractory)
Maximum instantaneous displacement rate (dpa/s)	$\sim 10^{-6}$	$3-10 \times 10^{-7}$ (mirrors and tokamaks)	$\sim 1-10$
		$1-10 \times 10^{-5}$ (theta pinch)	
Average [a] disp. rate (dpa/yr)	~ 50	10–30	10–30
Helium gas prod. (appm/yr)	~ 10	200–600 (Steel)	200–500 (Steel)
		25–150 (refractory)	25–150 (refractory)
Power cycles per year	~ 10	~ 10 (mirror)	$10^7 - 10^9$
		$10^3 - 10^5$ (tokamak)	
		3×10^6 (theta pinch)	
Stress Level (MPa)	60–120	60–120	100–200
		Desired Lifetime Conditions[b]	
Displacements per atom	100–150	300–1000	300–1000
Helium product (appm)	20–30	6,000–20,000	6,000–20,000
$\Delta V / V_0$ (%)	<5	<10	<10
Creep	<1	<1	<1

[a] PF = 70%.
[b] 30-yr lifetime.

temperature.) By way of contrast, spallation of wall material does not appear to be a serious problem since there will be no thermoelastic stress wave from ion energy deposition (because of the spread in arrival time). Also, if the X ray spectrum is harder than about 1 keV black body, only small amplitude transient stresses will be generated.

Neutron damage to structural materials will also play an important role in cavity design. The primary neutron damage mechanisms are atomic displacements and gas production (primarily helium). Displacement damage is expressed in terms of displacements per atom (dpa) and gas production is expressed as atomic parts per million (appm). The damage limits for type 316 SS at an operating temperature of 500°C are estimated to be 150 dpa and 500 appm helium. At a neutronic wall loading of 1 MW/m², an unprotected first wall of 316 SS would experience a displacement damage rate of roughly 10 dpa per full power year, and the helium production rate would be about 220 appm per full power year. A more detailed comparison of displacement rates and gas production for various reactor types is given in Table 10.3.

10.2.3.   REACTOR CAVITY DESIGNS

X-ray and charged particle debris damage to first wall materials is thought to be the most significant problem associated with inertial confinement fusion systems. A variety of reactor cavity designs have been proposed to deal with

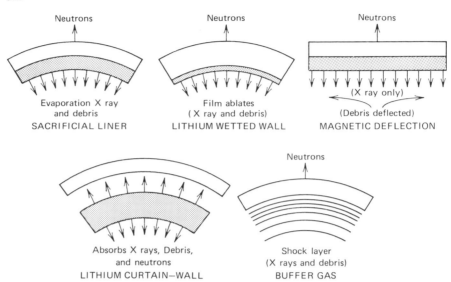

Figure 10.5. Various approaches to first wall protection in ICF reactors.

this problem, including[10,11]: (1) dry wall, sacrificial liners, (2) magnetically protected walls, (3) wetted walls, (4) fluid curtains or jets, and (5) gas-filled cavities. These approaches differ primarily in the way in which the inner surface of the first wall interacts with X rays and microexplosion debris. (These approaches are illustrated schematically in Figure 10.5.) In the dry wall[12] approach a sacrificial metal or ceramic liner is placed between the fusion chamber and the blanket. This wall would then be replaced periodically. The magnetic protection concept[13,14] uses a solenoid-generated magnetic field to divert the pellet debris away from the sides of a cylindrical blanket and into conical collectors at top and bottom. The wetted wall[15,16] approach features a thin layer of liquid metal that covers the metal wall and protects it from the blistering and structural ablation that would otherwise occur from the microexplosion debris. The fluid curtain[17,18] or jet[19] approach shields the first wall from X rays, neutrons, and debris with a thick falling region of liquid metal (lithium) or solid pellets. The gas-filled cavity[20] design fills the blast chamber with a buffer gas such as xenon at less than 1-torr pressure, sufficient to protect the first wall from the ions and soft X rays produced by the microexplosion. In this section we will discuss each of these approaches in detail.

Dry Wall/Sacrificial Liner Designs. We have noted that an unprotected or dry wall would experience such extensive surface damage from fusion reaction debris that it would not constitute an acceptable design by itself. However, it has been proposed[21] to place a sacrificial liner fabricated out of a material such as graphite that could protect the first wall. The sacrificial liner would experience thermal ablation and sputtering damage until it is reduced to a

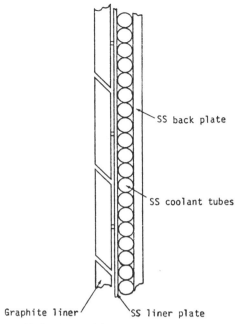

SS back plate

SS coolant tubes

Graphite liner   SS liner plate

Figure 10.6. A graphite-protected dry wall design.

minimum design thickness. At this point it would be replaced. A graphite protected dry wall design is shown in Figure 10.6 (where the first wall is fabricated out of stainless steel and contains channels for coolant flow). Other dry wall concepts which have been investigated have specified unprotected niobium walls.

Magnetically Protected Wall Designs. This design[22-24] utilizes an axial magnetic field produced by exterior coils to divert the charged particle debris out into conical energy sinks which are located on the ends of a cylindrical reactor cavity (see Figure 10.7). As designed, the graphite protected cylindrical first wall would see only the 10 to 20% X-ray yield plus the 0.1% reflected laser light flux. The energy sinks would be fabricated from refractory materials such as graphite and would be replaced periodically when radiation and material damage levels exceed operating tolerances. In principle, at least, the magnetically shielded first wall could be combined with direct conversion of the pellet debris kinetic energy into electricity (although this would only be attractive with advanced fuel schemes utilizing D-D or p-^{11}B reactions).

This approach suffers from several drawbacks, however. Although the alpha particles act as single particles and are quickly diverted to the sink cones, the slower debris plasma acts collectively in doing work against the magnetic field. Thus as the debris expands out the ends of the cylinder it first excludes and then compresses the magnetic field between the plasma and the cavity wall.

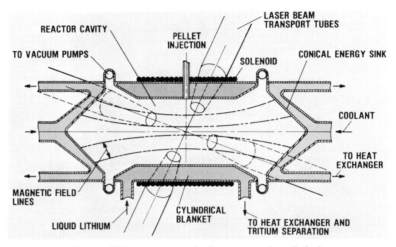

Figure 10.7. A magnetically protected wall design.

This can give rise to plasma instabilities that would cause particles to cross field lines and impact against the wall. Furthermore, magnetically shielded first walls present a disadvantage if liquid metal coolants are to be used because of the pumping power required to move the coolant across field lines. The blanket modules and the first wall would be more inaccessible than in the dry wall concepts.

Wetted Wall Designs. In wetted-wall designs,[15, 16, 25] the cavity wall is formed by a porous refractory metal through which coolants such as lithium flows to form a protective coating for the first wall surface (see Figure 10.8). This coating for the first wall will serve to absorb charged particles and reflected laser light while attenuating X rays. Typically, the coating will be about 1- to 2-mm thick, of which about 0.1 mm will be evaporated and ablated following each shot. Because of this ablation, such cavities would be limited to about one shot per second, the time delay associated with replenishing the protective layer and pumping the cavity back down to pressures of 10^{17} atoms/cm^3. The major disadvantages of this design are the large vacuum pumping loads required due to the high vapor pressure of the lithium flow and the complex first-wall designs which must allow the coolant to migrate from reservoirs to cover the first wall liners.

Lithium Curtain or Jets. This concept[5, 26-33] features a thick, continuously recyclable first wall of lithium jets that protect the first structural wall from direct exposure to the ICF microexplosions (see Figure 10.9). Each shot disassembles the "waterfall" or "forest" of lithium jets, which is then reestablished between shots. The lithium is continuously pumped to the top of the vacuum chamber through a reservoir region separating the first structural wall

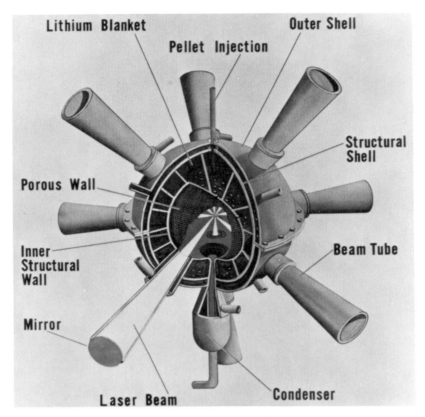

Figure 10.8.   A wetted wall design.

from the pressure vessel. A small fraction of the lithium flow circulates as the primary coolant to heat exchangers. The return flow from the heat exchangers is injected through a vortex generator to protect the top of the chamber.

The principal purpose of the fall is to reduce neutron damage in blanket structural materials, allowing them to survive the useful life of the plant. Besides moderating neutrons, the fall also absorbs photons (X rays and reflected laser light) and pellet debris (alpha particles, unburned fuel, and other pellet material). Because the fall is separated from the chamber wall, any shock wave produced in the fall would not be directly transmitted to the structural wall.

The falling lithium region contains enough lithium to significantly reduce neutron damage to the reactor structural materials from atomic displacements and gas production. Such a system could be operated at a wall loading of 4 MW/m^2 for the 30-year life of the plant without exceeding radiation damage limits. The liquid lithium waterfall or jet concept also appears to yield excellent

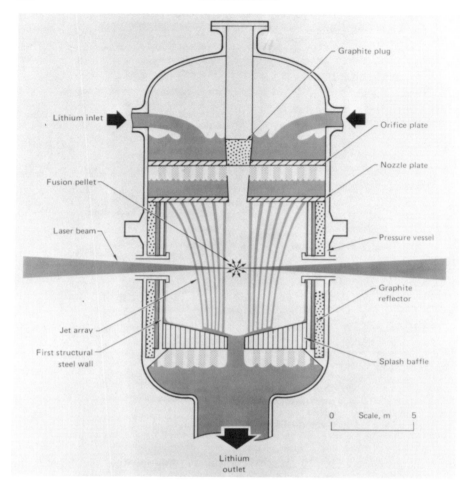

Labels on figure:
- Graphite plug
- Lithium inlet
- Orifice plate
- Nozzle plate
- Fusion pellet
- Laser beam
- Pressure vessel
- Graphite reflector
- Jet array
- First structural steel wall
- Splash baffle
- Lithium outlet

0   Scale, m   5

Figure 10.9.   First wall protection using lithium jets. (HYLIFE design.)

energy-conversion, energy-removal, and tritium-breeding characteristics. Nearly 99% of the total energy is deposited directly in the primary lithium coolant. This essentially eliminates cyclical thermal stresses in the structural walls.

The principal disadvantages appear to be the mechanical complexity of the design and the limitations of allowable partial atmospheres in the cavity to allow laser or heavy ion beam propagation. In addition, pumping power to maintain the waterfall or jet flow will add significantly to the recirculating power in the plant.

Gas-Filled Cavity Designs.[1, 20, 34–37]   In gas-filled cavity designs for laser drivers neon or xenon is included in the cavity at pressures of 0.5 to 1 torr to act as a buffer gas to prevent the charged particle debris from striking the first

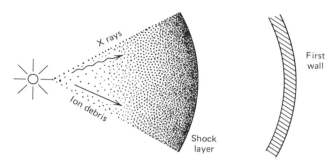

Figure 10.10.   First wall protection using a buffer gas.

wall. The energy deposited in the gas is radiated to the front wall over a relatively long time period (~ 1 ms) so that surface heating and thermal ablation of the wall become insignificant. The maximum overpressure generated at the first wall is only about 100 torr. The primary concern in this design is to keep the gas density sufficiently low to avoid laser beam defocusing and attenuation. An example of this design concept is the SOLASE laser fusion reactor system discussed later in this chapter.

In light ion beam driven systems, the gas pressure in the cavity must be between 5 to 50 torr.[38,39] This is because ionized channels are established in the gas between the diodes and the target by laser breakdown followed by a discharge current. The ion beam pulses are then injected into these z-pinch channels and propagate to the target. At this gas pressure, the target debris is quickly attenuated in the gas forming a fireball. The shock that is generated by this fireball propagates outward (see Figure 10.10) and reflects from the first

Table 10.4.   Current Status of Reactor Cavity Designs

Bare metal:	Very large cavity diameters (10's meters)
	Economically infeasible
Sacrificial liner:	Frequent liner replacement
	Probably not economic
Magnetic Deflection:	Periodic liner replacement (only from X-ray evaporation)
	Further analysis required
Gas fill:	Protection from X rays and ion debris
	Restoration of initial conditions uncertain
	Further analysis required
Wetted wall:	Protection from X rays and ion debris
	Significant wall impulse
	Repetition rate limited to about one shot per second
Thick lithium wall or jets:	Protection from X rays and ion debris
	Mitigation of structural radiation damage
	Significant wall impulse
	Restoration of initial conditions uncertain
	Further analysis required

wall. This shock can generate overpressures of 1 to 10 atmospheres at the wall. Hence in this design, the wall must be constructed of and supported by strong structural materials. A major difficulty in this concept involves the reestablishment of cavity conditions between pulses so that new propagation channels can be formed for the next shot.

We have compared various reactor cavity design concepts in Table 10.4.

10.3.   OTHER ASPECTS OF ICF REACTOR DESIGNS

10.3.1.   BLANKET DESIGNS

The blanket system of the reactor must perform several functions. It must convert the fusion energy into thermal energy, provide for the efficient removal of this thermal energy, and breed enough tritium to replace that which is burned in the fusion reaction. The blanket system must also maintain the required vacuum in the fusion chamber.

Perhaps the primary constraint on the design of most fusion reactor blankets is the requirement that tritium must be continually bred and processed from lithium. Since natural lithium is isotopically 7.4% ^{6}Li and 92.6% ^{7}Li, one can make use of two reactions:

$$^7\text{Li} + n(2.5 \text{ MeV}) \rightarrow {}^4\text{He} + {}^3\text{T} + n(\text{slow})$$

$$^6\text{Li} + n(\text{slow}) \rightarrow {}^4\text{He} + {}^3\text{T} + 4.8 \text{ MeV}$$

Note that the ^{7}Li reaction is a threshold reaction that is neutron conserving, and that the ^{6}Li reaction has a large thermal cross section and is exothermic. Thus, in a very ideal case, the best we can hope to do with a single 14.1-MeV fusion neutron and a pure lithium system is to cause a reaction first with ^{7}Li and have the resulting slow neutron absorbed in ^{6}Li. The result would be one surplus tritium atom (i.e., a breeding ratio of 2) and about 16.4 MeV of energy $(14.1 - 2.5 + 4.8)$.

In most designs, liquid lithium is used both as a breeding medium and as the blanket coolant. However, other tritium breeding compounds which may prove compatible with fusion reactor designs include molten salts (Li_2Be_4 or "flibe" and LiF), ceramic compounds (Li_2O and Li_2C_2) and aluminum compounds ($LiA1, LiA1O_2$).

Liquid-Lithium Cooled Stainless Steel Manifold. The conventional approach requiring the least in sophisticated technology would be a cylindrical annulus of stainless steel into which vertical coolant channels are drilled to form a manifold (see Figure 10.11). The stainless steel manifold concept[21] is compatible with either a dry or wetted first-wall cavity design. In the dry wall approach, one could use a graphite liner that is supported by stainless steel and cooled with liquid lithium. The graphite liner might be designed for an operational lifetime of one year. In the wetted-wall approach, one could

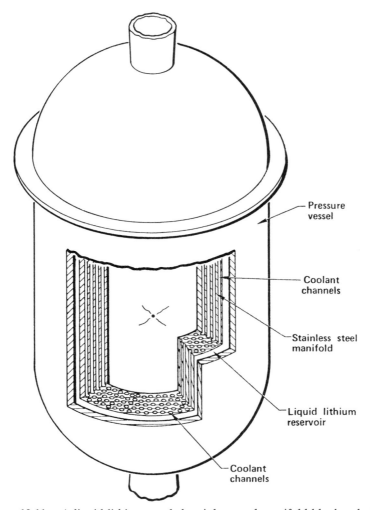

Figure 10.11. A liquid-lithium–cooled stainless steel manifold blanket design.

maintain a thin (3-mm) film of liquid lithium on the inner surface of the manifold. Tritium breeding considerations limit the thickness of a structural wall of solid stainless steel to less than 10 cm. The stainless steel manifold would operate at a neutronic first wall loading of about 1 to 2 MW/m², and this would require a relatively large chamber radius (10 to 15 m for a 4000 MWt system). For use with stainless steel, lithium temperatures must be limited to about 500°C to avoid excessive corrosion.

Gas-Cooled Graphite Manifold. A graphite manifold[21,40] would be similar to the stainless steel manifold, except that the vertical coolant channels are drilled into an array of graphite blocks that make up the fusion chamber. One could adapt high temperature gas-cooled fission reactor technology (HTGR) to

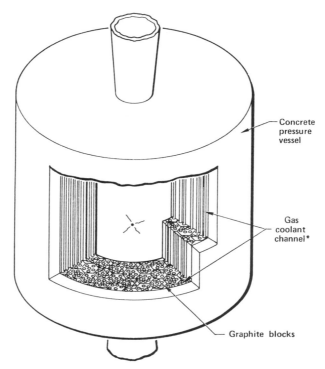

*Coolant channels filled with ceramic lithium

Figure 10.12.   A gas-cooled graphite manifold blanket design.

design a vacuum vessel of reinforced, prestressed concrete as shown in Figure 10.12. High pressure helium gas would then be pumped through the coolant channels, some or all of which are filled with pellets of a lithium ceramic. Tritium is removed from these channels by the gas coolant as it diffuses out of the lithium compound in which it is bred. The graphite manifold design is a reactor concept that exhibits low activation and low tritium inventories.

Fluidized Wall and Jet Concepts.   Fluidized wall concepts provide protection to the first metallic wall from high energy neutrons in addition to the X rays and debris. These concepts appear to be less dependent on materials development because radiation damage is significantly reduced. Designs using both a liquid lithium waterfall and an array of lithium jets have been proposed and studied. The jet arrangement is shown in Figure 10.13.

These designs feature a thick fall of liquid lithium that protects the first structural wall, allowing it to last for the useful life of the plant. By keeping the fall off the chamber wall, shock waves generated in the fall are not directly transmitted to the structural wall. The majority of the fusion energy is thus deposited in the liquid lithium, which serves as the primary coolant, fertile

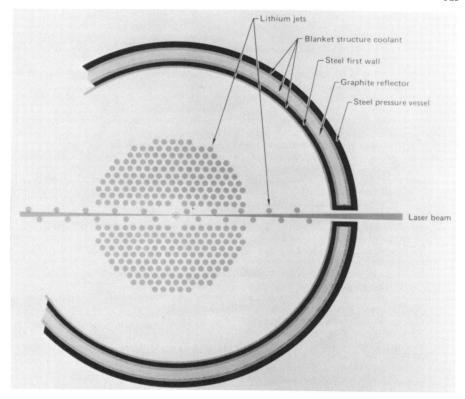

Figure 10.13. Lithium wall or jet blanket design.

material for tritium breeding, and first wall. Hence the system does not have to rely on conduction of heat through structural materials to remove thermal energy.

As one might expect, fluidized wall or jet concepts have excellent tritium breeding characteristics.[21,41] With no structural material between the fusion neutrons and the lithium fall, the design takes full advantage of the high energy $^7\text{Li}(n, n'T)$ reaction.

Gas-Filled Cavity Designs. The prototype gas-filled cavity design is the SOLASE system studied at the University of Wisconsin (see Figure 10.14). The SOLASE blanket is constructed primarily with graphite, either nuclear grade graphite or chopped-fiber type graphite composite. Lithium oxide particles, 100 to 200 nm in diameter, flow under gravity through the blanket and serve the dual purpose of tritium breeding and heat transfer. This design eliminates the need for metallic first walls and liners since the buffer gas attenuates the radiation to a level compatible with first wall thermal and mechanical tolerances. The estimated lifetime of the blanket structure at 5 MW/m² wall

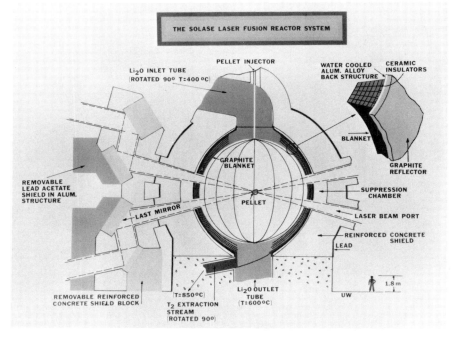

Figure 10.14. A buffer gas protected cavity design. (SOLASE design.)

loading is approximately 1 year (corresponding to 30 dpa). Hence the design utilizes a blanket structure that can be periodically replaced in segments.

10.3.2. TARGET INJECTION

Injection of ICF targets into a power reactor chamber will be quite different from the static mounting procedures used in experimental research today.[42,43] Electrostatic and pneumatic injection are the two injection modes of most interest. Electrostatic methods appear to be more susceptible to damage in the intense radiation and temperature environment of the reactor chamber. Hence most interest rests with pneumatic injection.

For example, one could inject the pellet with a pneumatically operated launching device that injects the pellet along a vertical trajectory. The state of the art of pneumatic injection is quite well developed, with present high precision air guns being able to deliver projectiles into an area of radius less than 1 mm at a distance of 12 m with a horizontal trajectory.

A more difficult problem involves tracking the pellets on their path to the focal point and reaiming the driver beam. For this, laser doppler velocimetry has been proposed along with adaptive optics in the beam lines to steer the beams.

10.3.3. DRIVER DEVELOPMENT

We have summarized the status of high power laser and particle beam driver development in Chapters 7 and 8. We suspect that driver development will be the most critical link in the achievement of ICF power generation systems. The complexity of developing reliable and efficient drivers with the required energy, power, and beam qualities necessary for ICF applications is formidable.

No presently known class of laser will obviously meet the requirements for laser fusion. However, several lasers such as the CO_2 and KrF can probably be scaled to meet minimum requirements for energy (1 to 4 MJ), power (100 to 1000 TW), and pulse width (1 to 10 ns). There are few lasers that have the potential to achieve an efficiency greater than 1 or 2%. If a power plant can be designed to operate with a laser of this efficiency, then the number of potential laser candidates increases substantially, and the chance of finding a suitable laser should also improve. However, there is still considerable uncertainty as to whether the laser-target coupling processes can yield pellet gains in the range from 100 to 1000 required by reactor applications.

While the requirements of efficiency and energy per pulse can be easily achieved with charged particle beams, and while, at least for ion beams, the driver-target interaction process seems adequate, there is still considerable doubt as to whether this type of driver can ever be developed for commercial applications. The problems in producing a beam of sufficient pulse length and power level and delivering it to a target look quite formidable.

Finally, it is important to keep in mind that the lifetime requirements for driver and power supply components must be about 10^9 shots for commercial applications. This is several orders of magnitude beyond the present state of the art and represents a serious challenge to successful driver development.

10.3.4. BEAM OPTICS

An important facet of laser fusion reactor design involves beam optics. To transport and focus intense laser beams into the target chamber, and to protect the necessary lenses and mirrors from the radiation produced in the microexplosion presents a serious challenge. Of particular concern are the last mirrors of the optics chain since these mirrors will be in a direct line of sight to the microexplosion.

The University of Wisconsin SOLASE study determined that uniform pellet illumination is not truly compatible with reactor requirements.[44] They therefore used nonuniform illumination with six beams on each side to yield two-sided illumination (much as in the Shiva and Nova systems).

Surface damage to final mirror components by X rays and pellet debris is a significant problem. Large quantities of pellet debris on mirror surfaces cannot be tolerated. Physical, and more importantly, chemical reactions at the mirror surface will degrade beam optical quality. A method to protect the mirrors from the pellet debris is to flow a gas, such as xenon or neon, in front of the mirror surface.

It is felt that neutron induced damage is not a serious problem for uncoated mirrors. Neutron damage in a Cu on Al substrate mirror at 15 m from a 150-MJ yield pellet at the rate of 20 Hz is moderate, about 10^{-7} dpa/s, and the neutron heating is less than 10 W/cm^3. This provides little incentive to move the mirror farther than 15 m from the target.[45]

However, mirror damage may be quite different if coated mirrors are required. Dielectric coatings are required to increase reflectivity for laser wavelengths of less than about 1 μm (see Ref. 20). These coatings may be susceptible to color center formation induced by neutron and X-ray damage. If this is the case, then the final optics in short wavelength laser driven reactors might require positioning as much as 100 m from the target. This seriously complicates pointing errors, but it does not seem to rule out this option.

10.3.5. FUEL PELLETS

Although detailed pellet designs are still highly speculative, one might well assume a generic pellet consisting of a frozen D-T fuel encapsulated by a glass or polyvinyl alcohol shell. This shell is then surrounded by a high-Z (or possibly low-Z) layer and a final low-Z, low density ablative zone.

The fabrication of these targets involves three main processes: filling, cryogenic processing, and layer deposition techniques. All techniques are widely used today and appear to be capable of meeting ICF target requirements, although fabrication costs may be a serious problem.

Storage of fuel pellets will be required because the manufacturing process is a batch process and because a store is needed to allow plant operation in the event of a failure. The storage is likely to be at cryogenic temperatures to minimize the outdiffusion of the D-T fuel. Furthermore, the total plant tritium inventory is highly sensitive to storage methods.

A particularly important aspect of ICF fuel pellet design will be the compatibility of pellet materials with cavity walls. Pellet designs must avoid the use of reactive materials and minimize oxygen and hydrogen content.

10.3.6. BALANCE OF PLANT

The design of ICF generating stations will vary considerably with the different first wall protection. For example, an early Los Alamos design [13-15] based on the wetted wall cavity protection system proposed incorporating up to 26 reactor cavities (each generating about 120 MWt) with pairs of cavities served by common heat transfer loops, steam generators, and fuel processing systems. Two eight-beam laser systems would each have the capability to drive all of the cavities via a rotating mirror that would direct the beam to each cavity, respectively. Each laser would have a redundant partner to achieve high reliability.

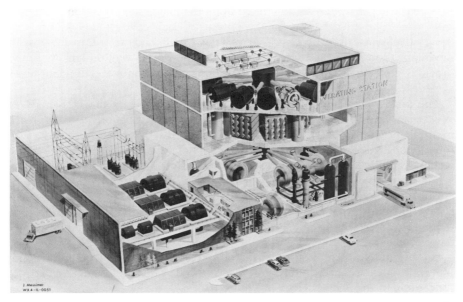

Figure 10.15. A schematic of an electric generating station based on a magnetically protected reactor cavity design. (Courtesy of Los Alamos National Laboratory.)

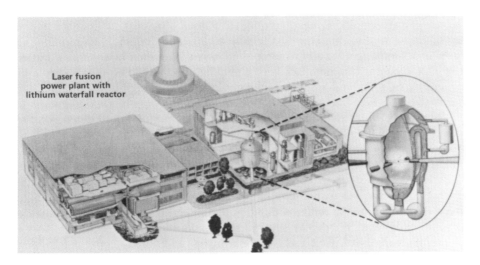

Laser fusion
power plant with
lithium waterfall reactor

Figure 10.16. An electric generating station based on the HYLIFE reactor design. (Courtesy of Lawrence Livermore Laboratory.)

An electric generation station based upon the magnetic protection concept is shown in Figure 10.15. Due to the output of each cavity (1249 MWt), this design is characterized by only four cavities which leads to a lesser degree of modularization and shared components than the wetted wall design. As in the previous concept, there would be a rotating mirror and a redundant laser system.

The use of such multiple cavities was briefly studied by the University of Wisconsin Fusion Project group. Although the primary motivation for such multiplicity was to increase the potential availability time of the reactor, they found that the most reliability-sensitive subsystem was the laser rather than the reactor chamber. Multiple laser and power supplies, while not economically attractive, would be preferable to multiple cavities. Furthermore, the use of multiple cavities increases the complexity of the beam line system and reduces accessibility. Multiple cavities would appear to be necessary only if the time required to reestablish the chamber environment (e.g., due to vacuum pumping) becomes too long.

An alternative plant design of some interest is that based on the lithium waterfall or jet concept, referred to as the HYLIFE design. An artist's conception of the plant is given in Figure 10.16.

10.3.7. A DETAILED EXAMPLE: SOLASE

The SOLASE reactor[20] designed by the University of Wisconsin Fusion Engineering Program is shown in Figures 10.14 and 10.17. This reactor is designed to produce 965 MWe at a net efficiency of 29% from laser driven ICF fuel pellets with a gain of 150. The laser energy on target is taken as 1 MJ, and 20 targets are imploded each second. The laser efficiency is assumed to be 6.7%, including multipassing of the next-to-last and last laser amplifier. The laser is designed, generically where possible, as a gas phase laser modeled after the CO_2 system, but no laser wavelength is specified. The optimistic laser efficiency of 6.7% still implies relatively large power supply needs, and the recirculating power fraction is 28%.

Thermonuclear burn dynamics calculations were performed to determine the pellet debris spectra for cavity design analysis.[46] A buffer gas of neon at 0.5- to 1-torr pressure is used to stop the ions. Multilayered cryogenic targets are produced in a batch process. Target delivery is by pneumatic guns, although trajectory diagnostic and correction techniques must be developed. The last mirrors are diamond turned copper on an aluminum structure located 15 m from the reactor cavity center with f-number 7.5. Heating of the mirror surface is minor so long as the debris ions are stopped in the buffer gas.

The reactor cavity itself is a sphere 6 m in radius. It is constructed from graphite designed to guide the gravitation flow of lithium oxide (Li_2O) pellets serving both as the tritium breeding and heat transport medium. The breeding ratio is 1.33, and the maximum Li_2O flow velocity is only about 1 m/s. The neutron wall loading is 5 MW/m^2 so that SOLASE presents a reasonably

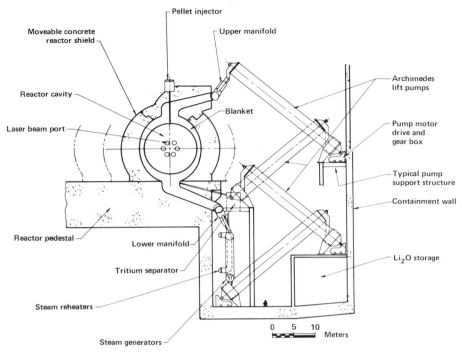

Figure 10.17. The SOLASE ICF reactor.

compact system given the net power produced. The blanket back structure is made from an aluminum alloy, and the shield can be either concrete or lead acetate solution. Thus, the overall levels of neutron induced activity decay very rapidly following shutdown. It appears that limited hands-on maintenance is possible after just one week.

This spherical system is highly accessible from the outside provided that two-sided target illumination (six beams on each side) is acceptable. A procedure has been developed for annual blanket replacement that is fast and simple. The expected downtime to replace graphite blanket segments is two weeks. The design philosophy is that blanket maintenance per se will be avoided; after draining the Li_2O, the graphite will simply be discarded.

In the steam power cycle, Li_2O pellets transport heat directly to steam generators that then drive turbines producing 1349 MWe gross.[47] The laser recirculating power requirement is 300 MWe, and other internal plant power requirements lead to a net plant electrical output of 965 MWe and a net thermal efficiency of 29%. The large recirculating laser power fraction appears typical of laser fusion systems unless gains much larger than 150 or laser efficiencies much greater than 5 to 10% can be achieved.

A cost analysis of the SOLASE design has been performed by United Engineers and Construction, Inc., by building up the cost, item by item, for all

Table 10.5.   Capital Cost Summaries for SOLASE Design (1979 Dollars).

Item	Cost (Millions of Dollars)
Direct cost	
Land and land rights	2.4
Structures and site facilities	131.2
Reactor plant equipment	879.0
Turbine plant equipment	206.0
Electric plant equipment	69.5
Miscellaneous plant equipment	14.6
Special materials	175.6
Subtotal	1478.3
Total design allowance (acct. 22 only)	87.9
Total spare parts allowance	27.5
Total contingency allowance	159.1
Indirect cost	
Construction facilities, equipment, and services	263.0
Engineering and construction management services	263.0
Other owner's costs	87.7
Total Direct and Indirect Cost	2366.5
Interest during construction (9 years)	1479.2
Total Capital Cost	3845.7
Total Construction Cost	$3985/kWe

reactor-related and balance of plant systems. A capital cost account summary is given in Table 10.5. When annual fuel and operational costs are added to the construction cost, the total busbar energy cost is 66.1 mils/kWh. The three largest cost items are the pellet fabrication factory, the Li_2O heat transfer system, and the laser system.

10.3.8.   SOME CONCLUSIONS

As we have noted, laser driven fusion power plants will have a large recirculating power fraction (30% or so) unless very high gain targets ($\gtrsim 500$) or high efficiency lasers (~ 10 to 20%) can be developed. The development of targets with even modest gains (~ 100) and lasers with modest efficiencies (~ 1%) is highly uncertain at present. The development of power supplies and pulsed power switching with high reliability (10^8 to 10^9 shots) and modest costs is also a major concern.

The constraints of few beams and pellet physics have led to a variety of reactor cavity concepts. There appears to be no essential constraint on cavity geometry (except cavity diameter), unlike magnetic fusion systems. Furthermore, the background density of cavity fill and debris gas appears to be low enough to permit beam propagation.

Most reactor design considerations are dominated by protection of the first wall and the achievement of high repetition rates. Preliminary reactor design studies appear to rule out the use of dry or bare metal walls. The tendency is to move toward either fluid walls or sacrificial (replaceable) liners with limited life. The success and choice of a cavity design will strongly depend on pellet materials and output characteristics.

Another important issue specific to ICF applications has to do with security classification. The relationship of ICF to technical ideas and information related to thermonuclear weapons could severely impede the commercialization of this technology due to extensive classification of pellet design ideas. It seems highly unlikely that ICF power generation could be developed for commercial implementation in the face of the present classification restrictions. In particular, it seems doubtful that an electric utility would add the extra burden of classification to the already overwhelming hurdles it faces in power plant licensing and regulation.

10.4. HYBRID FUSION/FISSION SYSTEMS

An alternative application of any D-T fusion system, whether based on magnetic confinement or inertial confinement fusion, is the utilization of the 14 MeV neutrons to produce fissile fuel (Pu or U-233) by neutron transmutation of fertile materials (U-238 or Th-232). This is made possible by the fact that high energy neutrons will undergo many neutron mutiplying reactions in the reactor blanket such as $(n, 2n)$, $(n, 3n)$, and $(n, \text{fission})$. This will allow the breeding of fissile fuel as well as sufficient quantities of tritium to complete the fusion fuel cycle.[50-56] In fact, there are roughly 2 to 4 neutron capture events per fusion neutron in most hybrid reactor blanket designs. In these systems, the blanket operates as a subcritical fission assembly driven by the fusion neutron source.

The blanket can be designed to achieve a number of different goals. If the neutron spectrum in the blanket remains hard, then the system will be a net producer of fissile fuel which can be periodically removed, reprocessed, and burned in conventional fission reactors. If the neutron spectrum is softened by including moderating materials in the blanket, then the bred fuel will be burned in situ. This results in a large power multiplication in the blanket, in some cases as high as $M=50$ to 100. Or the system can be designed to operate somewhere in between these two extremes.

The principal advantages of fusion/fission hybrid systems include the following: First, they produce prolific quantities of fissile fuel that can be burned in conventional fission reactors. This has several implications. First, the ultimate energy released per D-T fusion neutron is now roughly 20 MeV from fusion plus 200 MeV from the eventual fission of the bred fuel. This has the potential of providing more "bang for the buck" than a pure fusion reactor. Second, the majority of the power is produced in conventional fission reactors. Hence there need be no new technology development for this part of the energy production system.

The power multiplication in the blanket can significantly reduce the fusion performance requirements for the hybrid reactor. Recall that the recirculating power fraction is given by the expression

$$RPF = \frac{1}{\eta_L Q \eta_t [1 + f_n(M-1)]}$$

where M is the blanket multiplication. Hence the fusion gain can be reduced proportionally to the blanket multiplication. This is important for several reasons. First, the reduced fusion performance may allow an earlier introduction date for the hybrid than for a pure fusion system due to technological simplifications. Second, this earlier introduction date combined with the substantial fissile fuel production gives the hybrid reactor a large potential impact on the energy market.

Of course, hybrid reactors are not without their own special problems. In addition to combining the best features of neutron rich fusion reactions and energy rich fission reactions, they also combine several of the worst aspects of reactors based on either process alone. ICF fusion reactors at best will be very complicated devices with large amounts of advanced technology, including lasers or accelerators, target fabrication factories, tritium removal systems, and so on. To this, the hybrid adds the specter of fissionable material with the always present fission products and actinides. The technical problems presented by these materials may be overshadowed only by the sociopolitical implications of producing fissile fuel.

A number of hybrid reactor designs based on ICF reactors have been performed over the past few years.[57-62] The "fusion part" of these designs has been quite similar to that characterizing pure fusion ICF systems although with some relaxation of the target gain or the driver efficiency. An example of a hybrid reactor based on the lithium jet scheme discussed in section 10.3.1 is shown in Figure 10.18. The blankets in these designs have received the greatest attention. These have shown a striking similarity to liquid metal cooled fast breeder reactor (LMFBR) core designs (i.e., stainless steel clad fuel, sodium coolants, etc.).

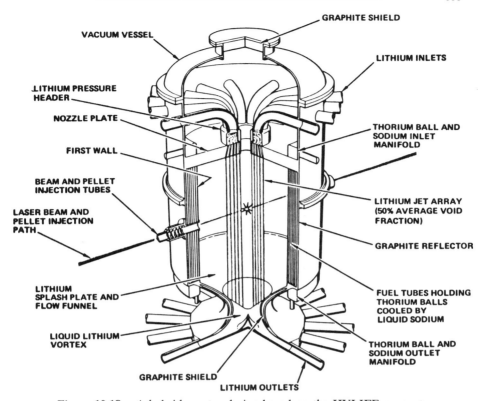

GRAPHITE SHIELD

VACUUM VESSEL

LITHIUM INLETS

LITHIUM PRESSURE HEADER

NOZZLE PLATE

THORIUM BALL AND SODIUM INLET MANIFOLD

FIRST WALL

BEAM AND PELLET INJECTION TUBES

LASER BEAM AND PELLET INJECTION PATH

LITHIUM JET ARRAY (50% AVERAGE VOID FRACTION)

GRAPHITE REFLECTOR

LITHIUM SPLASH PLATE AND FLOW FUNNEL

FUEL TUBES HOLDING THORIUM BALLS COOLED BY LIQUID SODIUM

LIQUID LITHIUM VORTEX

THORIUM BALL AND SODIUM OUTLET MANIFOLD

GRAPHITE SHIELD

LITHIUM OUTLETS

Figure 10.18. A hybrid reactor design based on the HYLIFE concept.

The neutrons produced by the D-T fusion reaction can also be used to transmute long-lived fission waste products (primarily actinides) to short-lived or stable isotopes.[63-65] This method could serve as a mechanism to relax requirements for long-term geological storage of high level radioactive wastes. One particularly interesting study, the laser fusion-driven actinide waste burner (LDAB) uses partitioned fission reactor generated actinide wastes dissolved in a molten tin alloy as fuel. A novel fuel-processing concept is used which involves the high-temperature precipitation of actinide nitrides from the liquid tin solution. This design allows for fission product removal at high burnup.

10.5. PROCESS HEAT AND SYNTHETIC FUEL PRODUCTION

Other applications of ICF reactors have been proposed. The temperatures in a laser fusion reactor blanket will be limited only by the properties of refractory materials in the blanket. Thus, temperatures above the HTGR limit of 1650°K can be achieved. Such temperatures are attractive for producing process heat for industrial applications.[66] (Indeed, about 28% of the energy consumption in the United States is used to generate process heat.) ICF reactors seem to be unique in this respect, since they do not suffer from the temperature limitations of magnetic fusion systems or fuel melting of fission reactors.

Los Alamos has designed several ICF blankets suitable for process heat production. Using two zone blankets, composed of pure carbon and a 90% carbon–10% boron carbide mixture, spherical ICF reactors have been designed that would supply from 20 to 100 MW of thermal power at about 2100°K.

ICF reactors could also supply the energy necessary to produce synthetic fuels. A variety of approaches have been proposed, including thermochemical, electrolytic, and radiolytic processes.[68,69]

The thermal energy from the ICF reactor could support a sequence of high temperature chemical reactions in various hydrogen-producing thermochemical processes. While these thermochemical processes generally do not consume the chemical reactants, they often use large quantities of hazardous and corrosive chemicals. Practical energy to 65% of the fusion energy is recoverable as thermal energy by burning the hydrogen produced.

Such reactors could also generate electricity for subsequent electrolysis of water to produce hydrogen. For existing electrolysis plants, the combustion energy of the hydrogen produced is 60 to 100% of the electrical energy input. Hence if we assume a 40% plant thermal efficiency, we find that this scheme

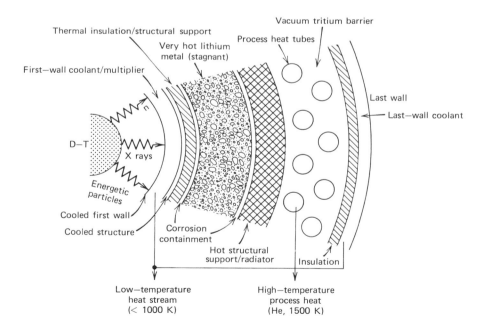

SIMPLE RADIATOR BLANKET

Figure 10.19. An ICF blanket design for synthetic fuel production.

would produce hydrogen with an overall energy conversion efficiency of 25 to 40%.

Hydrogen production by radiolysis can be achieved in several ways. The penetrating neutron radiation of an ICF reactor could be utilized by incorporating sufficient quantities of H_2O in the blanket regions surrounding the fusion vacuum chamber. In this sense, fusion reactors are quite attractive, since they produce an intense fast neutron flux in a surrounding blanket region rather that in the core proper (as with a fission reactor).

A variety of synthetic fuel production schemes have been studied which use an ICF reactor as an energy or radiation source. The blanket design for one such scheme is shown in Figure 10.19.

10.6. PROPULSION

ICF reactors have been proposed as energy sources for propulsion of marine vessels, aircraft, and spacecraft.[4,70,71] Actually, we can probably discard aircraft propulsion immediately since the power density of the overall ICF system would be less than that of chemical jets or rockets. ICF ship propulsion would appear to be characterized by features very similar to that of nuclear fission propulsion, and therefore this application would also not appear to offer any significant advantages over existing nuclear propulsion technology.

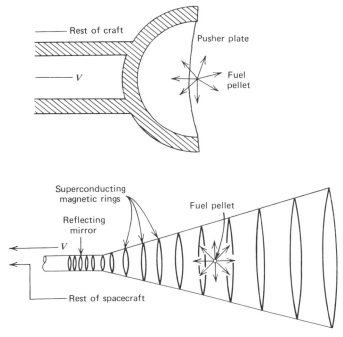

Figure 10.20. Two ICF propulsion schemes.

On the other hand, the very high velocities of fusion reaction products (10^7 m/s) suggest that rockets using this debris as propellant would be characterized by very high specific impulses (exhaust velocity divided by the gravitational acceleration g). For example, ICF systems might make possible specific impulses as high as 10^6 s in contrast to those of chemical fuels with specific impulses of 450 s or less and nuclear fission rockets with 2000 s.

Two schemes have been proposed (see Figure 10.20). The first would use the debris produced in an ICF microexplosion to collide with a pusher plate, thereby transferring some of its momentum to the spacecraft. This particular approach is in effect a microscopic approach to the Orion project in which the explosions were originally intended to be fission or fusion bombs.

A more attractive option would be to implode the pellet in a magnetic mirror that is reflecting at one end and open at the other. The charged particle reaction products would then be directed by the mirror field out the exhaust of the rocket. This latter approach would place a premium on fuels such as D-^{3}He or p-^{11}B that produce primarily charged fusion products.

Such ICF propulsion systems appear best suited for deep space missions. For example, Lawrence Livermore Laboratory has studied a laser fusion system using a 1-MJ laser pulse to produce a 260-MJ pellet yield that translates into 120-MJ producing thrust. For a pulse repetition frequency of

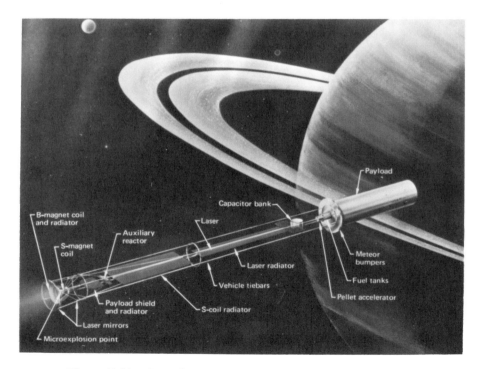

Figure 10.21. An artists conception of an ICF-powered spacecraft.

500/s, this system would develop a thrust of 2.2 tons at a specific impulse of 5.5×10^5 s (corresponding to an exhaust velocity of 6×10^6 m/s). For a spacecraft with a 300-ton propulsion system and a 200-ton deuterium fuel load, a payload of 100 tons could make a round trip to any point in the solar system in one year utilizing such an ICF drive (see Figure 10.21).

REFERENCES

1. G. A. Moses, R. W. Conn, and S. I. Abdel-Khalik, "The SOLASE Conceptual Laser Fusion Reactor Study," in *Proc. Third Topical Meeting on the Technology of Controlled Nuclear Fusion*, Santa Fe, NM (1978).

2. J. A. Maniscalco, J. A. Blink, J. Hovinsh, W. R. Meier, M. J. Monsler, and P. E. Walker, "A Laser Fusion Power Plant Based on a Fluid Wall Reactor Concept," in *Proc. Third Topical Meeting on the Technology of Controlled Nuclear Fusion*, Santa Fe, NM (1978).

3. H. J. Gomberg and W. W. Meinke, "Production of Synthetic Fuels: An Important Civilian Application of Laser Fusion," in *Miami International Conference on Alternative Energy Sources*, ed. by T. N. Veziroglu (1977).

4. Laser Program Annual Report-1977, Lawrence Livermore Laboratory Report UCRL-50021-77 (1977), p. 8–78.

5. J. A. Maniscalco, "Inertial Confinement Fusion," *Ann. Rev. Energy* 5, 33 (1980).

6. R. Gillette, "Laser Fusion: An Energy Option, but Weapons Simulation Is First," *Science* 188, 30 (1975).

7. A. J. Toepfer and L. D. Posey, "The Application of Inertial Confinement Fusion to Weapons Technology," Sandia Laboratories Report 77-0913 (1978).

8. J. Maniscalco, J. Blink, R. Buntzen, J. Hovingh, W. Meier, M. Monsler, and P. Walker, "Civilian Applications of Laser Fusion," Lawrence Livermore Laboratory Report UCRL-52349 (1977).

9. R. W. Conn, "First Wall and Divertor Plate Material Selection in Fusion Reactors," *J. Nucl. Mat.* 76, 103 (1978).

10. G. L. Kulcinski, "First Wall Protection Schemes for Inertial Confinement Fusion Reactors," in *Proc. First Topical Meeting on Fusion Reactor Mat.*, Miami, January, 1979.

11. S. I. Abdel-Khalik, R. W. Conn, and G. A. Moses, "Engineering Problems of Laser Driven Fusion Reactors," *Nucl. Technol.* 43, 4 (1979).

12. S. G. Varnado and G. A. Carlson, "Considerations in the Design of Electron Beam Induced Fusion Reactor Systems," *Nucl. Technol.* 29, 415 (1976).

13. L. A. Booth and T. G. Frank, "A Technology Assessment of Laser Fusion Power Development," Los Alamos Scientific Laboratory Report LA-UR-76-2060 (1976).

14. D. A. Freiwald, T. G. Frank, E. A. Kern, and L. A. Booth, "Laser Fusion Generating Stations Based on the Magnetically Protected Reactor Cavity," Los Alamos Scientific Laboratory Report LA-UR-75-2035; *Trans. Am. Nucl. Soc.* 22, 68 (1975).

15. L. B. Booth, "Central Station Power Generation by Laser Driven Fusion," *Nucl. Eng. Des.* 24, 263 (1973).

16. J. M. Williams, T. Merson, F. Finch, F. Schilling, and T. Frank, "A Conceptual Laser Controlled Thermonuclear Reactor Power Plant," *Proc. 1st Top. Meeting Technol. Controlled Nucl. Fusion*, San Diego (1974).

17. J. A. Maniscalco and W. R. Meier, "Liquid-Lithium "Waterfall" Inertial Confinement Fusion Reactor Concept," *Trans. Am. Nucl. Soc.* 26, 62 (1977).

18. J. Powell et al., "A Liquid-Wall Boiler and Moderator (BAM) for Heavy Ion-Pellet Fusion Reactors," *Trans. Am. Nucl. Soc.* 26, 64 (1977).

19. Laser Program Annual Report-1978, Lawrence Livermore Laboratory Report UCRL-50021-78 (1978).

20. R. W. Conn et al., "SOLASE, A Laser Fusion Reactor Study," University of Wisconsin Fusion Engineering Program Report UWFDM-220 (1977).

21. J. Hovingh, "First Wall Studies of a Laser Fusion Hybrid Reactor Design," Lawrence Livermore Laboratory Report UCRL-78090 (1976).

22. T. G. Frank, D. A. Freiwald, T. Merson, and J. J. Devaney, "A Laser Fusion Concept Utilizing Magnetic Fields for Cavity Wall Protection," *Proc. of the First Topical Meeting on the Technology of Controlled Nuclear Fusion*, San Diego (1974)

23. J. J. Devaney, "Magnetically Protected First Wall for a Laser Induced Thermonuclear Reaction," Los Alamos Scientific Laboratory Report LA-5699-MS (1974).

24. A. Freiwald, D. O. Dickman, and J. C. Goldstein, "Computer Simulation of a DT Pellet Microexplosion in a Magnetically Protected Laser Fusion Reactor," *Bull. Am. Phys. Soc.* **20**, 1238 (1975).

25. I. O. Bohachevsky, L. A. Booth, and J. F. Hafer, "Lithium Flow on the Inside of a Spherical Fusion Reactor Cavity," Los Alamos Scientific Laboratory Report LA-6362-MS (1976).

26. J. A. Maniscalco, W. R. Meier, M. J. Monsler, "Design Studies of a Laser Fusion Power Plant," Lawrence Livermore Laboratory Report UCRL-80071 (1977).

27. J. A. Maniscalco, W. R. Meier, and M. J. Monsler, "Conceptual Design of a Laser Fusion Power Plant," Lawrence Livermore Laboratory Report UCRL-79652 (1977).

28. W. R. Meier and J. A. Maniscalco, "Liquid Metal Requirements for Inertial Confinement Fusion," Lawrence Livermore Laboratory Report UCRL-80424 (1977).

29. P. E. Walker, "Environmental and Safety Features of a Lawrence Livermore Laboratory Laser Fusion Reactor Design," *Proc. Third Topical Meeting on the Technology of Controlled Nuclear Fusion, Santa Fe, NM* (1978).

30. M. Monsler, et al., "Electric Power from Laser Fusion: The HYLIFE Concept," Lawrence Livermore Laboratory Report UCRL-81259 (1978).

31. P. E. Walker, "Remote Systems Requirements of the High Yield Lithium Injection Fusion Energy (HYLIFE) Converter Concept," Lawrence Livermore Laboratory Report UCRL-81309 (1978).

32. W. R. Meier and W. R. Thomson, "Conceptual Design and Neutronics of Lithium Fall Laser Fusion Target Chambers," Lawrence Livermore Laboratory Report UCRL-80782 (1978).

33. J. H. Pitts et al., "Potential Design Modifications for the HYLIFE Reactor Chamber," *Proc. Eighth Symposium on Engineering Problems of Fusion Research, San Francisco* (1979); Lawrence Livermore Laboratory Report UCRL-82895 (1979).

34. J. Howard, "First Wall Protection Scheme for the SOLASE Conceptual Laser Fusion Reactor," *Topical Meeting on Inertial Confinement Fusion*, OSA, San Diego (1978).

35. S. I. Abdel-Khalik, G. A. Moses, and R. R. Peterson, "Inertial Confinement Fusion Reactors Based on the Gas Protection Concept," *Nucl. Eng. and Design* **63**, 315 (1981).

36. R. R. Peterson, G. W. Cooper, and G. A. Moses, "Cavity Gas Analysis for Light Ion Beam Fusion Reactors," *Nucl Tech/Fusion* **1**, 377 (1981).

37. G. A. Moses and R. R. Peterson, "First Wall Protection in Particle Beam Fusion Reactors," *Nucl. Fusion* **20**, 849 (1980).

38. D. Cook and M. A. Sweeney, "Design of Compact Particle-Beam-Driven Inertial Confinement Fusion Reactors," *Proc. 3rd Topical Meeting on the Technology of Controlled Nuclear Fusion*, Santa Fe, NM (1978).

39. D. Cook, M. Sweeney, M. Buttram, K. Prestwich, G. Moses, R. Peterson, E. Lovell, R. Engelstad, "Light Ion Driven Inertial Fusion Reactor Concepts," *Proc. of 4th ANS Top. Mtg. on Tech. of Controlled Nucl. Fusion*, Oct. 1980, Valley Forge, PA.

40. W. G. Wolfer and R. D. Watson, "Structural Performance of a Graphite Blanket in ICTRs," *Proc. Third Topical Meeting on the Technology of Controlled Nuclear Fusion*, Santa Fe, NM (1978).

41. W. R. Meier, "Two-Dimensional Neutronics Calculation for the HYLIFE Convertor," Lawrence Livermore Laboratory Report UCRL-83595 (1979).

42. M. Monsler, "Laser Fusion: An Assessment of Pellet Injection, Tracking, and Beam Pointing," *Proc. Third Topical Meeting on the Technology Controlled Nuclear Fusion*, Santa Fe, NM (1978).

43. R. G. Tomlinson, L. R. Boedeker, D. H. Polk, G. E. Palma, and R. W. Guile, "Pellet and Laser Beam Space-Time Interaction System Study," United Tech. Res. Center Report UTRC R78-954373-1 (Nov 1978).

44. J. E. Howard, "Uniform Illumination of Spherical Laser Fusion Targets," *Appl. Opt.* **16**, 2764 (1977).

45. M. M. H. Ragheb, A. C. Klein, and C. W. Maynard, "Three Dimensional Neutronics Analysis of the Mirror-Beam-Duct-Shield System for a Laser Driven Power Reactor," University of Wisconsin Fusion Engineering Program Report UWFDM-239 (1978).

46. G. R. Magelssen and G. A. Moses, "Pellet X-Ray Spectra for Laser Fusion Reactor Designs," *Nucl. Fusion*, **19**, 301 (1979).

47. G. Pavlenco, "SOLASE-Balance of Plant Analysis," United Engineers Report (March, 1980).

48. B. R. Leonard, "A Review of Fission-Fusion Hybrid Concepts," *Nucl. Technol.* **20**, 161 (1973).

49. L. M. Lidsky, "Fission-Fusion Systems: Hybrid, Symbiotic, and Augean," *Nucl. Fusion* **15**, 151 (1975).

50. A. G. Cook and J. A. Maniscalco, "Uranium-233 Breeding and Neutron Multiplying Blankets for Fusion Reactors," *Nucl. Technol.* **30**, 5 (1976).

51. J. Maniscalco, "Fusion-Fission Hybrid Concepts for Laser Induced Fusion," *Nucl. Technol.* **28**, 98 (1976).

52. U. P. Jenquin, B. R. Leonard, D. H. Thomsen, and W. C. Wolkenhauer, "A Fusion-Fission Parametric Study," Annual Controlled Thermonuclear Reactor Technology Report, Pacific Northwest Laboratory Report BNWL-1604 (1971).

53. B. R. Leonard, "A Hybrid Neutronics Analysis," Annual Controlled Thermonuclear Reactor Technology Report, Pacific Northwest Laboratory Report BNWL-1685 (1972), p. 18.

54. B. R. Leonard and W. C. Wolkenhauer, "Fusion-Fission Hybrids: A Subcritical Thermal Fission Lattice for a DT Reactor," Pacific Northwest Laboratory Report BNWL-SA-4390 (1972).

55. R. P. Rose, "Fusion Driven Breeder Reactor Design Study," Westinghouse Electric Corporation Report WFPS-TME-043 (1977).

56. M. M. H. Ragheb, M. Z. Youssef, S. I. Abdel-Khalik, and C. W. Maynard, "Three-Dimensional Neutronics Analysis of the SOLASE-H Laser Reactor Fissile Enrichment Fuel Factory," University of Wisconsin Fusion Engineering Program Report UWFDM-266 (1978).

57. "Laser Fusion-Fission Reactor Systems Study-4000 MW Laser Fusion Hybrid Reactor," Bechtel Corporation Research and Engineering, Lawrence Livermore Laboratory Report UCRL-13796 (1977).

58. J. A. Maniscalco, "A Conceptual Design Study for a Laser-Fusion Hybrid," *Proc. 2nd Top. Meeting Technol. Controlled Nucl. Fusion*, Richland, Wash. (1976).

59. W. P. Kovacik, "Laser Fusion Power Reactor Systems (LFPRS), Conceptual Design," Westinghouse Electric Corporation Report WFPS-RME-070 (1977).

60. R. W. Conn, S. I. Abdel-Khalik, and G. A. Moses, "The Laser Fusion Hybrid," *Nucl. Eng. Des.* **63**, 357 (1981).

61. R. J. Barrett and R. W. Hardie, "The Fusion-Fission Hybrid as an Alternative to the Fast Breeder Reactor," Los Alamos Scientific Laboratory Report LA-8503-MS (1980).

62. D. H. Berwald et al., "Parametric Systems Analysis for ICF Hybrid Reactors," *Proc. 4th Top. Meeting Technol. of Controlled Nucl. Fusion*, King of Prussia, Pa. (1980).

63. D. H. Berwald and J. J. Duderstadt, "Preliminary Design and Neutronic Analysis of a Laser Fusion Driven Actinide Burning Hybrid Reactor," *Nucl. Technol.* **42**, 34 (1978).

64. R. P. Rose et al., "Fusion Driven Actinide Burner Design Study," Electric Power Research Institute Report EPRI-ER-451 (1977).

65. W. Bocola et al., "Considerations on Nuclear Transmutation for the Elimination of Actinides," presented at *the Int. Symp. Manage. Radioact. Wastes Nucl. Fuel Cycle*, Vienna (1976).

66. H. I. Avci, K. D. Kok, R. G. Jung, and R. C. Dykheizer, "Production of High Temperature Process Heat in Pebble Beds in ICTR Blankets," *Trans. Am. Nucl. Soc.* **32**, 39 (1979).

67. D. R. Peterson, J. H. Pendergrass, G. E. Cort, and R. A. Krakowski, "A Tritium Self-Sufficient 1600 K Process Heat Reactor Blanket Concept," *Trans. Am. Nucl. Soc.* **33**, 74 (1979).

68. H. J. Gomberg and W. W. Meinke, "Production of Synthetic Fuels: An Important Civilian Application of Laser Fusion," *Miami International Conference on Alternative Energy Sources*, edited by T. N. Verziroglu (1977).

69. J. D. Fish, "Radiolytic Production of Chemical Fuels in Fusion Reactor Systems," Princeton University Ph.D. dissertation (1977).

70. R. Hyde, L. Wood, and J. Nuckolls, "Propulsion Applications of Laser Induced Fusion Microexplosions," *Proc. First Topical Meeting on the Technology of Controlled Nuclear Fusion*, San Diego, 1974, p. 159.

71. R. Hyde, L. Wood, and J. Nuckolls, "Prospects for Rocket Propulsion with Laser-Induced Fusion Microexplosions," *AIAA/SAE 8th Joint Propulsion Special. Conf. Paper*, New Orleans, 1972.

Index